BLACK TIDE

The Devastating Impact
of the Gulf Oil Spill

ANTONIA JUHASZ

WILEY
John Wiley & Sons, Inc.

Published by John Wiley & Sons, Inc., Hoboken, New Jersey
Published simultaneously in Canada

Design and composition by Forty-five Degree Design LLC

For general information about our other products and services, please contact our Customer Care Department within the United States at (800) 762-2974, outside the United States at (317) 572-3993 or fax (317) 572-4002.

Wiley also publishes its books in a variety of electronic formats. Some content that appears in print may not be available in electronic books. For more information about Wiley products, visit our web site at www.wiley.com.

ISBN 978-0-470-94337-3; ISBN 978-1-118-06761-1 (ebk.),
ISBN 978-1-118-06773-4 (ebk.), ISBN 978-1-118-06774-1 (ebk.)

Printed in the United States of America

10 9 8 7 6 5 4 3 2 1

To the men of the *Deepwater Horizon* who gave their lives so that their crewmates might live: Gordon Jones, Dewey Revette, Jason Anderson, Shane Roshto, Stephen Curtis, Blair Manuel, Karl Kleppinger, Adam Weise, Don Clark, Roy Kemp, and Aaron Dale Burkeen.

To those who opened their lives to me so that I would tell this story with depth, honesty, and to the most lasting effect.

To my family: Joseph, Suzanne, Alex, Jenny, Christina, Linda, Emma, Simone, Gabriel, Eliza, Zoe, Paul, Branny, Skip, and Lucky.

We have gone to a different planet in going to the deepwater. An alien environment. And what do you know from every science fiction movie? The aliens can kill us.

—Byron King, oil industry analyst, May 2010[1]

Contents

Introduction

On April 20, 2010, the *Deepwater Horizon* drilling rig exploded fifty miles from the coast of Louisiana, killing eleven men and setting off the largest oil disaster in U.S. history. It would have been the world's largest oil spill had Saddam Hussein not ordered his troops to impede the advance of U.S. soldiers in 1991 by opening the valves of Kuwait's oil wells and pipelines.

I went to Louisiana after the explosion to measure the disaster for myself and to learn from those for whom time is now forever marked as "before" and "after" the BP oil spill.

Having written two books investigating the oil industry and the government agencies that are both at the heart of this catastrophe, and having also worked for advocacy organizations and for the federal government on the issues most relevant to the cause of the spill, I hoped to find answers to four central questions: Why did this disaster take place? Why was each failure compounded to create a tragedy of such massive proportion? What are the disaster's impacts? Can we prevent another catastrophe of this magnitude from ever happening again?

My first trip to the Gulf ended with little in the way of answers. Instead, I came away with the realization that understanding and documenting this event would require much more than one visit or one article. It was clear that the tragedy involved far more than one company, BP, and was not limited to just one government agency or even one administration. Its impacts, moreover, would reverberate well beyond the *Deepwater Horizon* and the families of the eleven men who died, and even beyond the people and places of the Gulf of Mexico. I also learned that finding answers was going to be made all the more difficult by the lack of transparency—on the part of both corporations and the government—that kept many of the unfolding events out of the public eye and outside the reach of journalists.

To get to the truth would require time, deep investigation, and, most important, the voices, experiences, and knowledge of those at the heart of the Gulf oil disaster.

During the next eight months, I traveled countless miles across Texas, Louisiana, Mississippi, Alabama, Florida, and Georgia and then to Washington, D.C. I was awed by the incredible generosity of people whose lives had been forever changed by these events and who, at the time of greatest personal tragedy and turmoil, opened their hearts and homes to a stranger. They trusted me with their pain, sorrow, rage, exhaustion, anger, secrets, research, insights, revelations, frustrations, and heartbreak. They introduced me to their children, wives, husbands, and colleagues—and to their enemies. They invited me to their churches and workplaces, onto their boats and into their homes, to their protests and festivals, and to their most cherished beaches, wetlands, and waterways.

I interviewed hundreds of people, including family members of those who died aboard the *Deepwater Horizon,* as well as oil workers, fishers, crabbers, beach cleaners, oceanographers, government officials, lawyers, environmentalists, doctors, veterans, and countless others. They shared their lives with me so that I might share this story with you.

These people were my guides as I uncovered who set the oil loose and multiplied its deadly effects and as I tracked the oil on its rampage through the Gulf.

I was also led by the heroism of those working to stem this black tide and ensure that such a tragedy never occurs again. For this was not an isolated incident, and BP was not a lone actor. Rather, this tragedy was the predictable outcome of an industry that has pushed well beyond its own technological capacity and beyond the government's ability to regulate it. Nor is the oil gone or the cleanup complete. BP's oil monster continues to move through the ocean, sometimes washing up on beaches and sometimes content to lurk on the ocean floor, waiting for the next best moment to strike. It hides in the eggs of shrimp, the habitats of pelicans, and even the blood cells of people all along the coast.

The first time I held BP's oil in my hands, I was standing on a beach on Dauphin Island, Alabama. The oil had mixed with sand on its long journey from deep below the seafloor to form thick, brown gooey patties that covered the beach for as far as the eye could see. A warning immediately came to mind, uttered long ago by Carlos Andres Perez, the oil minister and then president of Venezuela and a father of the modern oil industry: "Oil is the devil's excrement."

Unless both the oil and those who crave it can be tamed, hell will surely once again come not only to the Gulf but also to the many other places where oil lies just beneath the earth's surface.

1

The Explosion of the
Deepwater Horizon

This event was set in motion years ago by these companies need-
lessly rushing to make money faster, while cutting corners to save
money. When these companies put their savings over our safety,
they gambled with our lives. They gambled with my life. They
gambled with the lives of 11 of my crew members who will never
see their families or loved ones again.

—Stephen Lane Stone, Transocean roustabout,
Deepwater Horizon survivor, May 27, 2010.[1]

At 9:49 p.m. on April 20, 2010, as a gas bubble raced up 18,360 feet of steel pipe on a deadly collision course with the *Deepwater Horizon*, a group of BP executives were at its helm, playing captain. The rig had arrived earlier that day with two Transocean executives for what was supposed to be a twenty-four-hour visit. BP would later tell reporters that the trip was arranged to celebrate a major achievement, that the "Deepwater Horizon was the first rig to go seven years without a lost-time accident."[2] Testimony would later reveal not only that the rig had not gone seven years without a lost-time incident but also that the trip had the far less celebratory purpose of applying senior-level pressure on the crew to finally complete the well.[3]

BP, the fourth-largest corporation in the world, was spending $2 million a day for a job that was more than fifty days behind schedule and nearly $100 million over budget and counting. At a rate of approximately $500,000 a day, BP leased the *Deepwater Horizon* from Transocean, the rig owner and operator. BP then paid an additional $500,000 a day in operating costs. BP's internal costs were another $1 million a day.[4]

Transocean is the largest deepwater driller in the world. In the Gulf of Mexico, it operates nearly half of all the rigs that work in more than 3,000 feet of water. All the major oil companies use its services: contracting out rigs, related equipment, and work crews at a day rate to drill oil and gas wells. According to the company, "We specialize in technically demanding segments of the offshore drilling business with a particular focus on deepwater and harsh environment drilling services."[5]

Of the 126 people on board the *Deepwater Horizon* on April 20, 79 worked for Transocean, and only 8 were BP employees.

The visiting executives were BP's vice president for drilling and completions for the entire Gulf of Mexico, Patrick O'Bryan, who had never been on an offshore rig before, and managers for performance and operations from BP and Transocean. They spent the afternoon and evening of April 20 in sometimes heated meetings and on tours of the rig. Around 9 p.m. they were ready to call it a night but instead decided to head up to the bridge, where they were greeted by Transocean captain Curt Kuchta, master of the *Deepwater Horizon*.

Master Captain Kuchta is not quite as imposing as his title. It may be the slight stutter when he speaks or his small and somewhat pudgy frame. But when the executives arrived, Captain Kuchta became more tour guide than master and "showed us around a lot of stuff up there," recalled O'Bryan.[6] Captain Kuchta asked if they wanted to use the bridge simulator. A joystick at a computer station allows the user to imitate steering the entire rig. The executives eagerly agreed and began taking turns.

"They were basically playing a video game," Captain Kuchta later explained.[7] And what a video game it must have been!

At 320 feet high, the *Deepwater Horizon* had an additional 50 feet on the Brooklyn Bridge. It was 369 feet long and 256 feet wide—the size of an entire NFL football field, including end zones, coaching boxes, and team areas. There was enough space to house 130 people, along with a gymnasium, movie theater, lounge, laundry, kitchen, and helicopter pad, with room left over for the actual work of the rig: drilling for oil.

It was also one of the most famous offshore rigs in the world. Just six months earlier, in September 2009, it had set the record for drilling the deepest oil and gas well in history at BP's Gulf of Mexico Tiber field. At nearly 40,000 feet below the ocean surface, the well is farther down than Mount Everest is up.

"It was an honor to be chosen again for that rig. Gordon was proud." When Keith Jones speaks about his son Gordon, his love is obvious. But when he talks about Gordon's job as a mud engineer, the only word that adequately describes his emotion is *restrained*. Gordon was part of the *Deepwater Horizon* drill team, but he wasn't supposed to be. He got into the oil business because of his communication skills and scratch golf game, both of which lent themselves to sales, the job Gordon was headed for after putting in his time on the rigs.

His father looks bewildered as he recounts the story. The Joneses are lawyers and teachers, not oilmen. Even more bewildering is how Gordon ended up in the business: Keith got his son the job through a lucky encounter with an old fraternity brother.

When I ask if Gordon liked his job, Keith replies, "He liked being home." It is a common response. Offshore work is typically "21 on, 21 off"—twenty-one days working on the rig and twenty-one days off work at home. Rig workers talk about having two families and two lives: one on the rig, one off. On the rig the work is intense: twelve-hour around-the-clock shifts, called tours (pronounced "towers"). While one team is on tour, the other sleeps; while the first team sleeps, the other is working. The payoff for many, in addition to the sizable paycheck, is the long period of downtime at home.

As on all offshore rigs, work on the *Deepwater Horizon* was exhausting. It was also uniquely complex. In contrast to older rigs and those in shallower waters that attach to the seafloor with cables or anchors, the *Deepwater Horizon*, built in 2001, was dynamically positioned. A satellite provided the coordinates used by Transocean dynamic positioning officers, Yancy Keplinger and Andrea Fleytas, to determine the rig's location. Using computers, they then directed the twenty-four-hour-a-day, seven-day-a-week efforts of eight 7,375-horsepower thrusters that kept the rig floating on two giant pontoons on the water precisely above the wellhead some four miles below.

On April 20, the two officers held the ship steady while the drill team ran pipe through 5,000 feet of ocean, 13,360 feet of earth, and into the Macondo oil well. The only thing attaching the rig to the floor was the drill pipe and the blowout preventer.

As offshore drilling has gone ever deeper in the last twenty years, experts have increasingly come to liken it, and the dangers involved, to space travel. "We have gone to a different planet in going to the deepwater. An alien environment," oil industry analyst Byron King told the *Washington Post*. "And what do you know from every science fiction movie? The aliens can kill us."[8]

We can only imagine the thrill for the visiting executives who got to "steer" the rig. They even tried to intensify the experience by simulating increasingly rough conditions. "We loaded into the simulator about 70-knot winds and 30-foot seas and two thrusters down and then you switch it into the manual mode and see if the individuals can maintain the rig on location," explained visiting Transocean executive Daun Winslow, operations manager-performance for the North American division. For BP vice president O'Bryan, the "newbie" on the rig, "we loaded up with the most environment," Winslow said.[9]

The thrill of the game very quickly turned to real-life terror around 9:50 p.m., just as O'Bryan took his turn. "All of a sudden, the rig started shaking," he later testified. "And I remember I believe it was Captain Kuchta that went over to the door and opened it. . . . He closed the door and said everybody stay inside. And then soon thereafter I don't know how quickly but pretty quick I heard a hissing sound. And that's when I heard the first explosion. . . . And you could actually see the rig floor, and the rig floor was on fire. And it was just right there, after there was a larger explosion. And that's when all the power on the bridge went out."[10]

The Doghouse

If the rig floor is the operational heart of a drilling rig, then the "doghouse," the driller's shack, is its soul. The doghouse looks like a giant airplane cockpit. Two large chairs dominate the room, each with its own joystick, telephone, microphone, and set of two computer screens. A large window above the screens provides the crew with a view of the rig floor, where the giant 242-foot-tall derrick resides.

Work on a drilling rig largely breaks down into two categories: that done on a computer and that done in the mud. The doghouse is the cord that ties the two together. Inside the doghouse, the crew is comfortably dressed in T-shirts and pants. Outside, they are in bright yellow coveralls, thick rubber boots, and hard hats. Using their computers, those in the driller's shack direct all the drilling operations, while those in the mud make sure that the pipes and the rest of the physical equipment behave as they should on their many-miles-long journey down to the well.

The mud is the lubricant that keeps the equipment flowing. It's called mud because that's what it looks like, but it is actually a toxic chemical cocktail more formally known as drilling fluid. The chiefs on the *Horizon* are the mud engineers: Gordon Jones, Blair Manuel, Greg Meche, and Leo Linder. They work for contractor M-I SWACO. Instead of working with the mud, they were, in Keith Jones's words, "technicians who slipped off to the lab, came up with a formula, and handed off the final product to the floor hands." The mud is composed of rock that is ground to a fine powder on shore, which is then mixed with chemicals and water on board, depending on the frequently changing needs of the rig and the well. Among its other functions, the mud provides a weight to counter the pressure down in the well, preventing gas or water from coming up through the pipe.

The *Deepwater Horizon*'s job, like that of any drill ship, was to look for a good hole in a well and then prepare the hole for a production vessel to follow. The crew does this by drilling a pipe into the well, hitting oil, sealing the pipe with cement (like a cork on a champagne bottle), disconnecting from the pipe, and moving off. A production vessel then moves in, uncorks the pipe, and removes the oil. At 9:50 p.m. on April 20, the crew was nearly ready to "cork the pipe" and move on. This may sound simple; however, it is anything but.

Drilling in deep water is an acrobatic dance with pressure. The deeper you go, the greater the pressure. Imagine a giant bag of popcorn, fresh from a microwave oven, waiting to be opened. Now bury it under 5,000 feet of ocean and 13,500 feet of earth. Instead of ripping the bag open, you insert an 18,500-foot straw in the form of the pipe and then place your thumb over the top to keep the contents from exploding out. The air in the bag is the gas, the popcorn is the crude oil, and the thumb is the cement job temporarily sealing the hole.

Managing the pressure so that oil and gas can be brought to the surface in a controlled manner is called well control. A well-control event or incident—also known as a kick—occurs when there are problems with managing this pressure. A loss of well control is called a blowout.

Thirty-year oil industry veteran Bob Cavnar described the difficulty this way: "I often think of deepwater drilling as like driving a car from the backseat; you can reach the steering wheel, but it's hard to control and you can't get your foot on the brake pedal very easily. Because the distance between man and well is so far in the deepwater, the technology must be the link between those two. As we've been made painfully aware, when the technology fails, or people fail, the consequences are catastrophic."[11]

The technology on board the *Horizon* was far from perfect. Just one example of hundreds of unattended repair issues on the rig was described as "the blue screen of death" in testimony by Transocean chief electronics technician for the *Horizon*, Mike Williams. There are three drilling chairs: A and B in the drill shack and C in the assistant drill shack. The computer screens locked up—regularly on the A chair, and occasionally on the B chair, according to Williams. When a screen locked up, it "would just turn to a blue screen," he testified. "You would have no data coming through." While waiting for a replacement system, the *Horizon* crew members "were limping along with what we had."

In questioning, Williams was asked, "Well, if the driller is sitting there trying to manage the well, and the blue screen of death shows up, how is the driller supposed to be able to manage the well?" Williams answered, "He's going to have to go to B Chair."

"And if B Chair isn't functioning properly?"

"He's got to go to C chair."

"And if that's not functioning properly?"
"Abandon ship."[12]

Although corporate executives are obsessed with oil, everything on a rig is focused on gas: how to keep the gas out while bringing the oil in. The Gulf of Mexico has many unique attributes, including being one of the most methane-rich production areas in the world, which also makes it one of the most dangerous places on earth to drill.

Gas kicks are routine. Even blowouts occur far more often than the industry would have us believe, and with increasing frequency. From 2005 to 2010, twenty-eight blowouts occurred in the Gulf of Mexico, four of which took place in the eighteen months preceding the blowout of the Macondo well.[13] From 1999 to 2004, there were twenty blowouts, and from 1993 to 1998 there were just eleven.[14]

Most of these blowouts were in shallow waters, because that's where the vast amount of drilling takes place. Of the 565 productive wells in the Gulf of Mexico, 488 are in water depths of less than 300 feet; only 37 wells are located at depths greater than 1,000 feet, but these wells account for nearly 70 percent of the oil produced in the Gulf.[15] As more drilling moves to ever deeper waters in search of richer fields, we can expect more deepwater blowouts.

Blowouts in deep water can be far more serious than those in shallow water because of the increased difficulty of reaching the blown-out well; the oil industry's lack of experience, technology, and equipment in deep water; and the potential for much greater amounts of oil to be spilled.

Each misstep made by BP, Transocean, Halliburton, and others on the *Deepwater Horizon*'s fateful journey opened a new door that allowed the gas from the Macondo well to enter and ultimately explode the rig.

The Macondo Well

The *Deepwater Horizon* wasn't supposed to be at the Macondo well. It was a replacement, taking over for Transocean's *Marianas*.

The U.S. government owns the land where the Macondo well resides, Mississippi Canyon Block 252, and the hydrocarbons within it. In 2008, BP, with minority partners Anadarko Petroleum (25 percent) and MOEX Offshore (10 percent), paid $34 million for a ten-year lease to the area. In exchange for the lease payment and the royalty fees on each barrel recovered, the hydrocarbons essentially became BP's. Estimates of the amount of oil in the well vary greatly, from 50 million to 1 billion barrels, worth as much as $86 billion in today's market.[16]

BP originally leased Transocean's *Marianas* to work the well, and it began drilling on October 6, 2009. Mechanical failures on the *Marianas* were intensified by the kicking well, which kicked so hard that it literally blew out the rig's blowout preventer. The blowout preventer shuts in a well in the event of a serious kick, in order to prevent a kick from becoming a blowout.

On November 1, the *Marianas* was forced to unlatch from its blowout preventer in "one of the singularly most costly events in the drilling of a well if it occurs," according to rig experts WEST Engineering Services.[17] When Hurricane Ida came through on November 7, the *Marianas* was already sitting idle.[18] It then sustained further electrical wiring damage from the storm, was sent to shore for repairs, and was later decommissioned.

Such problems on a Transocean rig are not unique. In fact, since 2008, 73 percent of incidents that triggered federal investigations into safety and other problems on deepwater drilling rigs in the Gulf have been on rigs operated by Transocean. This rate is out of proportion to the percentage of rigs the company operates in the Gulf: less than half.[19]

The *Deepwater Horizon* arrived at the Macondo well on January 31, 2010. It began drilling on February 15 and was supposed to be done just three weeks later, on March 8, for a total cost of $96 million. From the start, the well was kicking, and the crew lost control of it several times. In March, the drill pipe became so irrevocably stuck that it was left there while the crew moved on to try a new spot.

The reason, explained Mike Williams, was pressure to get the job done fast. A BP manager ordered a faster pace, telling the driller, "Hey, let's bump it up. Let's bump it up." Going faster caused the bottom of the well to split open. "We got stuck so bad we had to send tools down into the drill pipe and sever the pipe," Williams said. BP told the crew that it lost two weeks and some $25 million in the process. "And you always kind of knew that in the back of your mind when they start throwing these big numbers around that there was gonna be a push coming, you know? A push to pick up production and pick up the pace," Williams explained.[20]

As the operation grew further behind schedule and over budget, BP was cutting costs, Transocean's and other subcontractors' shoddy practices were put into heightened relief, and government regulators rubber-stamped each cumulatively more dangerous move.

Unfortunately, none of these events was unique to the Macondo well or the *Deepwater Horizon.* It was simply an unlucky cumulative impact that led this rig and not others to such catastrophic disaster. The enormity, extreme difficulty, and relative newness of deepwater drilling regularly puts production wells off schedule, which always means mammoth cost overruns. In response, companies cut corners.

I spoke with many offshore oil workers for this book; all of them are currently employed, so few would speak on the record. A typical response was provided by a completions engineer who has worked on many Transocean rigs as a subcontractor and is currently at work on one of BP's Macondo relief wells.

"When you start a new job, it's all about safety," he tells me. "But after you're out there for so long, you find you can save $2 million here or $3 million there, and they start cutting corners, save whatever money they can. I've seen 'em all cut corners they shouldn't cut. They all do it. Shell, Chevron, they all do it. They hit a comfort zone, get complacent, could do it a little cheaper, a little differently, and, yeah, eventually the worst happens. But, you know, we're paid to go out there, and you have confidence in the decisions they make. Doesn't do no good to worry and stress about it, because there's nothing you can do about it, anyway."

The Well from Hell

With seventeen years of experience in the oil industry, assistant driller Stephen Curtis knew a tough well when he worked one. He christened the Macondo "the well from hell" because it reminded him of another wicked well, Devil's Tower.

The Macondo well is not unique. It is, rather, a tragedy of foreseeable errors left uncorrected that has set its place in history. Mike

Williams worked both wells with Curtis. The wells "exhibited a lot of [the] same characteristics," Williams explained, "where we lost circulation, we're getting tons of gas back all the time, we got stuck, we had to sever the pipe. It was just—it was déjà vu all over again."[21] With Devil's Tower, however, the well had its way without loss of life. The oil company in charge, Italy's ENI, gave in, severed the pipe, and drilled in a different spot.

Floor hand Adam Weise was calling his girlfriend, Cindy Shelton, before and after every shift—unusual for him. He was frustrated with the problems on the project. "Everything that could go wrong was going wrong," she said. "Every time he'd call me, he'd say, 'This is a well from hell.'"[22]

Shane Roshto shared the same worries. Just twenty-two years old, he tried to explain the dangers of offshore drilling to his twenty-one-year-old wife, Natalie. "Baby," Shane told her, "the earth is . . . like blowing up a red balloon and taking a pin and just pushing it and pushing it and pushing it as far as it could go and it just blowing." On his last trip home before heading back to the Macondo, Shane told Natalie, "Mother Nature just doesn't want to be drilled here."[23]

"Dad, I know what they're trying to do is unsafe, and I know it's not right," thirty-five-year-old Jason Anderson, the leader of the *Deepwater Horizon* drill team, told his father on his last trip home before the disaster. Like many other members of the crew, Jason had been with the *Horizon* since its birth, bringing the rig to the United States from Korea, where it was built. He was proud to be on the team that drilled the world's deepest well.

Nonetheless, he was worried. Before he left, the father of two young children, ages five and one, gave his wife, Shelley, "a will and a list of things that he wanted handled if something happened to him."[24] This was his last tour on the *Horizon*. He had recently been promoted, and his new job as senior toolpusher on Transocean's *Discoverer Spirit* was to start on April 14, but the manager of the *Horizon* asked him to stay on for an extra week.[25]

On April 14, BP drilling engineer Brian Morel e-mailed a colleague, saying, "this has been [a] nightmare well which has everyone all over the place."[26]

Final Days

On April 15, the *Deepwater Horizon* was some forty-five days behind schedule and $90 million over budget. In response, BP was cutting corners. BP changed its well design three times within one twenty-four-hour period. Each new design was approved by government regulators, sometimes within minutes of the request being submitted. The design ultimately utilized was, according to BP, $7 to $10 million cheaper than an earlier plan, the "tie-back."[27] On March 30, BP's Brian Morel wrote, "not running the tie-back saves a good deal of time/money."[28]

The final design was one long tube running through the center of the well. The design was doable, according to Halliburton, the subcontractor hired to do the cement job to secure the pipe, as long as BP used twenty-one devices called centralizers to help hold the pipe in place while the cement set. The *Horizon* had six such centralizers on board, so BP drilling operations engineer Brett Cocales ordered an additional fifteen out to the rig.

On April 16, the centralizers arrived, but BP well team leader, John Guide, chose not to use them. In an e-mail that day, Guide argued that the new centralizers were not ideal for the job, and "it will take ten hours to install them."[29]

On April 18, Halliburton ran a new computer model of a cement job using fewer than seven centralizers. It showed that such a job entailed a "severe risk of gas flow."[30] Nonetheless, the decision to go with six centralizers stuck.

On April 19, Halliburton completed the job to BP's specifications. Not only did Halliburton perform a substandard job, it also used the wrong cement. Halliburton used a nitrogen foam, common in offshore drilling generally, but not for deep high-temperature, high-pressure zones such as the Macondo well.[31] An investigation by the National Commission on the BP *Deepwater Horizon* Oil Spill and Offshore Drilling, known as the National Oil Spill Commission, found that Halliburton and BP both had results in March showing that the cement mixture would be unstable, but neither acted on that data.[32]

On April 20, more than fifty days late and $100 million over budget, the *Horizon* was running a series of tests to make sure that Halliburton's cement job was holding. Tensions were high, and disputes appear to have been taking place all over the rig. One such fight took place at the 11 a.m. pretour meeting.

Douglas Brown, Transocean chief mechanic, described "a scrimmage taking place" at the meeting among Robert Kaluza, the so-called company man, the highest-ranking BP employee assigned to the rig; Jimmy Harrell, the offshore installation manager (OIM), the most senior Transocean position on the rig; and Dewey Revette, the toolpusher.[33]

Forty-eight-year-old Dewey Revette had been working offshore for nearly thirty years. He was on the *Horizon* when it broke its world record, but he nearly missed it. About four years earlier, he told his wife, Sheri, that he wanted to quit offshore and maybe start consulting. As Sheri recounted the story to me, she said no. The economy was in a tailspin. They had one daughter in high school and another in college. There was too much security in sticking with Transocean.

"I told him he could wait 'til it was time to retire, get his full benefits, and *then* go into consultin' if he wanted," Sheri tells me in her heavy Mississippi accent. "It was the last [work] decision we made. He would keep goin', then retire." Sheri's voice trails off. Dewey wasn't tired, she emphasizes. "He just liked bein' home. He liked huntin', fishin', playin' on his tractor. . . . He wanted to do 'round here and be with his family." Her voice trails off again. "We were gonna get my son-in-law offshore; this was about a year ago, but my daughter, she said no."

At the 11 a.m. meeting, Dewey was outlining his plans for the displacement test to be run that evening. The test, which displaces heavy mud in the drill pipe with lighter seawater to test the stability of the cement job, was one of the final procedures before moving off from the well. BP's company man, Robert Kaluza, stood up and said, "No, we have some changes to that." The changes were even more shortcuts and cost savings. The displacement would be faster and cheaper. Two tasks would also be performed at once: the displacement and the second negative test to measure upward pressure from the shut-in well.

According to Douglas Brown, "the company man was basically saying, 'Well, this is how it's going to be.' And the toolpusher and the OIM reluctantly agreed."

Kaluza later told investigators that he was just following BP's orders. "They decided we should do displacement and the negative test together; I don't know why," he said. "Maybe they were trying to save time. At the end of the well sometimes they think about speeding up."[34]

The risks involved appear not to have been lost on Jimmy Harrell, the Transocean OIM, who seemed to think that all was now resting on the rig's last line of defense, the blowout preventer. Brown testified that as Harrell left the meeting, he grumbled, "Well, I guess that's what we have those pinchers for," referring to the blowout preventer's pinchers that close in on the pipe, cutting it off from the rig and shutting in the well in case of a blowout.[35]

Weighing some 325 tons and standing nearly fifty feet high, the blowout preventer (BOP) sits at the bottom of the ocean on top of the wellhead. The riser is the portion of the drill pipe that descends from the rig into the top of the BOP. The pipe then passes through the BOP and descends into the well. Inside are a series of mechanisms that can, when activated, sever the pipe, thereby locking in the well. Activating the BOP is like sucking on a straw and then pinching the middle of it with your fingers to stop the fluid from moving, breaking off the top portion of the straw, and sticking a cork in what's left.

Harrell, who is facing possible criminal charges, says the disagreement happened the day before and was about something else. Other accounts support Brown's version and shed more light on the skirmish, including on Dewey's role. I shared one such report with his wife, Sheri, when I visited her home in State Line, Mississippi, a small rural community of about five hundred people.

Sheri is a petite woman. Her appearance is that of an attractive suburban American "everywoman." Sheri is from Fairborn, Ohio, which she describes as "the city" because "the mall was just twenty minutes away and the Kmart was within walking distance." By comparison, when she moved to State Line, "Dewey's parents didn't even have a telephone." But she was willing to make the sacrifice for Dewey. Sheri's smile and laughter are contagious as she retells the story of their whirlwind romance, engagement, and marriage—all in one year. She was just eighteen and he was twenty-one. "Everyone was lookin' for a baby," she laughs. "There wasn't one. We were just that in love." They met when Dewey drove up to the local Kerr gas station where Sheri worked. "We had the old-time Coca-Cola coolers. He reached in for one and he was sittin' there and we were talkin' and that was it." The extra emphasis on the last "t" conveys how Sheri never had a chance. "Dewey had this smile. It would make you melt." When Sheri says, "It was love at first sight," she means it.

Thirty years in rural Mississippi changed Sheri in many ways. And, while she looks much younger than her forty-five years, it remains very difficult to imagine her riding mud hills in Dewey's jacked-up four-wheel-drive Toyota truck, although she claims, "It was a blast!"

Sheri is also a calm woman, but it is an active calm, an intentional and self-imposed serenity. It is how she copes. It takes a jolt for the calm to be disrupted and the "mud trucker" to come out, which is just what happened when we sat in her kitchen and read the following account from *Esquire* magazine together:

> When [Kaluza] proposes a procedure that runs counter to the procedures the drilling team has in place, Dewey Revette, the driller, fresh from his circuit around the deck, begins to argue with him. Revette thinks that what Kaluza is proposing is reckless and premature, and when the argument grows heated, what the various crewmembers witnessing it remember is the passion and anger of an inherently careful man. "Dewey got pretty hot," one says.[36]

It's the last line that hits Sheri the hardest. "Dewey never—never—ever loses his temper. Never, ever, ever." The thought of him being that upset is not only shocking, it hurts. What perhaps hurts even more is that she wasn't able to be there for him. Dewey and Sheri usually talk around 9 a.m. every day before Dewey's tour, but "I missed his phone call that morning." Dewey called, but Sheri's phone didn't ring. "He left me a message, but I deleted it because I knew he was coming home, you know. Normally I always keep them. But for some reason I deleted that one."

Sheri had not seen the *Esquire* article, but she'd heard the story of the fight. If Dewey knew that something was wrong, "he wouldn't have done it unless they really, really made him buckle," Sherri tells me. He wouldn't have backed down "unless they really put him in his place and said—BP or whoever said—'We're going to do it like this because I said so.' I just don't know, you know. I would figure Dewey would tell them to take that job and shove it first. You know, he's just that headstrong, and if he really believed that any of this could have happened, he'd have never, never let them do it."

Dewey may, in fact, have had several reasons to be upset that morning. He may have learned not only that the displacement test was being messed with but also that the subcontractors hired to perform a key test on the well's stability were on a helicopter leaving the *Deepwater Horizon*.

Kill This Well

Schlumberger (pronounced "slumberjay," in the oil patch), an oil field service contractor, was hired by BP to perform a cement bond log, a final test on the cement job performed by Halliburton. Called "the only test that can really determine the actual effectiveness" of the well's seal, the test was a critical final step in ensuring the stability of the well.[37] Instead of running the test, the Schlumberger employees left the rig at 11 a.m. that day.

There are two explanations for why they left. One is that they knew the well was going to blow, could do nothing to stop it, and evacuated. The other is that BP decided at 7:30 a.m. on April 20 that it didn't need the test.

Sherrill Boega was the first person to alert me to the competing accounts. I met Sherrill at the Shrimp and Petroleum Festival in Morgan City, Louisiana—"after Mardi Gras, the largest festival in Louisiana." Like his father before him, Sherrill had spent his life working for Shell. While in college, he spent his summers working for the company and then, after a brief stint as a high school teacher, went to work for Shell full-time in operations offshore. He moved into the safety specialization and in 1999 retired as an offshore health safety environment coordinator after thirty years with the company. Since 1999, he has continued working for a variety of companies, including Exxon, BP, Chevron, and Shell.

The story he told me about Schlumberger was really about the gas. When Schlumberger arrived, "gas was showing, and the well was trying to come in. I know it sounds like the well has a personality, but what it means technically is that whatever was in there to prevent it from coming up was not doing its job. . . . The well was trying to come in. Supposedly, when [the Schlumberger employee] reported that to the BP guy and said, 'You need to kill this well,' the BP guy said, 'I ain't doing nothing; you go to work.' That's when [Schlumberger] asked for the helicopter. He was told that the helicopter wasn't available. That's when he called his supervisor on the beach, and I don't know what transpired, but I do know that Schlumberger sent a helicopter. . . . This guy ought to get an award, because he did what his company told him to do: if you can't stop the job, leave. That's what he was trying to do, have them kill the well, and the guy wouldn't have anything to do with it."

Virtually identical accounts appeared on several industry-insider Internet discussion boards, including the most often quoted account on the *Oil Drum*, posted on May 14. "AlanfromBigEasy" writes that Schlumberger (SLB) got out on the *Horizon* to run the test and "they find the well still kicking heavily, which it should not be that late in the operation." Schlumberger orders the well killed, the company man refuses, and "SLB gets on the horn to shore, calls SLB's corporate HQ, and gets a helo [helicopter] flown out there at SLB's expense and takes all SLB [personnel] to shore. . . . Pick your jaw up off the floor now."

In testimony and in its internal analysis, BP contends that its executives chose not to do the test and therefore sent Schlumberger home on a regularly scheduled BP flight. Oddly, no one from Schlumberger has testified at any of the dozens of investigations held by Congress, the Marine Safety Board, or the National Oil Spill Commission. All we have is a statement from Schlumberger spokesman Stephen Harris, who said that BP had a Schlumberger team on standby from April 18 to 20. "But BP never asked the Schlumberger crew to perform the acoustic test and sent its members back to Louisiana on a regularly scheduled helicopter flight at 11 a.m."[38]

Many have argued that BP chose not to do the test because of time and expense. The New Orleans *Times-Picayune* reported that "the test would have cost $100,000 or more, taken time and required a month of remedial work if it found problems, like an uneven cement job, at a likely additional cost of $30 million."[39] When asked, "Which option is cheapest: not running the cement bond log or running it?", BP's John Guide responded, "It's cheaper not to run it."[40]

Professor Robert Bea, a former Shell Oil Company engineer, heads the *Deepwater Horizon* Study Group at the Center for Catastrophic Risk Management at the University of California. The group was asked to provide monthly reports and cumulative findings to President Obama's

National Oil Spill Commission. Professor Bea told me that he believes it will be impossible to know what really happened with Schlumberger until "someone has to give sworn testimony" in a court case, "and we might not even know then."

Whatever the reason, the test was never run. Had the Schlumberger crew stayed on the *Deepwater Horizon* to run the test, they would surely have found that Halliburton's cement job was faulty; the findings of the negative tests performed later that day by the drill team, which showed the well to be stable, were wrong; and the displacement test that Dewey, Stephen, Jason, and the rest of the drill team were to perform later that night was doomed to catastrophic failure.

Blowout

At 9:20 p.m., Dewey was in the C chair in the assistant driller's shack, and Jason and Stephen were in the driller's shack in chairs A and B, respectively. All were following orders, performing the second negative test and the displacement at the same time, as designed by BP. It was one of the final necessary steps before the *Deepwater Horizon* could say good-bye to the Macondo well forever.

Why did these three men, and the rest of the crew, keep following what they knew to be bad orders? People who work offshore are, by and large, an order-abiding group. The vast majority are men, many are former military, and life on a rig is frequently described with military terminology: they are a unit at war against the sea, the weather, the gas, the oil, the government, the environmentalists—you name it. Whereas the Schlumberger crew was visiting, the Transocean crew was part of a unit, many of whom had been together since the birth of the rig nearly ten years earlier.

Although executives from BP and Transocean stressed in testimony that any crew member can at any time halt any operation on a rig, it rarely, if ever, happens. "Yeah it's in the manual," lawyer Kurt Arnold tells me, but from the perspective of his clients, more than a dozen of whom worked on the *Horizon*, "from a practical perspective it's a myth."

The crew not only believe in their unit, they also believe in keeping their jobs. Even though most have just high school educations, they easily earn from $50,000 to $100,000 working offshore. When you live in places like State Line, Mississippi, there aren't many other choices. "Dewey went offshore when he was nineteen," Sheri tells me. "It was either that or cuttin' timber." Not only have times not changed, they've gotten worse.

Thus, the crew did what they were told to do and then fought like hell to save the rig from the consequences of the bad orders they were given to follow.

There were several problems with the test they were performing. It was the second negative test that day; both tests were premature, adding to the already unstable well. The foamed cement used by Halliburton required forty-eight hours to strongly solidify. Nonetheless, the first negative test was performed just sixteen and a half hours after the cement was pumped, and the second test was just twenty-one hours later. The cement would have "almost no functional strength at [those] time[s] and would still be in a very fragile state," concludes industry analyst Paul Parsons.

The displacement test occurred at a much lower depth than is standard practice. Whereas government regulations are set at 1,000 feet, BP requested approval for 3,300 feet. The reason for doing so may have been further cost savings. As Paul Parsons wrote, it is possible that BP was "interested in the fact that setting the plug further down allows the removal of more expensive mud from the well for reuse elsewhere (all mud above the plug would be removed)." Parsons argued, "A deep negative test as BP performed (3,300 feet below the mudline) plus a full riser displacement would not only expose a problem with the cement but could also create or worsen a problem if performed too early."[41]

As Jason, Stephen, and Dewey were removing the mud and replacing it with lighter seawater, there would have been less pressure on the gas holding it in place, allowing more to escape into the pipes and up into the rig.

At 9:20 p.m., Randy Ezell, Transocean senior toolpusher, called Jason to ask how things were going. Jason had just completed the second negative test and was displacing the mud with seawater. "Do you need any help from me?" Randy asked. "No, man. . . . I've got this," Jason reassured. "Go to bed. I've got it."[42]

At 9:40 p.m., mud began overflowing onto the rig floor and then shot up through the derrick.

At 9:41 p.m., the blowout preventer was activated, but it did not act.

At 9:44 p.m., either Dewey or Jason called BP's well site leader, Donald Vidrine, and said, "We're getting mud back."

Randy had taken Jason up on his offer. He went to his quarters, called his wife, and got into bed. He had just turned off his overhead light when the phone rang. This time, the drill shack was calling him. BP's clock has the call at 9:45 p.m.; Randy testified that it was 9:50 p.m.

"The well is blown out."

It was Stephen Curtis.

"Do y'all have it shut in?" Randy asked.

"Jason is shutting it in now. Randy, we need your help."

Stephen was a big man, a former marine and a deer hunter who got married in full camouflage gear. For Stephen to ask for help meant that something was seriously wrong. Randy was horrified. "I put my coveralls on; they were hanging on the hook. I put my socks on. My boots and my hard hat were right across that hall . . . in the toolpushers office," he recalled in testimony. Just as he opened the door, "a tremendous explosion occurred. It blew me probably twenty feet against a bulkhead, against the wall in that office. And I remember then that the lights went out, power went out. I could hear everything deathly calm."[43]

The next morning, Sheri was expecting her husband to come home. Instead, her sister called at 5:15 a.m., telling her to turn on the television. She knew right away that her husband was dead. "My heart sunk, and I lost it," she says, choking on the words. "I knew where he would have been at that time because he was on tour. I said, 'I don't need to wait for a phone call, he's gone.'"

The Gas Attacks

We do not know exactly what happened in the drilling area at the time of the explosion, because everyone who was down there is dead. We do know this: before a blowout occurs, there are generally warning signs sent by the well and safeguards taken on the vessel. In the case of the Macondo, there were numerous warnings that the well was unstable before it blew. There were also many lines of defense that were supposed to stop the worst-case scenario from happening.

When the well blew, it evacuated the material thrust down the pipes: first the mud and then the cement. With the debris out of its way, the gas exploded into the rig, ignited two explosions, and sent a massive fireball up through the length of the derrick followed by oil.

"All of the workers that died during the blowout would have known they were in a dangerous situation and had time to flee," Parsons reminds us. "Instead, they elected to stay on position to try to regain control of the well."[44]

They should, however, have gotten a lot more warning and assistance in their efforts. The rig was equipped throughout with both visual and auditory alarms that would alert the crew if gas enters the rig in dangerous proportions. It also came equipped with control panels that automatically shut down operations in specific areas if gas is detected. The rooms in the drilling area were airtight, so if gas made it into one room, it would be isolated and not threaten the entire ship.

BP's internal investigation into the *Deepwater Horizon* explosion reveals that the ship's computer began to report gas approximately forty minutes prior to the explosion while Dewey, Jason, Stephen, and the rest of the drill team were running the negative test and displacement. As the gas entered the rig, they, and the rest of the crew, should have been alerted. They should have heard an alarm. They should have seen lights flashing. They should also have seen control panels automatically shutting down operations in their areas. They neither heard nor saw any such thing. Nor did anyone else on the rig. The automatic gas alarms were intentionally inhibited, set to record information on the computers but not to trigger alarms automatically.

Transocean chief electrician Mike Williams testified that more than a year earlier, he had asked why the *Deepwater Horizon* alarms were inhibited. From the OIM on down, he was told, "They did not want people woke up at three o'clock in the morning due to false alarms." When he tried to fix the alarms, Transocean subsea supervisor Mark Hay told him, "The damn thing's been in bypass for five years. Why did you even mess with it?" Hay said, "Matter of fact, the entire fleet runs them in bypass." Williams explained that Hay meant the entire Transocean fleet.[45]

Because the system was inhibited, the automatic shutdowns did not happen. The drill rooms were not shut in. And those elsewhere in the vessel were left unaware of gas until it was too late and the gas, now fully unleashed into the rig, was winding its way around the vessel like a wraith, heralding impending doom.

Transocean manager Daun Winslow was on his way to the bridge when "there was kind of a warm breeze, a whoosh of kind of high-pressure air went by,"[46] followed by the first explosion. For Douglas Brown, "the room was dark and the hiss of escaping gas was deafening."[47] BP manager David Sims heard "what sounded like a hissing sound. I would describe it as gas escaping something."[48]

Chris Pleasant, a Transocean supervisor, heard the hissing in his office. Over on the *Damon B. Bankston*, the cargo ship that gathered the mud from the *Horizon*, Captain Alwin Landry "heard a high pressure release of air or gas . . . right as the mud was flying," and then the first explosion hit.[49]

The first to die were Gordon Jones, Blair Manuel, and Karl Kleppinger. Just as Keith Jones would suggest that his son wasn't born to be on the rig, neither was he supposed to be in the mud room when the explosion occurred. Gordon was supposed to be asleep in his quarters. He wasn't on for another three hours, but his replacement, Leo Linder, seemed as though he could use more sleep, so Gordon offered to stay on.

Gordon, age twenty-eight, had a two-year-old son and a pregnant wife, Michelle, eagerly waiting for him at home. He was on the *Horizon* for just a one-week tour before heading home for seven weeks to be with Michelle for the birth. But he had "already eaten, he'd said goodnight to Michelle, he had the night shift, and Leo looked tired to him, so he said, 'Look, I've got this' and sent Leo off to bed. That's the last thing I know that Gordon said," Keith recalls.

His father knows the story because a few days later Leo visited the family. "He came to tell Michelle that he knew her baby was going to be a boy because Gordon knew and talked a lot about it." Leo told Michelle, "When your little boys get big enough, you tell them that their daddy's kindness saved my life."

"I know what my fervent prayer is. It's that Gordon never felt it when the explosion happened. But I know who Gordon was and that he had to be well aware of what was happening." Keith is looking away while he tells me this. It is not easy for him to talk about that night.

"He knew when stuff is gushing out of the well head . . ." Keith excuses himself. He's begun to cry. "He knew that the worst is about to happen." He pauses again and looks away. The tears flow down his face. "Gordon knew what a blowout was. Everybody does. I've seen BP's application to the MMS [Mineral Management Service] in which they describe that the worst-case scenario is a blowout. The worst thing that can happen is a blowout. Everyone knows that. Gordon knew that."

Gordon died with Blair Manuel. At fifty-six years old, Blair was one of the oldest men on the rig and the oldest to die. The father of three adult daughters was getting married in July. The hundreds of mourners who attended his memorial services wore a tie or a swath of purple and gold, the colors of Blair's beloved Louisiana State University.[50]

Karl, age thirty-eight, was in the shaker room, where the mud is separated from the liquids. His wife, Tracy, knew something was wrong when Karl, who never spoke about the rig unless he was worried, was talking about it incessantly on his last trip home. "He [was] losing sleep over it, because not only [was] it a 'bad hole,' his drilling team [was] getting pushed to drill it. They [were] way over budget and way behind schedule."[51]

The next in line were likely Shane Roshto, Adam Weise, and Roy Kemp, all of whom were in the pump room. At ages twenty-two, twenty-four, and twenty-seven, respectively, they were the youngest to die.

Five days earlier, on April 15, Shane wrote on his MySpace page: "Chillin out on the rig. . . . Ready to go home but gonna work over on the stack. . . . Missin Nat and Blaine . . ." The accompanying photo is of Blaine, Shane and Nat's three-year-old son, sitting on his father's lap, "driving" a very muddied jacked-up jeep. Shane was scheduled to go home on April 21.

Adam's grandmother told a local reporter, "I don't remember when we actually found out he was one of the ones in the pump room when the explosion happened. They said they believed they did not suffer, that it was almost instant. That's comforting."[52]

Roy was from Jonesville, Louisiana. The father of two young children was remembered as a devoted husband, father, and son with a "unique" sense of humor. He loved hunting with his dog, Ellie, but "most of all," he was "a man who loved the Lord."[53]

As the mud and oil was raining down from the derrick, drenching everyone in a black rain, floor hand Caleb Holloway said to Dan Barron,

"I smell gas. Run." They ran out the door from the drill floor to the deck just as assistant driller Don Clark was running in. He got through the open door and was greeted immediately by the second blast.[54]

Don was forty-eight years old and one of the few black men on the rig. He was a soybean farmer turned municipal worker. His wife, Sheila, never wanted him to go offshore, because he loved being home so much. But she was the woman of his dreams, and he wanted to earn enough money to support her and her two young children.

Don was on his way to help Jason, Stephen, and Dewey when he died.

They had all left the drillers' shacks and were seen on the drill floor "frantically [trying] to stop the natural gas before the rig blew."[55]

The Reverend Clyde Grier, who performed Jason's memorial, told a local newspaper, "Someone from the rig called and said a couple of minutes before the explosion, he had seen Jason on the drilling floor near the drill head. He was basically on ground zero."[56]

The eleventh and final man to die was likely the only man to die alone. Crane operator Aaron Dale Burkeen, a thirty-seven-year-old father of two, was celebrating his eighth wedding anniversary on April 20. He was on deck in his crane 100 feet aboveground when the first explosion happened. Crew member Daniel Barron saw Burkeen struggle to exit the crane's cabin. In the light of the flames, he watched as Burkeen made it out of the cabin and halfway down the stairs. Then the second explosion hit. It "literally picked him up . . . like a child would throw a toy" and slammed him into the deck fifty feet below.[57]

Chris Choy, a twenty-three-year-old roustabout, saw the fire engulf Burkeen. He tried to reach him, but a fireball erupted and blocked his path. "It just killed me that I knew I couldn't get to him. That's probably the hardest thing I've ever had to do in my life—and the hardest decision I've ever had to make in my life was I either can go over there and, you know, I might not make it back. That's, by far, the hardest decision I've made in my life, that we had to leave that and leave him there."[58]

Burkeen's supervisor, Bill Johnson, was already in the water in a lifeboat and "watched helplessly as the whole starboard side of the rig erupted in a cloud of smoke and flame. Burkeen just vanished."[59]

In February 2009, BP had submitted its fifty-two-page *Initial Exploration Plan* for the Macondo well to the U.S. Department of the Interior's

Mineral Management Service (MMS). The "worst-case scenario" was a blowout of 162,000 barrels of oil per day. BP certified that were the worst to happen, the company had the capability to respond to such a blowout "to the maximum extent practicable."[60]

Professor Robert Bea of the *Deepwater Horizon* Study Group is seventy-three years old and bald with a sharp white beard. His voice was ravaged by years spent in New Orleans investigating the engineering failures associated with Hurricane Katrina's devastation. When he recounts for me the testimonials of men who survived the rig floor at the time of the explosions, he sounds like the narrator of a 1940s horror movie.

"Each man described this strange transparent fluid moving across the drill deck. That fluid was gas. When the gas reached the mud pits . . . they were the first to die, that's where the first explosion happened. Now the gas continued to ignite back to the drill floor. The second explosion was the drill floor, that's where those people got incinerated instantly. . . . It was that damn gas that got loose. That's the killer. Once that fire happened and then they couldn't get loose from the well, that's when the *Deepwater Horizon* becomes toast."

Chaos

The two explosions, just seconds apart, rocked the *Deepwater Horizon*. The first explosion hit at 9:49 p.m., followed ten seconds later by the second one.

The explosions should not have happened, nor should they have destroyed the rig once they did occur. First, operations in the mud pit should have been automatically halted when the gas entered the area.

Second, the engine that sparked the second explosion should have automatically shut down. The second explosion is believed to have occurred when gas entered engine 3, causing it to overspeed, which most likely created a spark that led to an explosion. Engine 3 was located immediately below the living quarters where at least half of the crew was preparing for bed. The engine should, however, have shut off automatically as soon as the gas was detected. It should also have shut off automatically once it began to overspeed.[61]

Third, the blowout preventer should have prevented the blowout. Fourth, the emergency disconnect should have separated the rig from the well. Finally, the automatic mode function should have had fresh batteries.

In a thick Louisiana accent, Chad Murray, chief electrician aboard the *Deepwater Horizon*, described the fear and chaos that ensued as the explosions rocked the rig. "It's—it's absolutely a disaster. . . . I mean, we're talkin' 'bout alotta noise, alotta fire, and a tremendous amount a'heat, people panickin', and the fear a'losin' your life. At that time, I mean, you do what you can to survive. It's just a bad deal."[62]

Jimmy Harrell, the rig's OIM, described the scene on deck. "It was very chaotic. A lot of people were jumping over the side during this explosion."[63]

In an interview with the *Wall Street Journal*, crew member Carlos Ramos said, "People were in a state of panic. Flames were shooting out of the well hole to a height of 250 feet or more. Debris was falling. One crane boom on the rig melted from the heat and folded over. Injured workers were scattered around the deck. Others were yelling that the rig was going to blow up."[64]

Oleander Benton was one of just six women on board the *Horizon*. A cook, she was part of the catering crew. April 20 was her fifty-second birthday. She and another woman were in the laundry when the explosions occurred. The Associated Press reported that she "hit the floor as ceiling tiles and light fixtures came crashing down on her head and back. The concussion had blown a door off its hinges and pinned her friend to the floor. 'My leg! My leg!' the woman screamed." Benton couldn't move the door. "She told her friend to lie flat and slide herself out, and the two made their way into the darkened hallway." A man yelled to them, "Come on, Miss O! Go this way. This is the real deal! This is the *real deal!*" They made their way in the dark clogged halls past "dazed and injured people" and emerged safely on the deck.[65]

Steve Bertone, a Transocean chief engineer, had just finished smoking a cigarette and taking a shower. He was getting into bed to read a book when what sounded like a freight train coming through his room jolted him to attention. "There was a thumping sound that consecutively got much faster and with each thump, I felt the rig actually shake," he later testified. "Then there was an initial boom. The lights went out. I jumped out of the bed, ran to my door. . . . When I turned to go grab my clothing, the second explosion occurred, which threw me across my room."

Bertone made it out to a stairway only to find a group of people standing there, frozen; the stairway was blocked with debris. "I hollered out to head to the port forward or starboard forward spiral staircase and go to your emergency stations. I ran to the port spiral staircase and made my way to the bridge."[66]

The second explosion buried Randy Ezell, Transocean senior tool-pusher, under a mountain of debris. "I tried two different times to get up, but whatever it was, it was a substantial weight. The third time it was something like adrenaline had kicked in and I told myself 'Either you get up or you're going to lay here and die.'"

As he recounted in testimony, Ezell tried to stand up but ended with a face full of smoke. On hands and knees he crawled in the direction of air. "The living quarters was pretty well demolished. Debris everywhere. But I made it to the doorway and what I thought was air was actually methane and I could actually feel like droplets. It was moist on the side of my face." He kept crawling till he put his hand on the body of an injured Wyman Wheeler. He heard another voice calling out in the dark, "God help me. Somebody please help me." He looked to where the maintenance office had been and saw "a pair of feet sticking out from underneath a bunch of wreckage and debris." It was Buddy Trahan, "one of the visiting Transocean dignitaries."[67]

Trahan had left the bridge after his turn at the video game simulator. The initial explosion had hurled him thirty feet through a wall, burning most of the clothing off his back in an instant. He regained conscious-ness just in time to get Ezell's attention. Ezell not only dug Trahan out from beneath the rubble but also yanked away a steel door whose hinge was stuck in Trahan's neck a half-inch from his carotid artery.[68]

Wheeler and Trahan were put onto stretchers, but getting them through the living quarters was treacherous; there was "debris hanging from the ceiling, the wall was jutted out, the floor was jutted up," Ezell recounted. "I mean it was just total chaos."[69]

The first explosion threw chief mechanic Douglas Brown up against the control panel. A hole opened up in the floor beneath him, and down he fell. "I was confused. I was hurting. I was dazed, and I proceeded to try to get up and the second explosion happened. And I ended up falling back down in the hole and the ceiling caved in on top of me. . . . I started hearing people screaming and calling for help, that they were hurt, they needed to get out of here."[70]

Mayday

Confusion reigned on the bridge, where most of the people seemed to be the last on board to realize the full extent of the disaster. BP's Pat O'Bryan was running the video game simulator when the first explosion hit. With him was BP's David Sims. Daun Winslow from Transocean arrived shortly thereafter, following his encounter with the gas in the hallway. With so many executives in the room, the bridge crew might have been distracted, and they might also have been unclear about who was actually in charge.

Steve Bertone arrived on the bridge and went to his station only to discover that there was no more power on the rig. "There were no engines, no thrusters, no power whatsoever," he recalled. He tried calling the engine room. "There was no dial tone whatsoever. . . . I hollered out, 'We have no coms!'" With the communication system down, he ran to the starboard window of the bridge and looked back to the derrick.

He later recounted that "prior to this, for whatever reason, the second explosion and everything had not registered with me. . . . I was fully expecting to see steel and pipe and everything on the rig floor. When I looked out the window, I saw fire from derrick leg to derrick leg and as high as I could see. At that point, I realized that we had just had a blowout."

Bertone heard a voice behind him saying that the engine room, the engine control room, and the pump room were gone. "They are all gone." He turned around but didn't recognize Mike Williams, whose face was covered in blood from a deep head wound. Bertone asked, "What do you mean gone?" The voice behind the blood responded, "They've blown up. They're all gone. They've blown up."[71]

There was at least one person on the bridge who was paying attention: twenty-three-year-old dynamic positioning officer Andrea Fleytas. It was Fleytas, one of just three women on the crew, who manually set off the ship's general alarm at 9:47 p.m. when the automatic alarms failed to signal that combustible gas had entered parts of the rig where crews were working. It was Fleytas who, at approximately 9:53 p.m., four minutes after the explosions—with power out, communication out, engines down, fires throughout the rig, and men throwing themselves overboard—noticed that no one had sent a distress signal to the outside world. It was she who activated the distress button—"grabbed the radio and began calling over a signal monitored by the Coast Guard and other vessels. 'Mayday, Mayday. This is *Deepwater Horizon*. We have an uncontrollable fire.'"

When Captain Kuchta realized what Fleytas had done, the *Wall Street Journal* reported, "he reprimanded her. 'I didn't give you authority to do that,' he said, according to Ms. Fleytas, who says she responded: 'I'm sorry.'"[72] In a statement, Steven Bertone wrote that the captain's response was even stronger, that "the Captain was screaming at Andrea for pushing the distress button."[73]

Disconnect

Meanwhile, Bertone realized that no one had activated the emergency disconnect system (EDS). The EDS triggers the blowout preventer (BOP) and separates the rig from the wellhead. Bertone immediately yelled to Chris Pleasant, the subsea supervisor, "Have you EDSed?!" Pleasant said he needed permission to do so. Bertone turned to see Transocean's Daun Winslow standing right next to him. "Can we EDS?" Winslow said yes, and Bertone turned back to Pleasant when "somebody on the bridge hollered out, 'He cannot EDS without the OIM's approval!'"

Bertone then spun around and saw the OIM, Jimmy Harrell, running across the bridge. "I hollered out for Jimmy, 'Can we EDS?!' 'Yes, EDS, EDS!' Harrell replied. "When I turned back to Chris," Bertone continued, "he was in the panel pushing a button. I hollered to Chris, 'I need confirmation that we have EDSed!' He said, 'Yes, we've EDSed!'"[74]

Records indicate that Pleasant activated the EDS at 9:56 p.m.

Nothing happened.

•   •   •

Federal regulations require BOPs to be recertified every five years. The *Deepwater Horizon* BOP had been in use for nearly ten years and had never been recertified. Getting it recertified would have required Transocean to take the rig out of use for months while the four-story stack was disassembled.[75]

There are two control panels on the rig to operate the BOP, one on the rig floor close to the driller and the toolpusher and one on the bridge close to the OIM and the master captain. These rig control panels link to two control pods on the BOP, called the yellow pod and the blue pod. These pods respond to signals by activating hydraulic valves that channel hydraulic pressure to open and close the BOP devices. The BOP has the capacity to store enough pressure to fully close all valves one time in the event that pressure is lost from the riser.[76]

As noted earlier, BP's records indicate that at 9:41 p.m. the BOP was activated, most likely by the drill team. Nothing happened.

At 9:56 p.m., the bridge again tried to activate the BOP, this time using the EDS. Again nothing happened. The well continued to blow, and the *Deepwater Horizon* remained tethered to the well by the riser pipe.

There were several problems with the BOP that were well known on the rig and that had been reported in the BP Daily Operations Reports as early as March 10. Both BP and Transocean officials knew that the yellow control pod had a hydraulic leak. They also knew that federal regulations required that if "a BOP control station or pod . . . does not function properly," the rig must "suspend further drilling operations" until it's fixed. When asked if regulations were followed and operations were suspended, Ronnie Sepulvado, the BP company man on the *Horizon* until April 16, answered, "Well, no, it wasn't. And the reason it wasn't, I guess we assumed that everything was okay since I reported it to the team leader and he should have reported it to the MMS."[77] The team leader was BP's John Guide, who admitted to knowing about the problems but failing to report them.[78]

Chief electronics technician Mike Williams also reported that the BOP was damaged just four weeks before the explosion. The annular is a key component at the top of the BOP. It is one of the devices that closes off the drill pipe in the case of a blowout. While testing the annular, Williams reported, a crewman on deck accidentally nudged a joystick, applying hundreds of thousands of pounds of force and moving fifteen feet of drill pipe through the closed BOP. Later, while he was monitoring drilling fluid rising to the top, the man "discovered chunks of rubber in the drilling fluid. He thought it was important enough to gather this double handful of chunks of rubber and bring them into the driller shack." Williams asked the supervisor if this was out of the ordinary and was told, "Oh, it's no big deal." He recalls thinking, "How can it be not a big deal? There's chunks of our seal now missing."[79]

When the BOP failed to activate from the floor and from the bridge, there should have been one more backup, the automatic mode function (AMF), but it failed, too. The BOP has a built-in battery-powered AMF device, commonly called the "deadman switch," that would activate the blind shear rams that close in the well if both the hydraulic pressure and the electrical communication with the rig were lost.

According to BP's internal investigation, "insufficient charge was discovered on the 27-volt AMF battery bank in the blue pod, and a failed solenoid valve 103 was discovered in the yellow pod."[80] In other words, the batteries had been allowed to run down.

All across the rig, the technology on which everything on the *Deepwater Horizon* so dearly depended was failing, and with catastrophic results.

Escape

In the dark of night and fifty miles from shore, the *Deepwater Horizon* was inextricably locked to an exploding oil well. Gas and fuel met to ignite new explosions that kept rocking the ship like the aftershocks of two giant earthquakes. Without power, the water pumps would not function, and the crew was without means to fight the spreading fires. As the crew crawled out of collapsed living quarters and ran toward life rafts, chaos prevailed.

Some, but not all, of the chaos was preventable. The rig was equipped with two main lifeboats, each with a capacity of seventy-five people, and two smaller life rafts, with a capacity of about twenty-five. Although the crew had been required to practice mustering to lifeboats, the drills were so routine that they were nearly useless. The drills were conducted at the same time and on the same day every week: Sunday between 10 a.m.

and noon. People at work were excused from the drills, which did not include lowering the lifeboats into the water. There were never any surprise drills, nor were any drills conducted under distress scenarios.

On the evening of April 20, the lack of adequate drills was nearly catastrophic. Rather than wait for lifeboats, ten people threw themselves from the rig—an estimated 100-foot drop. Meanwhile, one lifeboat was dangerously overcrowded when lowered while the other was not even full. One life raft was never deployed, whereas the other was lowered while tethered to the ship, leaving it dangling at a ninety-degree angle, attached to the burning rig. Moreover, according to the testimony of several on board, the captain abandoned the ship before all of the crew was off.[81]

"There was no chain of command. Nobody in charge," Carlos Ramos said. "People were just coming out of nowhere and just trying to get on the lifeboats," said Darin Rupinski, one of the operators of the rig's positioning system. "One guy was actually hanging off the railing. . . . People were saying that we needed to get out of there."[82]

Douglas Brown went to the lifeboats and was waiting to receive orders. "It was just complete mayhem, chaos, people were scared, they were crying. I heard later that people were jumping overboard."[83]

Crane operator Micah Sandell described pure pandemonium. "A lot of screaming, just a lot of screaming, a lot of hollering, a lot of scared people, including me." While he was trying to get people on the boats, others were yelling, "Drop the boat, drop the boat!" even though it was not yet full. "We was still trying to get people on the boat and trying to calm them down enough to—trying to calm them down enough to get everybody on the boat. And there was people jumping off the side. . . . We were trying to get people to count '1, 2, 3' . . . people couldn't even count right because they was too scared."[84]

Greg Meche, a compliance specialist with M-I SWACO, is one of those who jumped. He had an understandable difficulty trying to explain the decision to an investigatory committee. He cited the chaos and slow speed of the mustering. He also expressed his pure fear and quick, five-minute decision to jump rather than wait for a lifeboat.[85]

Roustabout Chris Choy made it into a lifeboat, but "there he found chaos. Men, including some with broken bones, open wounds and burning flesh pushed onto the lifeboat." The lifeboat, an enclosed fiberglass capsule, "was supposed to save them but now it was looking like a hot and smoky tomb. Choy had seen other men jump off the rig. . . . They had jumped to get away from the heat of the flames. Now, Choy thought about jumping too. 'And there was a couple people, you know, yelling, trying to get everybody to calm down and just, you know, stay as calm as we could so everybody could hear, you know, what we needed to do. Some people just refused to do it and they just kept screaming, you know: The derricks fixing to fall. Hollering, cussing: We've got to get out of here.'"[86]

The visiting executives—Patrick O'Bryan, Daun Winslow, and David Sims—made it safely off the rig on lifeboat number 2.

Transocean executive Buddy Trahan recounted his escape to a reporter. Rig workers Stan Cardin and Chad Murray had strapped him to a stretcher and put him into a lifeboat. He suffered twelve broken bones and was bleeding heavily from second-degree burns and gashes, including a nine-inch, bone-deep slash across his left thigh and a fist-sized hole in his neck. "While I was laying on the stretcher," Trahan recounted, "I could see the rig floor engulfed in flames, and that's when I knew we lost everybody on the rig floor, including three who were dear to me."[87]

After the two lifeboats deployed, about ten people were likely left on the rig. The last to leave the *Deepwater Horizon*, these were the crew from the bridge and a few others. But not all were lucky enough to leave on a raft. By this time the fire had engulfed much of the rig. The heat was unbearable, and smoke combined with the dark night sky made sight almost impossible. Because of the turmoil surrounding the escape, there is conflicting testimony of exactly how the events unfolded.

Steve Bertone, who made it into the life raft, described the terrifying scene. "There was a lot of explosions still going on, smaller explosions, and immense heat. All the flames and heat from the rig floor were coming down the forward part of that deck, as well as all of the flames and the heat from under the rig. They were meeting, I guess, in like a vortex or something right there at the life raft."[88]

Randy Ezell made it to the bow of the *Horizon* with the injured Wyman Wheeler on a stretcher just in time. The first thing he noticed was that both of the main lifeboats had already been deployed and were gone. He looked to his left and saw Captain Kuchta "and a few of his marine crew deploying a life raft."[89] Ezell made it to the raft, got Wheeler on, and the raft set off with approximately six people on board. At this point, there is disagreement over who cut the cord that lowered the raft into the water. Some say it was Captain Kuchta, others chief mate Dave Young. What we do know is that there were still people on the rig when this, the last raft, was lowered.

In an interview with *60 Minutes*, Mike Williams recounted standing on the burning rig, watching the final raft leave, and realizing that there were no choices left but to jump. "I remember looking at Andrea and seeing that look in her eyes. She had quit. She had given up. I remember her saying, 'I'm scared.' And I said, 'It's okay to be scared. I'm scared too.'" Andrea asked what they should do. "We're gonna burn up. Or we're gonna jump," Williams replied. He closed his eyes, said a prayer, and asked God to tell his wife and his little girl "that Daddy did everything he could and if, if I survive this, it's for a reason. I made those three steps, and I pushed off the end of the rig."[90]

It appears that Andrea Fleytas jumped at Williams's encouragement, but she jumped into the descending raft rather than into the water.[91]

When the raft hit the water, Bertone jumped out and swam beside it; with the help of others, he dragged it out of the way of the burning rig. He looked up into a "tremendous amount of smoke bellowing out from under the rig" just in time to see "a person's boots and his clothing and stuff come shooting through the smoke." It was Captain Kuchta. Within seconds, another pair of boots came flying out of the smoke. This time it was Yancy Keplinger.

Finally, Bertone looked up and saw a man running full speed across the deck. "When he jumped off the end of the [deck], he was still running. Just before he splashed into the water, he was actually looking over at us and that was Mike Williams."[92]

Williams crashed down into an ocean on fire. Burning fuel from the rig mixed with the gas and the heat was unbearable. Williams thought, "What have you done?! The fire's gonna come across the water, and you're gonna burn up!" So he swam as hard and as fast as he could until a hand reached down and scooped him into a life raft.[93]

Rescue

By sheer luck, the *Damon B. Bankston* was tethered by a hose to the *Deepwater Horizon* that day to collect mud for reuse on another BP well. Had it not been there, it is all but certain that there would have been far fewer survivors.

The 260-foot-long *Bankston* is a multipurpose supply vessel that looks like a barge. It provided a range of services to the *Horizon*, such as running supplies and cargo, and exported mud from the rig's drilling operations.

As the mud from the blown-out well rained down on his ship, Captain Alwin Landry ordered the hose untethered and the *Bankston* a safe distance away. As soon as the explosions hit, he began a search and recovery. Over the course of about one hour and twenty minutes, all surviving crew members of the *Horizon* were scooped out of the water and from lifeboats and life rafts onto the *Bankston*.

The *Bankston* was aided by another lucky stroke: four guys on a fishing trip. Marine biology student Albert Andry III and three high school buddies had come to the *Deepwater Horizon* for "a couple leisurely days of tuna fishing and beer drinking" in Andry's twenty-six-foot catamaran, the *Endorfin*. The men were fishing for bait under the lip of the platform when water began raining down from the rig's network of pipes. Luckily, one of the men, Wes Bourg, had worked offshore. "Go, go, go, go, go-o—o!" he shouted.

"With no radar and only the light of a crescent moon to see by, Andry pointed the bow north, gunned the twin 140-horsepower Suzuki outboards and hit the deck. They were about 100 yards from the *Deepwater Horizon* when the lights went out, and the first explosion hit."[94]

They stayed until they were nearly out of fuel, four hours later, pulling survivors out of the water and searching for the missing.

The Missing

On April 20, the heroic efforts of the brave crews on the *Deepwater Horizon* and the *Damon B. Bankston*, the fishermen on the *Endorfin*, and many others enabled 115 people to escape the *Deepwater Horizon*. The

Keith Jones in October 2010, whose son,
Gordon, died aboard the *Deepwater Horizon*.

U.S. Coast Guard arrived at approximately 11:30 p.m., just as the last of the survivors had been brought safely aboard the *Bankston*. By early the next morning, images of the burning vessel began to appear on television sets, and the families began receiving phone calls from their loved ones' employers.

When the Jones family learned of the explosions, they gathered at the home of Gordon and his wife, Michelle. There were rumors about a lifeboat that the coast guard had seen but lost sight of, and some were holding out hope that Gordon was on this craft. Keith didn't say it at the time, but he told me later that he knew it was nonsense.

"Gordon told me about those lifeboats. They have not just food and water, they have telephones. They have GPS. The coast guard doesn't lose track of those boats. I knew, I accepted before anyone else did, that Gordon was gone."

Sheri Revette listened politely to the people from Transocean and the coast guard who called to tell her to hold out hope. Dewey's parents were hoping, and so was Sheri's family. But she knew Dewey was dead.

Sheri Revette in September 2010, whose husband, Dewey, died aboard the *Deepwater Horizon*, sits with the commemorative bronze helmet Transocean gave her upon her husband's death.

The coast guard searched for the missing men for nearly seventy hours. With none of the bodies recovered, the hunt was officially ended at 5 p.m., April 23, 2010.

Several months later, I sit with Keith in Baton Rouge. It's our fifth interview and my second visit to his office. This time I notice the piles of paper stacked up in the corners, on the desk, and on the bookshelves. I notice how tired Keith is. We then talk about anger.

"I know I should feel anger," Keith tells me, "but I don't." When I ask why he should, he suddenly turns toward me, and he is angry.

"They murdered my boy, Antonia. Greed murdered my boy."

2

What You Can't See Can Kill You

The BP Macondo Oil Monster Escapes

A t 9:50 p.m. on April 20, 2010, as methane gas ripped a hole into the heart of the *Deepwater Horizon*, Professor Samantha Joye was 700 miles away in Athens, Georgia. Hard at work in her home office tucked away in the wooded hills near the University of Georgia, Dr. Joye was enjoying a typical Tuesday night. She was completing a manuscript on the fate of methane derived from gas hydrate dissolution, preparing for the upcoming release in *Nature Geoscience* of a paper describing the abiotic production of nitrous oxide via chemo-denitrification, and finalizing the proofs for a third paper in *Deep Sea Research* on deep Gulf of Mexico sediment microbial processes.

Dr. Joye, age forty-five, has been described as "a librarian version of Angelina Jolie" by CNN. Throw in a strong rural North Carolina accent and a constant state of sleep deprivation, and the comparison works. Dr. Joye's husband, Christof, certainly likes it. "I mean, that makes me Brad Pitt, right?" he jokes.

Dr. Samantha Joye on the deck of the *Oceanus* research vessel near the Macondo well in the Gulf of Mexico in September 2010.

Other than her obvious brains and clear tendency toward nerdiness, however, Dr. Joye, who prefers to be called Mandy, fits not a single librarian stereotype. She bears more of a resemblance to the superheroines that Jolie portrays in her films: she rides horses and does triathlons, travels to the bottom of the sea in tiny submarines, seems to be perpetually bruised or broken from her latest daredevil move on a ship, and suffers neither fools nor bullies lightly.

Think of a Mrs. Frizzle version of Angelina Jolie—the science teacher in the children's cartoon *The Magic School Bus*, who bravely takes her students on fantastic field trips from outer space to the inside of the human body, teaching important lessons all the while.

On April 20, Dr. Joye's colleagues from the University of Mississippi were sampling Gulf waters and sediments at a site less than ten miles from the Macondo well on board the *Pelican* research vessel on a cruise funded by the National Institute for Undersea Science and Technology. The next day, Dr. Joye received word that they could see a plume of smoke suggestive of a rig explosion but facts were scarce. Shortly thereafter, they spotted an oil slick off the side of their vessel. Dr. Joye got online to hunt for information, finally finding images of the burning rig on the *Sabine Times* Web site.

Within days, everyone across the country saw the same images. The photos were taken by the coast guard, provided to media outlets, and quickly seared into our collective psyche. Against a clear blue sky and on still blue waters, the *Deepwater Horizon* erupted in flames. A bright yellow fireball raged at the center of the rig, dwarfing the handful of vessels surrounding it that were trying to put out the fire. Their water cannons looked like squirt guns against the eruption of black smoke that engulfed the rig and then reached as far into the heavens as the images would allow us to see.

Quickly identified as a BP oil rig, the images went global. Even for people overly accustomed to seeing things explode on television, it was a shocking sight. Perhaps it was the isolation of the event: no cityscape, no people that we could see, and, in most cases, no sound. The comparisons most commonly made were to outer space, including the explosion of the space shuttle *Challenger*. But this did not suffice, as comparatively the *Challenger* looked like a small clean white pinprick against the blue sky. Comparisons to the destruction of the spacecraft in the movie *Avatar*, which was still in the midst of its historic box office run, were readily made. Of course, these explosions were fanciful, whereas the explosion of the *Horizon* was all too real.

When she saw the images, it was more than the knot in her stomach that told Dr. Joye that the trouble had only just begun. Dr. Joye is a biogeochemist oceanographer in the Department of Marine Sciences at the University of Georgia. As she explains, she is an expert in "the dynamics of chemosynthetic habitats" and "the cycling of carbon, nutrients, and bioactive metals and microbial ecology, metabolism, and physiology in Gulf of Mexico waters and sediments." In other words, she is one of a handful of people in the world who have dedicated their professional careers to understanding the role of oil and gas in the ecosystem of the Gulf of Mexico.

The look of the fire told her the rig would not survive the blaze. The location of the rig told her it was on top of a very rich well. And she knew that any subsequent oil spill would be "very, very serious." The only way to know how serious was to measure the amount of oil and gas that had spilled and to monitor where it went.

Dr. Joye had no idea, however, that her efforts to track the BP Macondo oil and gas monster as it burst forth on a deadly rampage through the Gulf and onto the shores of five states would put her in the middle of an ongoing global battle with BP and the U.S. government over exactly how much oil and gas escaped the "well from hell," where it went, and what further destruction it may yet bring upon us. The knot she felt in her stomach has yet to go away.

Going Home

As the world awoke to the explosion of the *Deepwater Horizon* on April 21, the ordeal for the rig's survivors and their families had only just begun. Shortly before midnight and just hours after the explosion, the survivors gathered on the deck of the *Damon B. Bankston*. Some were only half dressed, having been startled out of bed or even showers when the explosions hit. Many were suffering from gory injuries, seventeen seriously so. All were exhausted, scared, and traumatized.

Before the sun had set, it had been a clear day in the Gulf of Mexico. The colors of the *Deepwater Horizon*—yellow, blue, white, and red—shone brightly as the sun played against blue skies and still waters. As the survivors now looked out to the rig, however, they saw burning oil blazing black on the water's surface, gray steel stripped clean of paint, billowing black smoke, and searing red flames. The many young war veterans in the crew must have been experiencing their own particular hell.

As they gathered, a muster was taken. Some took a roll call. Others walked around to try to identify those who might be unable to hear or be in too much shock to respond. As the names of Gordon Jones, Dewey Revette, Jason Anderson, Shane Roshto, Karl Kleppinger, Adam Weise, Stephen Curtis, Blair Manuel, Roy Kemp, Don Clark, and Aaron

Burkeen were read, silence blanketed the ship. A head count was taken, then another one. Ultimately, the truth set in: eleven men were not there.

The *Bankston* stayed in place beside the burning rig for another eight hours, until 8:15 a.m. April 21. Captain Alwin Landry testified that it was his decision to stay and that he did so to help with the search and rescue efforts, including rumors of an overturned missing life raft.[1]

"That was hard," Christopher Choy recalled later. "We sat there and watched the rig burn. It makes you sick to your stomach, just watching that knowing that you're missing guys and that they're up there somewhere, not knowing if they're alive or dead or if they jumped off and somebody's looking for them in a boat."[2]

The survivors watched as the vessels arrived to fight the fire and search for the missing. Captain Michael Roberts, one of the ship's captains who responded to the Mayday, arrived around 2:30 a.m. on April 21 and saw the blaze emerge "as bright as the sun" over the horizon. The heat from the fire was so intense that it peeled paint, melted electrical equipment, and forced crew members to take shelter. The various vessels had to take turns firefighting, retreating from the scene to cool off before heading back in.[3]

Dennis Dewayne Martinez, age thirty and a supervisor on the *Deepwater Horizon*, later told the *New York Times* that as he stood on the *Bankston* and watched his home for most of the past eight years burn, he thought about a ring he had received from his father. "I lost my daddy when I was 23. He was 46," Martinez recalled. He only removed the ring when he was working. Though he had escaped, his ring had not. Another worker, startled by a memory, jammed his hand into his pocket. He pulled out a small photograph of his son back home. He caught his breath, stared at it, then exhaled."[4]

Worse than the wait was being forbidden to call their families. The crew was under a communications blackout that would last for more than thirty hours: at least, most of them were. While on the *Bankston*, Transocean offshore installation manager Jimmy Harrell was reportedly overhead yelling into a satellite telephone, "Are you f—king happy? Are you f—king happy? The rig's on fire! I told you this was gonna happen!" in a call to Houston.[5]

It is unclear whether the coast guard or Transocean called for the blackout, but the ban applied to all Transocean rigs out in the Gulf that night. Transocean told its employees that the blackout was to quash rumors, particularly those concerning who was alive or dead, before they could be confirmed and the families could be notified.

Others on board that day thought that the blackout was to ensure that the companies could get their stories straight and control the flow of information to the public. Whatever the reason, the pain and concern it caused spread to the families of hundreds of workers on the Gulf coast that night who were cut off from loved ones.

Sitting at a small table in the mini-mart of a Chevron gas station in Raleigh, Mississippi, a twenty-seven-year-old woman with a thick Alabama accent tells me what it was like to wait for that phone call from her husband, a subcontractor on deep offshore rigs. She and her husband have asked for anonymity for fear of jeopardizing his continued employment, so we'll call them Jane and John. As their one-month-old daughter sits contentedly in a rocker on the table, Jane describes waking up at 6 a.m. on April 21 to images of the exploding rig on television and phone calls from worried relatives.

She had wanted her husband to quit "offshore" ever since he started. The job terrifies her, even more than when he was a marine fighting in Fallujah. Even though everyone who lives in their small Louisiana community (they're in Mississippi visiting family) is "either an oil fielder or makes their living by the oil field," Jane maintains her sanity by keeping an "out of sight, out of mind" approach to John's work. She doesn't know any other "offshore wives." She doesn't go to the potlucks or other social events. She doesn't let John talk about the work, and she didn't even know until we were all sitting there how many times John had suffered broken bones on the job.

Even though she knew he was out in the Gulf on April 20 on a Transocean rig, she did not know the name of the vessel. The next morning, part of her did not want to know. Six months pregnant, with their five-year-old son sensing her worry and attaching himself to her hip, Jane tried to avoid both the images and the phone that would not stop ringing. There was only one call she was going to answer, but it would not come.

The call didn't come because John wasn't allowed to make it. Working on a rig about fifty miles away from the site of the disaster, he and the rest of his crew got word of the explosion two hours after it happened. At that time they were also told of the communications blackout: no Internet and no phones.

The Internet ban was the worse of the two. Most rigs come equipped with just two satellite telephones. This, and perhaps the relative youth of the crew, explains why Facebook is their dominant mode of communication with friends and family on shore. The blackout meant no communication other than the wave of rumors coming across the television: "All the workers are saved," "Fifteen are dead," "Eighteen are missing," "A lifeboat is overturned," and on and on throughout the night.

"I wasn't worried, though," John says, as his wife and I listen. Apparently trying to alleviate Jane's fears, he continues, "Rigs catch on fire all the time—that's just one of them things. . . . So, it wasn't until the next day that I fully realized the magnitude of what had happened." The realization came at 7 a.m. on April 22, when the blackout was partly lifted: phones, no Internet.

The more than 130 crew members quickly, but quietly and patiently, got in line to wait to call home. Those who were in special circumstances moved to the front, including those who "were kin to some of the ones that died," John explained, as well as those with pregnant and particularly anxious wives. "I was further back in line," John says, "and somebody saw me and said, 'Go over there! Get your wife on the phone and let her know you're okay!'"

The call was quick. "'Honey, I'm alive, it wasn't my rig. I love you, talk to you later.' Click. Next!"

About twenty-four hours earlier, at 8:15 a.m. on April 21, the crew of the *Deepwater Horizon* was finally heading home, but they were still not permitted to call their family members. "Everybody on the boat kept talking about was, man, I wish they'd just let me call my wife," recalled Christopher Choy. "I wish they would just let me call my wife, so she could know I'm okay."[6]

The most seriously injured members of the crew were transported by helicopter, whereas the rest took the nearly seventeen-hour-long journey back to shore on the *Bankston*, making two stops along the way. The first was a forty-minute stop at the *Ocean Endeavor* deepwater drilling rig. There, four crew members who ran remotely operated vehicles (ROVs) were offloaded onto the *Max Chouest*, a small supply craft, so that they could help with efforts to shut in the well while two medics were brought on board.[7]

Next, under the direction of the coast guard, they arrived three and a half hours later at the *Matterhorn* drilling rig. Investigators from the coast guard and the Interior Department were waiting there to board the *Bankston*, but the crew was forced to wait an extra forty minutes for the lawyers from Tidewater, the corporate parent of Transocean, to arrive.

With all the investigators and lawyers finally on board, the *Bankston* took off for the remaining nine-and-a-half-hour ride home to Port Fourchon, Louisiana. En route, the government investigators questioned some of the crew and had all fill out written statements.

At 1 a.m. on Thursday, April 22, twenty-seven hours since their ordeal began, the crew finally arrived at Port Fourchon. Many had been awake for more than forty hours. All were anxious to connect with the outside world, which was equally eager to connect with them. As they pulled up to shore, they expected to see excited family members and maybe even TV cameras, but instead they saw security guards and rows of toilets. As is standard practice for "incidents when one or more deaths or damage of greater than $100,000" occur, explained coast guard investigator Lieutenant Barbara Wilk, a drug test was administered.[8] Standard practice or not, to crane worker Micah Sandell, "It all felt like a nightmare. . . . And I still wasn't sure if I was awake."[9]

Still without access to their families, the press, or lawyers, the crew was "zipped into private buses," driven to a hotel, and then escorted to the back of the hotel, explained lawyer Steven Gordon. There, before they had a chance to see their family members or even, for many, to call them, they were told to sign a statement by Transocean.

The statement was a form letter. The crew members were to fill in the date, their names and addresses, and where they were at the time the evacuation was ordered. Two sentences at the end read "I was not a witness to the incident requiring the evacuation and have no firsthand or personal knowledge regarding the incident" and "I was not injured as a result of the incident or evacuation." The crew was asked—if they agreed—to initial those statements.

Those initials are now being used against the survivors as they file lawsuits seeking payment for emotional distress and other claims. Gordon said, "When we were hired by one of the survivors, we gave notice to Transocean's lawyers. And the immediate response was 'Wow, we're surprised. Here's a statement that says he's not hurt.'"[10]

Desperate to put the entire ordeal behind them, crewmembers signed the forms, got past the lawyers, and on into the arms of their awaiting family members. The survivors of the *Deepwater Horizon* were home.

Had the *Deepwater Horizon* itself been among the survivors, our story would now largely be over.

The Sinking of the *Deepwater Horizon*

As the survivors were finally reunited with their families, the fight to save the *Deepwater Horizon* entered its third day. It would not take long for it to be painfully obvious that no one involved—not BP, Transocean, or the federal government—was prepared for an oil rig fire or failure on this scale. All were ill equipped, unqualified, and unable to meet their obligations.

As a result of their actions, many people began to question whether it is even possible to save a burning oil rig fifty miles from shore and in 5,000 feet of water. Is it ever possible to stop an oil gusher in time that lies below that rig, when the oil itself is another 13,500 feet below the seafloor? If the largest oil producer and the largest oil rig operators in the Gulf of Mexico couldn't do it, then what companies could?

If the federal government could not be counted on to make them do it, then who could or should? The federal government not only allowed these failures by inadequately regulating their activities, it then lied to mislead the public on the scope of the disaster at hand and quash the outrage that is the only avenue toward meaningful policy change.

The stakes could not have been higher. Oil was released in the blowout and was already escaping from the well. A visible oil sheen was slowly growing on the water's surface—evidence of a lurking menace whose size, origin, and intent were a mystery. Because the *Deepwater Horizon* was unable to disconnect from its riser, the only thing holding the estimated 50 million to 1 billion barrels of oil and nearly half as much gas in the Macondo well was the riser and the rig itself, acting as a sort of giant cork on top of a clogged straw. If the rig sank and the riser broke, the mass of oil and gas would escape.

Three critical activities began simultaneously. Two were attempts to save the rig; the third, the only successful activity, was to shield the public from accurate information about the size of the oil spill. Above water, Good Samaritans sought to put out the blaze. Underwater, BP tried to close the blowout preventer to hold in the oil. On shore, BP and the U.S. government tried to stay one step ahead of the public.

The Fire

It is the responsibility of the rig owner and operator to fight fires on commercial offshore rigs in the United States. The rig leasee bears the ultimate responsibility for what happens on its rig. Transocean (the rig owner and operator) was the immediate, and BP (the leasee) the ultimate, responsible party for the *Deepwater Horizon*. One might assume that the federal government would also have a backup plan to protect federal waters from oil rig fires in case the commercial interests fail to act. As it turned out, none of the parties was prepared to address the catastrophic fire on the *Deepwater Horizon*.

The coast guard's guidelines state that its personnel are not to "actively engage in firefighting except in support of a regular firefighting agency under the supervision of a qualified fire officer." The qualified fire officer responsible for firefighting on the *Deepwater Horizon* was Transocean chief mate David Young. Coast guard representatives have also stressed since the explosion that the coast guard does not bear the responsibility for offshore oil rig firefighting, nor does it have the capacity or the expertise. Rather, coast guard vessels and aircraft focus solely on searching for and rescuing human survivors.[11]

Firefighting was not an option while on the *Deepwater Horizon* that night. When the crew abandoned ship, the rig itself was largely intact, but it was also engulfed in fire, attached to a blown-out burning oil well, without power or fire pumps, and unequipped to fight the fire itself. During the thirty-six hours of firefighting that took place, however, neither Young nor Captain Kuchta took control of the firefighting effort. Nor did anyone else.

One can imagine that having survived an explosion, they might not have been in any shape to guide firefighting efforts. Yet no one else from

Transocean or BP—or anyone else, for that matter—took control. There was no one in charge, no plan, and clearly no preparation. There was no team of Transocean or BP offshore firefighting vessels or personnel trained, ready, and deployed from some other location. There was no adequate, informed, or trained firefighting response. Instead, the crew sent out a call for help.

Four to six Good Samaritans who happened to be in the area arrived on the scene. Without anyone directing their activities, they sprayed salt water on the fire until the rig capsized and sank. Their efforts, though undeniably noble in intent, may have inadvertently contributed to the rig's sinking. This is because the *Deepwater Horizon* stayed afloat with a series of buoyancy chambers—large spaces filled with air and ocean water. It is possible that the thousands of pounds of water poured onto the rig flooded the buoyancy chambers and forced the rig to sink. There is also concern that salt water was the wrong response. Rather, Transocean should have ensured that foam was available and utilized to fight the fire.[12]

The Marine Board of Inquiry investigated the handling of the fire. At a hearing on May 11, coast guard captain Hung Nguyen questioned Kevin Robb, the coast guard watch commander on duty on April 20, asking if there was a certified fire marshal in charge of the firefighting scene, to which Robb responded, "To the best of my knowledge, sir, there was not."[13]

Captain Nguyen is the cochairman of what emerged as the most important federal investigation into the *Deepwater Horizon*: the United States Coast Guard/Bureau of Oceans and Environmental Management Marine Board of Investigation into the Marine Casualty, Explosion, Fire, Pollution, and Sinking of Mobile Offshore Drilling Unit *Deepwater Horizon*, with Loss of Life in the Gulf of Mexico, 21–22 April 2010— otherwise known as the Marine Safety Board Investigation.

A Marine Safety Board Investigation is the highest level of investigation by the coast guard involving death at sea or the loss of a vessel. As the primary regulatory body, the Department of the Interior was the co-convener. It launched the investigation by joint order with the Department of Homeland Security on April 27. Unlike the National Oil Spill Commission, the Marine Safety Board Investigation has subpoena power to force witnesses to appear and comply with information requests.

The eight-man panel, composed of four coast guard officials dressed in full uniform, three Interior Department personnel, and retired U.S. district judge Wayne Anderson, has interrogated nearly a hundred people from each government agency and private company that was involved in the disaster during twenty-one days of hearings in the six months from May 11 to December 9. The investigation is ongoing.

Witnesses, who thought the somewhat nerdish appearance of the lead lawyer, Interior's Jason Matthews, meant that they would have an easy go of it were quickly corrected. In his monotone voice and unflappable, calm bearing, Matthews presented witnesses with often-times shocking evidence of malfeasance and quietly pushed them to get the answers the panel required. That is, with the exception of those who refused to appear and exercised their Fifth Amendment rights against self-incrimination, including BP's well site leaders, Robert Kaluza and Donald Vidrine, and BP drilling engineers Brian Morel and Mark Hafle (the latter refused to testify when he was called for a second appearance).

The majority of the hearings took place in a nondescript stuffy conference room at the Radisson Hotel just down the road from the Louis Armstrong New Orleans International Airport. The attendees were primarily lawyers and a handful of reporters.

A single witness was often forced to spend five hours or more sitting before the panel, offering testimony and answering questions first from the board, then from each of the many lawyers representing both the companies and the individuals under investigation.

In this stale setting, Captain Nguyen stood out for many reasons. One was the juxtaposition of his thick Vietnamese accent against the preponderance of deep Louisiana drawls of those he was interrogating, and their difficulty understanding what he had to say. The other was his unwillingness to accept unsatisfactory answers.

It did not take much to get Captain Nguyen riled. He became especially worked up when confronted by what became increasingly clear was his view that, opposed to the strict guidelines that control the offshore operations with which he became accustomed in his quarter century with the coast guard, when it comes to offshore drilling, anarchy rules. He found that the oil industry's self-regulation consisted of a lax and unruly oversight in its operations. Captain Nguyen was also particularly frustrated when the coast guard was a party to these inadequacies. On this day, his exasperation was piqued by the disorder that reigned over the *Deepwater Horizon* firefighting effort.

"The purpose of this investigation is to obtain information to prevent or reduce recurrence of such an incident," Captain Nguyen told Commander Robb. "So what we're looking at here is maybe if there's no coordination out there, no direction out there, we may be throwing water onto a disabled vessel that may lead to this sinking; is that correct? Is that the potential?"

When Robb indicated that he didn't understand the question, the captain pressed, "Well, if the firefighting efforts are not coordinated and we're putting water onto a disabled vessel, there's the possibility that no coordinated action may result in the sinking of the vessel; is that correct, any vessel?"

"That is exactly correct, Captain," replied Robb.[14]

That no one was prepared to address this kind of fire is clear. In the case of the *Deepwater Horizon*, however, it is possible that there was no firefighting effort that could have saved the rig. This was Dr. Joye's initial instinct on seeing pictures of the fire, and it is the belief of many other experts in the field.

"I do not believe anyone thought they could put the fire out with foam or water—it was too big and too hot," said Paul Bommer, a twenty-five-year veteran of the oil industry and now a senior lecturer at the University of Texas. "Without putting the fire out—which was impossible—there was no way to save this vessel." But, he added, that is true only as long as the rig was connected to the riser and the well.

"If the emergency disconnect had worked . . . the fire would have gone out," possibly saving the rig. Saving the rig would most likely have stopped the subsequent oil spill. "It is sure that for as long as the vessel was floating the spill was greatly minimized; it is possible that it could still be floating and there would have been only minimal pollution."[15]

The Blowout Preventer and the Oil Leak

While the fire raged above, BP worked underwater in a feverish attempt to close the blowout preventer (BOP) and shut in the well. From the moment of the blowout, and for several months afterward, BP, not the federal government, had total control of wellhead operations. Oversight of these operations took place at BP's U.S. corporate headquarters in Houston. At the wellhead, unmanned underwater remotely controlled vehicles (ROVs) operated by people sitting at computer terminals in Houston did the work. BP did not have the necessary equipment immediately available, however, to close in the well, nor did it have accurate information about the BOP. Neither, for that matter, did the federal government. It is possible that BP's attempts to close the BOP, though again clearly meant to save the rig, also contributed to its sinking.

Few above water had any idea what was taking place below. BP kept those operations as far away from the public eye as possible for as long as it could. Even when it did provide information, it did not always live up to expectations.

For example, at nearly two hundred pages, BP's internal investigation into the disaster, released in September 2010, is an imposing-looking document. It is, however, like a book that should have been published as a ten-page paper: with the use of big print, huge spaces between lines, fat borders, and pictures, the contents fill up a book's worth of space. The BP report utilizes all of these tricks, but even more frustrating, it is filled with repetition: material presented first as bullet points, then repackaged as lists, tables, and text in paragraphs. There is useful information in the report, but it could easily have been provided in ten pages.

Two far more useful sources give us a clearer look into the last few hours of the *Deepwater Horizon*: BP's internal handwritten logs, acquired in June by the *New York Times*, and the coast guard's logs, released in June through a Freedom of Information Act request to the nonpartisan research organization the Center for Public Integrity.

There are contradictions between the logs, but two important points are made very clear: BP was stumbling in the dark, losing valuable time, and potentially increasing the size of the spill by trying to repair the irreparably broken BOP, and both BP and the U.S. government knew that oil was leaking from the Macondo well from the time of the blowout and long before they chose to share that information with the public.

As described earlier, the BOP has a built-in battery-powered automatic mode function device, commonly called the deadman switch, that will activate the blind shear rams to close in the well if both hydraulic pressure and the electrical communication with the rig are lost. Experts have testified that among the reasons the deadman switch did not activate are that the BOP was without adequate batteries and was leaking hydraulic fluid.

BP and Transocean knew of the latter problem at the time, but it appears that this information was not utilized by the engineers now working on the BOP. The *New York Times* reported from its review of BP's handwritten logs, "Within the first few days, engineers had already begun to wonder whether a leak of hydraulic fluid had crippled the ram. 'May have had leak & have lost pressure,' one entry reads."[16]

It is possible that the next ten days of efforts to close the BOP were not only fruitless but also potentially catastrophic in helping to trigger the collapse of the well. Efforts to close the BOP began on April 21. That day, the coast guard logs record that at 4 p.m. "the second attempt of the ROV to shut in the well has failed. Additional measures are planned." The logs also note the existing flow of oil from the well, stating the development by Transocean of "a plan to stop the flow/fire using an ROV."[17]

BP's formal investigation also reports ROV operations beginning on April 21, but somewhat later in the day. Regardless of what time they began, it is clear that the information gained by these underwater operations provided the basis for the coast guard's first recorded estimate of the "potential environmental threat," which was made that evening: "700,000 gallons of diesel on board the *Deepwater Horizon* and estimated potential of 8,000 barrels per day of crude oil, if the well were to completely blow out."[18]

Around 2 a.m. on April 22, BP tried to drive the BOP closed, using a robotic submersible equipped with its own hydraulic pump. "But the pump did not have nearly the needed strength," the *New York Times* reported. "It could not pump water fast enough to budge the blades." The paper then noted that BP should have been well warned that the ROV would not have the strength to accomplish this task, because "industry studies had highlighted the problem of submersibles without sufficient strength years earlier." Rather than have the appropriate vessels ready and waiting, BP and Transocean officials were now in a search around the globe for more powerful ones.[19] While they hunted for the necessary equipment, the engineers plotted their next course.

At 2:45 a.m., an underwater robot went down to cut the hydraulic

lines physically, simulating the loss of hydraulic power from the rig. As
the coast guard described the task, "ROV is making its 3rd attempt to
shut in the well by cutting the umbilical cord to activate the BOP ram."
But with insufficient hydraulic pressure and batteries in the BOP, this
too would be useless and potentially catastrophic, for within one hour
and fifteen minutes of this action, two explosions rocked the *Horizon*:
"2 explosions around 3:30–4:00 this morning & rig listing at about 35
degrees," BP's logs reveal. "High risk of sinking."

At 7:30 a.m. they tried again. This time a submersible cut a firing
pin on the blowout preventer, simulating the activation of the emergency
disconnect system (EDS) and the rig's pulling free. "This time, the
blowout preventer shuddered, as if struggling to come back to life.
'L.M.R.P. rocked & settled,' one note says, referring to the top half of
the blowout preventer. But after a few moments, *as oil continued to flow*,
it became clear that this, too, had failed" (emphasis added).[20]

Professor Bea's *Deepwater Horizon* Study Group concluded, "The
failure of the BOP, although foreseen and foreseeable . . . was ostensibly
never contemplated by BP—there were no contingency plans to address
the developing situation as there were no spill contingency plans to deal
with the scope of the disaster."[21]

Three hours later, the *Deepwater Horizon* sank.

A Killer Is Released

At 10:21 a.m. on April 22, 2010, the *Deepwater Horizon* capsized into
the Gulf of Mexico. As it fell, the riser pipe that had tethered it to the
ocean floor like a giant steel umbilical cord finally snapped. The *Horizon*
slowly sank through 5,000 feet of water to rest—beaten, broken, and
defeated—atop the "well from hell." With it were the remains of the
dead, including Stephen Curtis, who had given the well the wicked nickname in the first place.

When the rig sank and the riser came free, the bottom of the pipe
remained lodged like a spear puncturing deep into the well's core. It was
held in place by the BOP in the same way the base of a lightbulb screws
into a lamp socket. Severed in three places, the pipe gave the oil and
gas little choice. Up and out they went.

Of course, both the oil and gas would have been perfectly content to spend the next thirty million or so years below the ocean floor, just as they had for the previous multiple millennia, happily in balance with their earthen and ocean ecosystem. Occasionally, a bit of each would seep out into the water, where Dr. Joye and her colleagues would check them out. Over the course of so many millions of years, however, all affected parties had gotten pretty well accustomed to that. The well had certainly given the companies plenty of warning that it didn't want to be disturbed; that poking at it with a stick wasn't going to end up pretty for anyone. Instead of listening, the companies kept pushing as far as they could go, and government regulators kept letting them do it.

Maybe the fact that the oil and gas never really wanted to come out of the ground is why we all got lucky in the end. Even though it was the largest unintentional oil spill in world history, it could have been far worse. When the rig sank, it snapped the riser pipe, but it did not dislodge it. The result was a gusher strangled by a pipe, which then sprang several leaks, rather than a gusher exploding straight out of the bottom of the ocean. The riser pipe fell and snapped back upon itself, coming to rest on the ocean floor, kinked in two places. Each kink acted as a stop on the flow of the oil and gas out of the well.

The leaks came from the top of the riser—where the oil had been spewing since the time of the blowout—and from holes at each kink. We were also lucky because although the BOP failed to prevent the blowout and sever the pipe, it did close out other avenues through which the oil and gas could have escaped had nothing been plugging that hole. If there is a next time, we may not get so lucky. Even with the blocks slowing its path, with the well punctured, the oil and gas had nowhere to go but up and out.

Ripped out of the earth's core, forced out of the well, they exploded into the Gulf. The oil and gas created a raging black cloud pouring out of the earth with a terrible ferocity spewing forth in a torrent as much as 80,000 barrels of oil strong plus another 30,000 barrels' worth of gas a day.[22] For nearly three months they cut a path through the water, setting forth on a deadly rampage that would take them across the Gulf of Mexico and onto the beaches, wetlands, and shores of four states and then out again into the water's loop current, where they were let loose into the Atlantic Ocean.

It would be well over a month, however, before the public was let in on the secret of the full size of the monster that had been set loose. For many, the monster was not real until it was seen in person or brought into their homes through the release of the "spill cam" video footage on May 25. Until then, the only people who were in on the secret were a small circle of BP and government representatives and an even smaller handful of scientists who weren't buying their lies.

The Cover-Up

Was this a crisis that would be contained—limited to those whose lives and livelihoods were directly tied to the *Deepwater Horizon*? Or would an underwater menace be released, spreading across the waters of the Gulf and harming all those whose lives depended on it? This is the question that held the public's mind from the moment of the explosion. Our ability to answer this question, however, was far more limited than it should have been. The reasons range from innocent misstatements, misreporting, and the inherent confusion of a crisis of this scale to intentional lies, disinformation, and an at times compliant media.

BP's motivation for wanting to limit the release of information about the size of a spill is obvious. The fact and the size of a spill directly affect how much money BP will be forced to pay in restitution.

In 1972, Congress passed the Clean Water Act in response to (among other things) the 1969 blowout of a Unocal oil rig and subsequent oil spill off the coast of Santa Barbara. The act applies penalties for each barrel spilled of both oil and gas, from $1,000 for an accidental spill to $4,300 if the spill is a result of "willful negligence."

In 1990, Congress passed the Oil Pollution Act in response to the *Exxon Valdez* oil spill disaster. In 1989, the supertanker *Exxon Valdez* spilled more than 11 million gallons of crude oil into Prince William Sound, Alaska. Until the *Deepwater Horizon*, it was the largest oil spill in U.S. history. The act stipulates that in the event of an oil spill, the private company is responsible for plugging the well, cleaning up all of the pollution its creates, and compensating all of the victims. Thus, the more oil and gas spilled, the more BP pays.

For the entire length of the disaster, BP stifled the public's ability to measure the size of the monster it had released in what appears to be a crass attempt to limit the size of its ultimate financial payout.

BP knew that its operations on the Macondo well could result in a blowout that could release 162,000 barrels of oil per day. In February 2009, as noted in chapter 1, BP submitted its *Initial Exploration Plan* for the Macondo well to the Interior Department's Mineral Management Service (MMS). The worst-case scenario was a blowout of 162,000 barrels of oil per day. BP certified that were the worst to happen, the company had the capability to respond. In fact, it certified that it had the ability to handle an oil spill nearly double that size—as much as 300,000 barrels of oil per day—its worst-case scenario for any of its Gulf operations.

But, BP argued, it was "unlikely that an accidental surface or sub-surface oil spill would occur from the proposed activities." In the event an accident did take place, the plan asserted that due to the well being forty-eight miles from shore and the response capabilities that would be implemented, "no significant adverse impacts would be expected."[23] A few weeks later, the MMS exempted the operation from the detailed environmental impact study required under the National Environmental Policy Act after it too concluded that a massive oil spill was unlikely.[24]

In the event of an oil spill, while the company is responsible for fixing and cleaning up its mess, the federal government is in charge of overseeing and coordinating the response effort. The Obama administration immediately mobilized the resources at its disposal to address the disaster and stated throughout that it was operating on the worst-case estimates. It maintained its own complex set of motivations, however, for refusing to release these estimates and for misleading the public about the full scope of the disaster. Those motivations will be discussed in chapter 7 but can be most readily summarized as concern over pending legislation in Congress and national elections taking place eight short months later.

Regulation of offshore oil rigs falls to both the U.S. Coast Guard and the Interior Department. Simplifying a bit, the coast guard regulates the aspects of offshore drilling that move around in the ocean (for example, ships), whereas the Interior Department regulates everything to do with the actual oil and gas operations (for example, drills, pipes, and

BOPs). The coast guard also has the authority to investigate deaths and serious injuries at sea.

In the case of an offshore oil spill, the coast guard provides an on-site coordinator to manage all the elements—local, state, federal, and commercial—involved in the response. With eleven men missing at sea and an oil spill in the Gulf, Rear Admiral Mary Landry, commander of the Eighth Coast Guard District, headquartered in New Orleans, was put in charge of the federal government's early response.

Public attention to the disaster built over the course of the event: sparked by the explosion, heightened by the sinking, and captivated by oil hitting the shores and images of oil-soaked birds; then it became an obsession with the spill cam. In contrast, the amount of information released by BP and the federal government about the size and fate of the spill went in the opposite direction: most forthcoming in the first two days and significantly less thereafter, leading to outright disinformation and lies.

On April 21, an oil slick two miles by half a mile was visible in the water around the disaster site. Rear Admiral Landry led a joint press conference in New Orleans with Interior deputy secretary David J. Hayes, BP vice president David Rainey, and Transocean vice president Adrian Rose. In answer to questions posed following the conference, Rear Admiral Landry admitted that oil was already spewing forth from the Macondo well. She said that the coast guard was "estimating 13,000 gallons of crude might be emanating from the source per hour" (the equivalent of approximately 7,500 barrels per day).[25]

The next morning, shortly after the explosion, CNN quoted agency spokeswoman, coast guard petty officer Ashley Butler, essentially repeating the rear admiral's comments, saying that "oil was leaking from the rig at the rate of about 8,000 barrels of crude per day," and that the coast guard was preparing for possible leaks of up to 700,000 gallons of diesel fuel "but could do little to protect the environment until the fire was out." Coast guard senior chief petty officer Mike O'Berry also said of the explosion, "It obviously was a catastrophic event."[26] It is the last time for a long time that any official would again speak so freely about the size of the lurking BP oil monster.

Later that day, a second press conference was held in New Orleans, and a very different set of information was provided. A sheen of crude oil one by five miles was visible, spreading across the Gulf of Mexico's surface where the rig had exploded and sank. At the same time that BP's logs indicated that "oil continued to flow" from the well head, the message delivered for public consumption was that the oil was probably from the explosion, not the well.

CNN reported from the press conference, "Officials do not know whether oil or fuel [is] leaking from the sunken *Deepwater Horizon* rig and the well below." Rear Admiral Landry said that the sheen "probably is residual from the fire and the activity that was going on on this rig before it sank below the surface."[27]

New York Times reporter Campbell Robertson asked Landry, "How that could be—if the well was visibly spewing out hydrocarbons before it sank, and the flow was not stemmed, how could one say with any confidence that the oil on the surface was not fresh?" After being pressed on this point by Robertson and a reporter from Reuters, Landry responded that they did not know what was going on beneath the surface.[28]

On April 23, the wagons were circled and the media were blitzed. BP and the federal government established the *Deepwater Horizon* Joint Information Center in New Orleans to coordinate all media coverage related to the disaster. BP set up an incident command post and held an executive meeting attended by Tony Hayward, BP's chief executive officer, Andy Inglis, BP's chief of exploration and production, and senior management from Transocean, Cameron (which made the blowout preventer), and Wild Well Control (which had been brought in to help in the capping effort). The purpose of the meeting, according to the notes, was to review "what has been done and what the options are ahead."[29]

By 8 a.m., BP and the coast guard had determined privately that 64,000 to 110,000 barrels of oil per day could emanate from the well in the event of a full blowout.[30] That is precisely the range of the amount of oil that is currently known to have been emanating from the well at that time. It would be more than three months, however, before this information would be shared with the public.

In the early hours of April 23, Rear Admiral Landry hit the airwaves, making back-to-back appearances on *The Early Show*, *Good Morning America*, *Today*, and *American Morning*, among other news programs. Landry was perfect for the role. Her credentials were impeccable: a meritorious thirty-year career in the coast guard, the majority of which was in marine safety. Landry had well-honed skills as a spokeswoman, developed in her most recent posting as director of governmental and public affairs at coast guard headquarters in Washington, D.C.

On camera, Landry was both authoritative and comforting. Her gold-plated shoulder boards ostentatiously presenting her rank and a broad chest plate displaying rows of ribbons for the many activities in which she served with honor were juxtaposed with baby blue shirt sleeves, soft white pearl earrings, light pink lipstick, and "Carol Brady" blond hair.

The rear admiral came on the air not to warn the public of an impending threat but rather to herald the good news: the well was not leaking any oil. Zero! Zilch! *Nada!* Landry stressed two points repeatedly. First, there were extensive underwater operations taking place, and second, no oil was leaking from the well. The first point was clearly accurate, whereas the second was simply false.

Fourth months later, the White House director of the Office of Energy and Climate Change Policy, Carol Browner, made the identical round of early morning show appearances with an equally erroneous declaration, announcing that "the vast majority of the oil is gone" when it simply was not.[31]

Rear Admiral Landry's April 23 interview with NBC's *Today* host Matt Lauer was typical of her appearances that day. She explained, "At this point in time we're confident that the sheen we are working with is from the explosion—residual from the explosion. We have air assets, surface assets, and subsurface assets. A remotely operated vehicle with both visual and sonar capability is tracking any oil that might be coming from the well itself beneath the surface. We have not seen any visible signs of crude emanating from the well." Lauer did not push for any more information on the spill, and no more was offered.

It is possible that Landry was trying to get away with wordplay in these interviews. The oil was not *literally* emanating precisely from the wellhead. Rather, it was pouring from the wellhead into the riser, and the leaks were coming from the riser. All the interviewers followed an identical script, but only one journalist asked specifically about the riser.

CBS's *Early Show* host Harry Smith has spent years reporting on the oil industry and has never shied away from a tough question about the industry's practices. Following the same lead as the other reporters, he asked Landry, "What can you tell us, then, about what's going on with the environment?" However, he pushed further, continuing, "Because as this rig then goes under, there's oil coming out of the pipeline. At what rate do we know, and how difficult is it going to be to plug it up?"

Rear Admiral Landry did not mince words when she responded, "At this time there is no crude emanating from that wellhead at the ocean floor. But we have to constantly monitor that." A somewhat incredulous Smith started to interrupt but let her finish and then pressed, "So as best you know, there is no oil coming out of the ocean floor. But what about where the pipe would have broken off from the base of the rig itself?" Without hesitation the rear admiral stated, "There is no oil emanating from the riser, either." Smith responded, "Wow."[32] Wow, indeed.

These and similar interviews took place at dawn in New Orleans. Although it is conceivable that the rear admiral did not have the information about the revised spill estimate at that time, her interviews did not change throughout the day, nor did they change the next day. Rather, Landry went even further, taking on the issue of the riser directly, without being asked.

For example, on the afternoon of April 23, she appeared on WDSU-TV in New Orleans and stressed that both the government and the companies "had eyes" on the riser and that no oil was emanating from it.

"At this point we have a 24/7 visual using a remotely operated vehicle on the wellhead and we are not seeing crude emanating from the wellhead," she stated. "We are also not seeing crude emanating from the riser that is attached to the wellhead, but it's kinked and it's lying on the bottom adjacent to it. . . . We also have sonar on the wellhead and on the riser and on the subsurface. . . . These companies have brought very, very sophisticated tools to give us that capability to maintain that eyes on target. . . . There is no oil spilled other than that small amount we were able to work with, you know, that's a good thing."[33]

The coast guard logs from April 23 mirrored Admiral Landry's public statements that day that there was no oil emanating from the well. They even indicated the possibility that the BOP had worked that morning. However, they made no mention of the riser, and Landry made no mention of the BOP on television. The logs also did not mirror BP's logs—or common sense. As the rear admiral stressed repeatedly, ROVs had been operating at the wellhead for nearly three days. They had sonar imagery

and ROV imagery on the wellhead and on the riser. They had unsuc-
cessfully attempted to close the BOP multiple times. BP's logs asked
matter-of-factly on this day, "Is the well safe?"[34]

Both BP and the coast guard had stated several times in their logs
and in public that they knew oil was flowing from the well and they had
done nothing to stem the leaks. It is certainly possible that Rear Admiral
Landry was given misinformation and that there was simply a great deal
of uncertainty and confusion, but she reflected none of this on television; rather she stated her assertions as clear and uncontested fact. It is
also deeply disturbing that these contentions were not more forcefully
questioned by more of the press.

The disinformation at this point set the stage for months to follow
regarding the amount of oil and gas released into the Gulf and their fate.
As the days progressed and private estimates of the size of the spill
steadily grew, the White House and BP closed ranks on a much rosier,
albeit erroneous, vision of the spill for public consumption.

The National Oil Spill Commission concluded, "Throughout the
first month of the spill, government responders officially adhered to what
we now know were low and inaccurate estimates. Non-governmental
scientists, on the other hand, used the small amount of publicly available flow data to generate estimates that have proven to be much more
accurate."[35]

Put more bluntly in October 2010 by fisherman Clint Guidry of the
Louisiana Shrimp Association, "This has been an exercise in lessening
BP's liability from day one."[36]

The Reveal

If anything good can be said to have come from the BP Gulf oil disaster,
Dr. Samantha Joye contends, it is that "it has given academic scientists
a visible, actual, important role in a process that otherwise people would
not have been able to understand. And I think that's good." Christof, aka
"Brad Pitt," who is also a scientist and a professor of marine science,
nods his head in agreement. While the public and the press were largely
buying a line from BP and the Obama administration that could easily
be summarized as "Jump in, the water's fine!" while sharks circled, Dr.

Joye and her fellow marine scientists were hard at work trying to figure out how much oil and gas were actually released into the Gulf and how much more was yet to come.

From the first days of the explosion, Dr. Joye and her colleagues discussed a plan to get out to the spill site. It is not as easy as you might think to arrange a scientific research cruise fifty miles out into the Gulf of Mexico. There are very few craft available with the technical capacity to serve this purpose. It is incredibly competitive to gain access to them, much less obtain funding to operate them, and they are booked years in advance.

Fortunately, the R/V (research vessel) *Pelican* was already booked for the first two weeks of May by the National Institute for Undersea Science and Technology (NIUST). NIUST is a program of the University of Southern Mississippi and the University of Mississippi in partnership with the National Oceanic and Atmospheric Administration (NOAA). The R/V *Pelican* is a 116-foot research vessel that is owned by Louisiana Universities Marine Consortium. It features four labs with a variety of scientific equipment, and can hold up to sixteen scientists and five crew members.

Dr. Joye and her colleagues, Dr. Ray Highsmith, NIUST executive director; geologist Dr. Arne Diercks, chief scientist on the cruise; engineer Dr. Vernon Asper; and microbiologist Dr. Andreas Teske, "designed a sampling plan to obtain background and early spill sediment and water samples from around the area in about three days!" Dr. Joye is still shocked when she recalls how quickly the events unfolded. "During this time, I was also talking with Ian MacDonald [a biological oceanographer of "deep-ocean extreme communities" at Florida State University], about this because he and I have worked together in the GOM [Gulf of Mexico] for years, and Ian, of course, was similarly worried about the magnitude of the spill and how in the world it would be stopped." The cruise was ready to depart on May 5. It would be the first research cruise designed specifically to investigate the size of the *Deepwater Horizon* oil spill.

While the cruise was being planned to acquire original data, scientists were forced to work with the tidbits of information released to the public. From the beginning, Dr. Joye and her colleagues knew the oil estimates were simply too low to be accurate.

On April 24, the surface oil sheen had grown to a massive twenty miles by twenty miles—eighteen times the size of Manhattan or about 200,000 football fields. Public attention and concern were growing in tandem with the size of the sheen and its proximity to shore. On this day, the coast guard and BP held the first Joint Unified Command (the coast guard, MMS, and BP) press briefing. They were now ready to acknowledge to the public that there was, in fact, fresh oil pouring from the Macondo well rather than just a sheen growing from the initial explosion.

According to Rear Admiral Landry, new information had been gained the day before when ROVs traced the riser and discovered two leaks: one from a kink in the riser above the BOP, called "the kink leak," and the other, a primary leak, from the end of the riser, where it had broken off from the rig. They put out an estimate: up to 1,000 barrels of oil a day were flowing from the two leaks in the riser. Neither the coast guard nor BP divulged the data or methodology behind this estimate. Later investigation revealed that BP came up with the number and, without providing any supporting documentation, handed it off to the federal government, which then reported it to the public.[37]

The scientists were not convinced. "It was clear to everybody who worked on the GOM that this was not a thousand-barrel-a-day oil spill," Dr. Joye tells me. Among the many ways that Dr. Joye resembles Angelina Jolie are her very large eyes and slim face. When she is frustrated, flabbergasted, angry, or just wants to emphasize a point, her eyes widen even more. This happens a lot, but as the months progress, when discussing NOAA, her eyes truly bulge from her face.

"We were looking at that spill, and Ian was estimating that this was a twenty-five thousand barrels a day spill. It was driving Ian insane that they were lying about it. NOAA just believed BP—BP gave them a number and they just believed it. It wasn't until the scientific community started pushing that the numbers began to change."

It turns out that it was clear to BP at the time, as well, but—as BP would do throughout the next three months—it took the lowest spill estimate from a range of amounts. Thus, BP's internal documents show that at least as early as April 27, BP estimated the actual amount of oil spilling from the two leaks (a third leak would be identified later) in the range of 1,063 to 14,226 barrels per day.[38] The number that both BP and the coast guard went public with was the low end of the spectrum, rounded down to 1,000 barrels of oil per day (standard rounding would be 1,100 barrels).

In response to this obviously low estimate, independent experts began releasing their own analyses of the size of the spill that would prove to be far more accurate. The first came on April 27, 2010, when John Amos, a geologist who has worked as a consultant with companies such as BP, ExxonMobil, and Royal Dutch Shell on tracking and measuring oil spills from satellite data, used publicly available satellite images to estimate the leak. His estimate was five to twenty times larger— between 5,000 and 20,000 barrels of oil a day—than BP's and the government's estimate. Amos's estimate gained immediate attention and spread rapidly throughout the media.

Like clockwork, the next day the coast guard released a new estimate, five times larger than the last. The coast guard announced the discovery of the third leak in the riser, and that the spill was 5,000 barrels of oil a day, not 1,000.[39] Although the new estimate was still just a third of the size of BP's own high-end estimate prior to the discovery of the third leak, BP disagreed with the government rate.

At a joint press conference to announce the new figures, Doug Suttles, chief operating officer for BP, pointed "to a diagram that plotted the leaks" and said that "he did not believe the amount of oil spilling into the water was higher than earlier [1,000 barrels] approximations."[40]

This time the coast guard got its numbers not from BP but from an unsolicited, one-page document e-mailed to Admiral Landry's Scientific Support Coordinator by a concerned NOAA scientist.[41] The scientist (who has remained anonymous) knew the estimate that he was offering was very rough, but he nonetheless wanted to warn government officials that the 1,000 barrels of oil a day flow estimate was startlingly low. The coast guard claimed that the revised estimate was based on its new knowledge of the third leak. Quite to the contrary, the 5,000 barrels of oil a day estimate was based on the scientist's visual observation of the speed at which oil was leaking from the end of the riser alone.

The existence of the anonymous scientist was revealed in the findings of the National Oil Spill Commission. On May 21, President Obama signed an executive order establishing the National Commission on the BP *Deepwater Horizon* Oil Spill and Offshore Drilling. The seven-member independent bipartisan commission is cochaired by Democrat senator Bob Graham, who chaired the 9/11 Commission, and Republican William K. Reilly, who under President George H. W. Bush was EPA administrator at the time of the *Exxon Valdez* disaster. The commission's work has been extensive and invaluable, particularly due to its unique access to candid and illuminating interviews with key players within the Obama administration and BP.

In an October working paper, the commission staff concluded that the anonymous scientist's estimate was still many magnitudes too low. This is because it did not take into account the other leaks and his methodology for estimating the velocity of the oil was imprecise. Further, there is no indication that the scientist had expertise in estimating deep-sea flow velocity from video data or that he used an established or peer-reviewed methodology when doing so. "This is not a criticism of the scientist," the commission emphasizes, "who made clear his assumptions and that the 5,000 bbls/day figure was a 'very rough estimate.'" Rather, the commission was criticizing the government for using the erroneous number as its public estimate of the oil flow.

"Despite the acknowledged inaccuracies of the NOAA scientist's estimate, and despite the existence of other and potentially better methodologies for visually assessing flow rate," the commission concluded, "5,000 bbls/day was to remain the government's official flow-rate estimate for a full month, until May 27, 2010."[42]

Meanwhile, BP was still trying to close the ill-fated BOP. On April 26, six days after the explosion, BP sent down tanks of pressurized hydraulic fluid. The goal was to use the ROVs to inject the fluid directly into the ram and kick-start the device—a steel blade designed to cut through the drill pipe—to close in the well. Those involved in the operations were stunned when the blind shear ram's hydraulic system leaked, which meant that pressure could not be maintained on the shearing blades and they would not cut the pipe.

The *New York Times* reported, "This leak shocked engineers because the blowout preventer's hydraulic system was obsessively checked for leaks." The underwater robots tried to find and fix the leak, but by now, leaks were springing up on nearly every component of the BOP. "'Retighten leak,' reads a note from 4 a.m. on April 26. At 4:45: 'Retest & leak still present.' Fifteen minutes later: 'Retighten loose connection.'"

A moment of success is recorded on April 27, when BP managed to repair a leak on the ram and apply 5,000 pounds per square inch of hydraulic pressure on its blades—nearly double the pressure it typically takes to shear pipe. But it was quickly over. "A BP report tersely described the results: 'No indication of movement.'"[43]

A Spill of National Significance

The next few days would prove to be game changers in the history of the spill, bringing a rapid succession of groundbreaking "firsts." The first "first" was most likely the catalyst that triggered all the others when the unthinkable happened: on April 29, oil from the Macondo well made landfall, hitting Birds Foot Delta, a national wildlife refuge in Louisiana, and its critically vital barrier island marshes—the state's last line of defense against hurricanes. The oil oozed up on shore like gooey brown excrement.

Chemical dispersants had begun to be used at the source of the spill on April 22. Any onlooker could see that the material leaching onto the beaches and marshes was thoroughly unnatural: a deadly cocktail of greasy chemicals and far more oil than the waters could withstand. The oil-chemical mix could be seen moving in from the ocean, lapping up on waves and coming to shore like a creature on the prowl and springing to attack. From the first moment it made landfall, the attacker captured prey, and the images were caught on film.

The first images of oil-soaked birds hit the news media on April 30, followed on May 1 by what would become the iconic image of the Gulf oil disaster: brown pelicans buried, suffocating, and suffering in oil.[44] The brown pelican had come off the Endangered Species List only the year before, after having been made nearly extinct from chemical pesticides. The brown pelican is the state bird of Louisiana (known as the Pelican State), and its image appears on the state flag, state seal, and the official state painting. In its breeding season when the oil attacked, the pelican population was at its height. The pelicans tried to take flight to escape, but their wings were coated, their eyes blinded, and their hearing muffled by soaking oil. The collective air went out of the nation's lungs when we saw these pictures.

A brown pelican drowning in BP's Macondo well oil on East Grand Terre Island, Louisiana, in June 2010.

These images were also made unique by how quickly we would be prevented from seeing more of them. Shortly thereafter, BP, with the help of private security contractors, state and local police, and the coast guard, began barring reporters, authors, and the public at large from visiting many of the sites where the oil hit ground.

The rest of the "firsts" came in quick succession on April 29. Louisiana governor Bobby Jindal announced a state of emergency. ABC News reported that BP had made its first public request to the U.S. government "for help cleaning up the mess."[45] In a press release, BP made its first note of the newly revised flow rate, "5,000 barrels per day."[46] Homeland Security secretary Janet Napolitano declared the disaster "a spill of national significance."

Most significant, President Barack Obama addressed the nation for the first time about the disaster on this day. Speaking from the White House Rose Garden at a press conference originally called to honor the 2010 National Teacher of the Year, the president assured the public that "while BP is ultimately responsible for funding the cost of response and cleanup operations, my administration will continue to use every single available resource at our disposal, including potentially the Department of Defense, to address the incident."[47]

BP CEO Tony Hayward responded the next day, saying that the company would take full responsibility for the spill and would pay "all legitimate claims" and the cost of the cleanup.[48]

The federal government's oil spill response is laid out in the National Oil and Hazardous Substances Pollution Contingency Plan. The plan was established by Congress in 1968, and in later years was significantly expanded to address key problems that plagued the *Exxon Valdez* response when, for example, the George H. W. Bush administration declared that government involvement would be "counterproductive."[49] The changes gave the federal government greater authority over the response effort, including coordinating communications and deployment of equipment. The changes also addressed Exxon's underestimation of the potential size of the spill by requiring consideration of a "worst-case discharge scenario."[50] Now emboldened with a law that gives the government more authority and requires both the government and the corporation to consider the largest potential spill, the Obama administration should have been ready for action.

The plan, however, then sets up a complex multilayered quasimilitary command structure but lacks the military's clarity of duties. Much of the response is dependent on the formal designation of the disaster as a "spill of national significance," defined as "a spill which due to its severity, size, location, actual or potential impact on the public health and welfare or the environment, or the necessary response effort, is so complex that it requires extraordinary coordination of federal, state, local, and responsible party resources to contain and clean up the discharge."[51] The BP Macondo well blowout was the first time such a designation had been made.

Because the administration waited nine long days to make this designation, however, the amount of time that people had to mobilize was delayed, and the very steep learning curve that BP, the oil industry, and the federal government had to surmount in order to confront the BP oil monster was intensified.

Because the spill was in coastal waters, the coast guard provided the coordinator positions, including an on-scene coordinator who focused on efforts at the scene of the spill and a federal on-scene coordinator who had authority over the command system, consisting of a Unified Area Command with a regional focus and incident command posts for front-line responders.[52] With the "spill of national significance" determination, the commandant of the coast guard, in this case Admiral Thad Allen, designates a national incident commander to provide national-level support and a National Incident Command Center.

On April 29, Admiral Allen appointed himself national incident commander. Allen was "the Hulk" to Landry's "Carol Brady": a tall, large, imposing man who brought rugged authority to the job. From the outset, however, Admiral Allen and the Obama administration either directly partnered with or, more often than not, completely turned control over to BP throughout the formal command structure.

The organizational charts from the Unified Area Command and the incident command posts show BP employees scattered throughout the command structure, and in some command chains, a BP employee was at the top and a coast guard member would report up to BP. In the Unified Area Command at Houma, Louisiana, for example, most coast guard responder positions had a BP counterpart, and coast guard members and BP employees worked side by side. Doug Suttles, BP's chief operating officer of exploration and production, was at the Unified Area Command at Houma, and the federal on-scene coordinator, coast guard rear admiral James Watson, viewed Suttles as his counterpart.[53]

Moreover, two of the four command posts were operated directly out of BP offices. In the first days of the spill, responders established a Unified Area Command post at Robert, Louisiana, in a BP training facility. Coast guard responders adopted BP Houston headquarters as a formal incident command post.[54]

As described above, press conferences were held jointly between the administration and BP, and at times Transocean, for the first month and a half of the disaster, with the press often failing to distinguish between BP and federal employees. For example, Ron Rybarczyk, BP government and public affairs director, was identified on several occasions in the press as simply "a spokesman for the Joint Information Center in Louisiana," including when he downplayed the seriousness of the oil fast approaching Louisiana.[55] On April 27 he was identified as such by the Voice of America and quoted as saying, "If you can imagine putting a little bit of oil in a cup of water or a bathtub or when we see rainbow type sheen in a parking lot after a rain—it doesn't take much oil to cause a sheen that spreads a long way. So the fact that 97 percent of the surface area of this plume is sheen is a very positive thing."[56] Fortunately, not everyone was easily swayed by such descriptions.

As aerial and satellite images revealed an enormous orange swirling mass of thick oil heading out from the spill site, across the water, and onto the beaches of Louisiana, Dr. Joye's colleague Dr. Ian MacDonald released the second independent flow estimate on May 1. Dr. Mac-Donald's estimate was five times larger than the estimates provided to the public by BP and the federal government, although it was still less than half of what both BP and the government privately calculated. Using a publicly accessible coast guard map that tracked the spill's surface size and classified the color of the surface oil throughout, Dr. MacDonald estimated that 26,500 barrels of oil per day were gushing from the well.[57]

The estimate was shocking not only for its size but also because it gave the Gulf oil disaster yet another key first: it was larger than the *Exxon Valdez* oil spill, making the BP disaster the largest oil spill in U.S. history. Nonetheless, coast guard commander Thad Allen said on May 1, "Any exact estimate [of the flow rate] is probably impossible at this time."[58] In contrast, Dr. MacDonald told the *New York Times*, "The government has a responsibility to get good numbers. If it's beyond their technical capability, the whole world is ready to help them."[59]

His estimates were also the first to use an established scientific protocol—the Bonn Convention—for determining surface oil thickness. "Thus, while estimating volume from surface appearance may be inherently unreliable," concluded the National Oil Spill Commission, "the non-government scientists appeared to make greater efforts to be clear and rigorous in their methodologies, possibly leading their estimates to be closer to the actual flow rate."[60]

Publicly, BP was still indicating its grudging acceptance of the 5,000-barrel-a-day rate, while privately it admitted a potential spill more than ten times greater. In a press release, BP stated, "Accurate estimation of the rate of flow is difficult, but current estimates by [NOAA] suggest some 5,000 barrels (210,000 U.S. gallons) of oil per day are escaping from the well."[61]

At the same time, however, BP admitted privately in a closed-door briefing for members of Congress "that the ruptured oil well in the Gulf of Mexico could conceivably spill as much as 60,000 barrels a day of oil, more than 10 times the estimate of the current flow."[62] Again, this was based on the low-end estimate of BP's own numbers. Recall that on April 23, BP had estimated that the maximum likely rate of oil would be between 64,000 and 110,000 barrels a day. Once again, BP took its own lowest number and rounded that down, when standard rounding would have yielded 65,000 barrels a day.

BP based its newly revised estimate on the information gained from its underwater camera monitoring the flow from the wellhead twenty-four hours a day. Congress would later demand that these "spill cam" images be made public. It would take more than a week for BP to comply.

The National Oil Spill Commission found out that NOAA wanted to release worst-case discharge models to the public at this time but that the White House would not allow it. The commission staff learned that "in late April or early May 2010, NOAA wanted to make public some of its long-term, worst-case discharge models for the *Deepwater Horizon* spill and requested approval to do so from the White House's Office of Management and Budget. Staff was told that the Office of Management and Budget denied NOAA's request."[63]

The *Pelican* Cruise

It is appropriate that the research vessel that created the most controversy over the *Deepwater Horizon* oil disaster was named after the bird that would become the iconic image of the event. The *Pelican* cruise set sail on May 2. Although NOAA might have wanted to go public at this time with its real oil flow projections, it was not interested in conducting its own research or even in backing the research conducted by its funded scientists. The *Pelican* was the first research vessel organized by anyone to measure the size of the oil spill.

"There was no plan in place when this well blew," Dr. Joye explains of BP and the federal government. "There was no plan in place to measure the oil. No plan in place as to where even to start if there was a deepwater blowout. No one in charge of the blowout had a clue what to do to track the oil, to measure it and follow it. It was never part of their plan."

The *New York Times* reported on BP's approach to the size of the spill, "BP has repeatedly said that its highest priority is stopping the leak, not measuring it. 'There's just no way to measure it,' Kent Wells, a BP senior vice president, said in a recent briefing" in May.[64]

It is also somewhat ironic that the *Pelican's* funding came from NOAA, given the agency's subsequent sharp attacks on its research. Although the agency did, of course, accept and authorize the cruise's revised research plan, the funding was essentially a by-product of the agency's existing partnership with NIUST.

Dr. Joye served as a coordinator of the mission but was forced to work out of her lab in Athens, benched by a bad back. As a teenager, she had been thrown from a horse; the resulting persistent injury flares up at times of great stress, and this was certainly a tense moment. Yet even though the stress ultimately landed her in the hospital, Dr. Joye never again let her back keep her from personally participating in a Macondo well cruise.

Dr. Joye grew up poor on a farm in North Carolina. Her father wanted her to be a heart surgeon because that's what he had aspired to be. Two of his older brothers had been killed in World War II, and when his father died, his mother asked him to come home from the war and run the family farm. He bought Dr. Joye her first microscope when she was eight, gave her a copy of *Grey's Anatomy* when she was eleven, and refused to ever respond to a question that she could find the answer to herself. Had he not died when she was seventeen, she is sure that she would be a doctor today. Instead, her life's work was inspired by family time spent on the coast fishing, swimming, and canoeing. After a college professor exposed her to oceanography, her life was never the same.

Dr. Joye now lives and breathes with the Gulf of Mexico. "It is as if the Gulf of Mexico is my best friend. I love this system," she tells me. She is protective of it in the way that most people are protective of small children or animals—innocents often pummeled by forces over which they have no control. When her father died, she became, in many ways, the head of the household. Because of this, and perhaps also because she was regularly harassed as a child for being both "a geek" and poorer than those with whom she went to school, Dr. Joye developed a hard shell and a short fuse for bullies. As the *Pelican* departed, she still had little idea that she would need both attributes to take on BP and the federal government.

We now know that the actual amount of oil escaping the well was three times larger than the estimates made based on the size of the surface sheen alone. One reason the estimates were so far off is that such a large amount of the oil was either chemically dispersed (as discussed in the next chapter) or below the surface of the water.

Dr. Joye knew where to look for BP's oil and the *Pelican* crew found it lurking deep below the surface in giant plumes, like storm clouds building up before a massive cascade of driving rain.

As mentioned earlier, a unique and troubling attribute of the Gulf of Mexico is the large amount of methane that dwells beneath the sea. When the oil and gas exploded out of the riser, the force broke up the oil into small droplets that were then carried in a gaseous oil-water cloud by the methane. These plumes suspended the oil and the gas below the water's surface. They moved with the currents, changing locations and size, and Dr. Joye and her team were about to begin a many-months-long hunt to track, count, and measure them.

During its two weeks at sea, the *Pelican* identified three of these plumes; they were both deep and giant. They were found at roughly 2,200 feet, 3,280 feet, and 4,260 feet below the ocean's surface. The largest was 15 miles long, 5 miles wide, and 300 feet thick in spots.[65] That is 5,100 times the volume of the Superdome, or about half the volume of Utah's Great Salt Lake.

Oil plumes are an anticipated outcome of deepwater oil well blowouts. Both the National Research Council and the U.S. Minerals Management Service have identified the likelihood of their presence.[66] Thus, while those aboard the *Pelican* were surprised to see so much oil and gas permeating the deep sea, their findings were not especially shocking, nor did they suspect them to be particularly controversial.

At first, no one outside of the scientific community even took notice of the *Pelican*'s results due to the sizable distractions of the oil hitting land and the release of underwater oil spill footage, and the fact that Dr. Joye had yet to make a firm phone call to unsuspecting *New York Times* reporter Justin Gillis.

On May 12, in response to demands from two senators, Barbara Boxer (D-CA) and Bill Nelson (D-FL), BP released a thirty-second clip of the oil coming out of the end of the broken riser.[67] Once the video was available, it was obvious why BP had tried so hard to keep it from the public. For the first time, we could all see what an oil gusher at 5,000 feet below the ocean's surface looked like: a dark billowing mass roaring out of the pipe in a torrent that appeared to have no end in sight. Within twenty-four hours, at least three scientists using different methodologies derived not only massively larger flow rates but also more accurate ones than the government and BP were reporting.

Neither the White House nor BP had contacted the scientists; rather, National Public Radio had sought out their expertise.[68] The speed with which the scientists were able to turn the video into reliable data, and their eagerness to do so, made it painfully obvious yet again that had the White House only sought out the judgment of experts, we would all have known exactly how much oil was emanating from the wellhead in the earliest days of the disaster. The National Oil Spill Commission concluded that "the national response may have benefited early on from a greater sense of urgency, which public discussion of worst-case discharge figures may have generated."[69]

The scientists were still limited by the information available to them, for all they had was the video from the end of the riser. In early May, BP had succeeded in closing the third and smallest leak. Because this leak was so small, however, BP conceded that stopping it was not "expected to affect the overall rate of flow from the well."[70] Even so, their estimates were low, because they did not include the second remaining leak point.

Dr. Timothy Crone, a marine geophysicist at Columbia University's Lamont-Doherty Earth Observatory, estimated that 25,000 to 50,000 barrels of oil a day were flowing from the end of the riser, using a technique he developed called optical plume velocimetry.

Dr. Eugene Chiang, an astrophysicist at the University of California and an orders-of-magnitude estimation expert (that is, estimating size or scale from small amounts of data), estimated the flow of oil from the end of the riser to be 10,000 to 50,000 barrels of oil a day.

Dr. Steven Wereley, a mechanical engineer at Purdue University and an expert in fluid mechanics, estimated the flow between 29,000 and 43,000 barrels of oil a day, by using particle image velocimetry, which uses a computer program to identify and track distinct flow structures in the plume as it exits the riser (akin to the billows of a cloud).[71]

As outside scientists were inching ever closer to BP's own internal estimates of the spill, the company sought to dismiss the (accurate) measurements. BP spokesman Bill Salvin told NPR's Richard Harris, "We've said all along that there's no way to estimate the flow coming out of the pipe accurately." Instead, Harris reported, "BP prefers to rely on measurements of oil on the sea surface made by the Coast Guard and the National Oceanic and Atmospheric Administration."[72]

At the same time, BP experienced the final crushing blow in its attempts to close the ill-fated BOP. BP had no way to see into the BOP. The only way the engineers knew whether their operations were having an effect was by the change, or lack thereof, in the amount of oil escaping from the well. In mid-May, with the help of Department of Energy secretary Steven Chu and the scientists at Los Alamos National Laboratory, the ROVs were armed with cobalt-60, a radioactive isotope that generates gamma rays, and a picture came from inside the BOP. The images showed one wedge lock fully engaged, which meant that at least half of the shear ram had deployed. "I don't think anybody who saw the pictures thought it was ambiguous," Los Alamos scientist Scott Watson said.

Based on BP's internal logs, the New York Times reported, "It was a crushing moment. Engineers realized that all their efforts to revive the blowout preventer had probably never budged the critical component at the machine's core, the blind shear ram. They had assumed that at some point early on the blades had tried to close. They had hoped to close them all the way. But now, the gamma ray images showed that at least one blade was fully deployed, and they had run out of options for forcing the other one closed. Continuing to push on the ram's pistons with more hydraulic fluid would achieve nothing. The last line of defense was a useless carcass of steel."[73]

Another New York Times reporter, Justin Gillis, had been covering the disaster since its first days. On May 13, he wrote on the emerging controversy over the new spill estimates in which he cited several researchers who were critical of NOAA's response and the inability of scientists to go out in the water to acquire their own data.[74] Dr. Joye read the story, threw down her newspaper, and yelled, "They're doing it right now!"

At that very moment, Dr. Joye's colleagues were out in the Gulf of Mexico, conducting original research on a NOAA-funded trip, yet no one seemed to know about it. Incensed, Dr. Joye did something that she had never done before—and that several times since, she has regretted. She phoned the *New York Times*, and once through to Gillis yelled at him. Gillis took his beating, listened carefully, and immediately got on the story.

Months later, Dr. Joye sends me an e-mail while on her third cruise out to the Gulf of Mexico since the blowout. "Crazy thing is," she writes, "I called him to *defend* NOAA because the *Pelican* cruise was out there measuring impacts. How ironic is that??"

Gillis's front-page article ran on May 15 with the headline: GIANT PLUMES OF OIL FORMING UNDER THE GULF. It featured a photograph of an anxious and somewhat ill-looking Interior secretary Ken Salazar standing over an oil-soaked brown pelican. After writing that "the discovery is fresh evidence that the leak from the broken undersea well could be substantially worse than estimates that the government and BP have given," Gillis quoted Dr. Joye extensively, as well as the scientists on board the *Pelican*.

"There's a shocking amount of oil in the deep water, relative to what you see in the surface water," Dr. Joye told Gillis. "There's a tremendous amount of oil in multiple layers, three or four or five layers deep in the water column."[75]

Congressman Edward Markey (D-MA) read about the plumes in the *New York Times* story and was incensed. His staff immediately contacted Dr. Joye to learn more about the *Pelican*'s research and its findings. "These huge plumes of oil are like hidden mushroom clouds that indicate a larger spill than originally thought and portend more dangerous long-term fallout for the Gulf of Mexico's wildlife and economy," Congressman Markey said the next day in a widely reported statement.[76] Within weeks, Dr. Joye would appear before the congressman's committee to testify on the amount of oil and gas hidden below the ocean's surface in one of the dozens of hearings conducted by Congress about the disaster.

BP would later dismiss the existence of the plumes outright. In the story, Gillis quotes an earlier response by BP that knowing the size of the spill could in fact be dangerous to the response effort. Replying to entreaties from scientists that they be allowed to use instruments at the ocean floor to get a more accurate picture of how much oil was really gushing from the well, BP spokesman Tom Mueller said, "The answer is no to that. . . . We're not going to take any extra efforts now to calculate flow there at this point. It's not relevant to the response effort, and it might even detract from the response effort."

Gillis wrote, "The undersea plumes may go a long way toward explaining the discrepancy between the flow estimates, suggesting that much of the oil emerging from the well could be lingering far below the sea surface." Given the criticism of the *Pelican* scientists that came next, it is also important to note that Gillis specifically stated, "Much about the situation below the water is unclear, and the [*Pelican*] scientists stressed that their results were preliminary."[77]

The story went live online around 8:30 p.m., and that's when Dr. Joye's phone started ringing; it hasn't stopped yet. "Skype calls, calls at my home, calls at my office, TV crews showing up at my work!" she exclaims. "I called Justin, and he said he was getting the same reaction. It was insane." The story exploded in the media, answering the key question for the public's understanding of the crisis: Where is all of the oil?

The administration's response came just as rapidly and was equally intense. The news of the oil plumes came while the *Pelican* crew was out on-site. When it came back to shore on May 16, NOAA asked the scientists to stop talking to reporters.[78] On May 17, NOAA administrator Jane Lubchenco issued the following statement: "Media reports related to the research work conducted aboard the R/V Pelican included information that was misleading, premature and, in some cases, inaccurate."[79]

"But at that point," Dr. Joye stresses, "neither Dr. Lubchenco or anyone else from NOAA had seen the *Pelican* data. So how in the world could they make such a statement?"

On June 13, NOAA issued a research update stating that the samples taken by the *Pelican* did not comply with Environmental Protection Agency (EPA) guidelines, were rendered invalid, and "will not be used."[80] The statement still appears on NOAA's Web site.

The *Pelican* researchers immediately responded with letters, phone calls, and e-mails to NOAA defending their work. On July 29, a representative of NOAA responded by e-mail, agreeing with the scientists' complaints and promising that "NOAA is taking down the statement immediately (today) and will re-post with retractions later next week." Nonetheless, as of this writing, the statement still appears on NOAA's Web site.

Dr. Joye describes what should have happened immediately following the discovery of the plumes as a "man overboard" drill. "A man falls overboard, you point your finger at him, you don't take your finger off the person until they are on board. When this oil well blew, they should have put their finger on the plumes and kept it there." Rather than track the plumes, NOAA and BP denied their existence. "They took their finger off. I don't know why. But ever since, we've had to hunt it down for them."

The *Pelican* findings were supported by the work of other scientific teams, including that from the Woods Hole Oceanographic Institution, a highly respected private nonprofit organization in Massachusetts that conducted its own research cruise from June 19 to 28.

The team, led by Dr. Richard Camilli, found two oil plumes from the Macondo well. The largest was nearly 22 miles long, more than 656 feet high, and 1.24 miles wide, at a depth of some 3,600 feet. The second was "more diffuse" at about 164 to 1,640 feet in depth. The identification of these two plumes did not rule out the existence of others in other locations, it only meant that in the locations where this team searched on these days, these plumes were identified.

In order to avoid a showdown with the administration, however, the Woods Hole team did not go public with its results until after publishing them in the scholarly journal *Science*, on August 19.[81]

The very next day, Dr. Lubchenco said that NOAA is now "very concerned about the impact" of the oil below the surface and would begin more aggressive monitoring of it. NOAA estimated as much as 42 million gallons of oil may be lurking below the surface but preferred the phrase "an ephemeral cloud" to describe the oil, rather than "plume."[82] NOAA's terminology has thus far failed to catch on.

Perhaps feeling vindicated—the members of the *Pelican* team described publicly for the first time what had happened to them upon their return in May. "I got lambasted by the Coast Guard and NOAA when we said there was undersea oil," University of South Florida (USF) marine sciences dean William Hogarth told Florida's *St. Petersburg Times* in August. He was asked by NOAA to retract the public announcement, and Dr. Hogarth compared this to being "beat up" by federal officials.

University of Southern Mississippi oceanographer Vernon Asper said, "We expected that NOAA would be pleased because we found something very, very interesting. . . . NOAA instead responded by trying to discredit us. It was just a shock to us." Of Lubchenco's statement, Dr. Asper said, "She basically called us inept idiots. We took that very personally."

Administrator Lubchenco confirmed that she had instructed the researchers to stop speaking with the media. "Her agency told USF and other academic institutions involved in the study of undersea plumes that they should hold off talking so openly about it," the *St. Petersburg Times* reported.

"What we asked for," the administrator told the paper, "was for people to stop speculating before they had a chance to analyze what they were finding. . . . We just wanted to try to make sure that we knew something before we speculated about it."[83]

Until recently, Dr. Joye had never spoken to Lubchenco personally but describes being bullied by federal officials in response to the Gillis story. "Everyone involved in the *Pelican* plume discovery got their hands slapped and were asked to 'stand down' and not respond to media requests," she told me. She believes this is because NOAA "wanted to control the flow of information." Some scientists—Dr. Joye is not saying who—have even had their federal funding threatened, which is no small matter, given that this is a main source of funding for oceanographic research. "They told us, 'We're not trying to tell you what to say . . . we're just asking you to temper your statements, not be inflammatory.'"

"The *truth* is inflammatory," Dr. Joye tells me, and her large eyes widen again. "That's a shame. That's the situation, that the truth is considered inflammatory. It's a sad state of affairs, and I would never have expected that from this administration. It's pretty scary."

Reality Bites

At the same time that Dr. Joye and the other scientists were going head-to-head with NOAA in mid-May, the public estimates of the actual amount of oil being released from the Macondo well were finally catching up with reality.

Thirty seconds from the spill cam were enough to tell the world that we needed to see more. Congress put pressure on BP to release more video, and BP responded by releasing video of the kink leak. Testifying before Congress on May 19, Dr. Wereley estimated that the kink leak was producing a flow of roughly 20,000 to 30,000 barrels of oil and gas a day. Adding that figure to his previous estimate of the flow from the end of the riser, he arrived at a total flow rate of approximately 50,000 barrels of oil a day.[84] This same day, a small amount of oil from the Macondo well was reported to be entering the Gulf of Mexico Loop Current.[85]

Congress continued to press—particularly Congressman Edward Markey (D-MA), chairman of the House Energy and Environment subcommittee, who wanted BP to make its video footage available to all who wished to view it, whenever they wished to view it, twenty-four hours a day. On May 20, BP finally complied, and the spill cam went live.

The effect of the spill cam on the public psyche was profound. Within the first twenty-four hours of Congressman Markey's posting the link, hundreds of thousands of people visited the subcommittee's Web site. Perhaps because it just made such good television, the video footage opened morning news programs and closed out the nightly news. The public was riveted, and the demand for better information and action grew accordingly.

Within the National Incident Command, a Flow Rate Group was established to create analyses of the size of the flow. The group was broken up into teams: plume modeling, mass balance, and riser insertion tube tool. The group's initial estimates, however, proved chronically low, plagued by a lack of inclusion and transparency.

"The Group's estimates may have suffered from a failure to disclose enough information to enable other experts to assess the group's methodologies and findings," the National Oil Spill Commission found. "If more of the Group's data had been made public, its estimates may have evolved more rapidly with input from the broader scientific community."[86]

During the next three months, the group put out low estimates with little supporting information. Accurate assessments were finally made not by the Flow Rate Group but by two other teams: one led by the secretary of energy, Dr. Steven Chu, and one from the Woods Hole Oceanographic Institution. Dr. Chu's work depended on resources made available only in July when BP began its "top hat" process of lowering a cap on the wellhead to close in the well. At Chu's urging, the cap was armed with equipment to measure oil flow. The analysis of the Woods Hole team, however, was based on tools at its disposal, and the analysis was offered for use in early May.

A good example of the Flow Rate Group's work came on May 27, when the group published an estimate—without any supporting documentation—of a total flow of 12,000 to 25,000 barrels of oil and gas a day. About a week later, three pages of summary information were released, which noted that the team produced "a range of lower bounds" but did not provide data on how the numbers were reached, nor did it include documentation of the upper ranges of the estimate.

The National Oil Spill Commission staff learned that the lower-bound range "was simply a collection of the minimum estimates produced by each of the Plume Team members." The commission also learned that the team members had produced maximum estimates, several of which were in excess of 50,000 barrels of oil a day, but these amounts were not made public.

Further, the team's original report revealed some divergence of opinion, but these were omitted from the public version. On June 10, a revised estimate of 20,000 to 40,000 barrels of oil a day was publicly released.

The Flow Rate Group also released a finding by Dr. Richard Camilli and the Woods Hole team on June 10. The Woods Hole researchers had used an ROV mounted with sonar and acoustic sensors to determine the volume and velocity of the outflow from the end of the riser and kink leak. Their initial rough estimate, derived on May 31, was a flow rate of 65,000 to 125,000 barrels of oil and gas a day.[87]

The Woods Hole team had offered its services to BP weeks earlier, and BP had initially agreed, "for reasons related to broken equipment, not the environmental consequences," Dr. Camilli later testified. But BP then backed off. "Our team was thanked politely by BP representatives for our efforts, and I've had no further communication with BP since May 6," Camilli said.[88]

On August 2, the final "government estimate of discharge" was released, using Secretary Chu's analysis and data. The final finding was that between 55,800 and 68,200 barrels of oil a day flowed from the Macondo well from April 22 until the riser was cut on June 3—the equivalent of approximately one *Exxon Valdez* tanker disaster pouring into the Gulf every four days. After June 3, the rate declined to between 47,700 and 58,300 barrels of oil a day until the well was capped on July 15.[89]

The Woods Hole group ended with higher figures, concluding that the flow rate was initially 55,000 to 80,000 barrels of oil a day, declining to 44,000 to 67,600 barrels of oil a day from then on.[90]

The final consensus (although one disputed by BP) is that roughly 5 million barrels—210 million gallons—of oil were released by the Macondo well, making it the largest environmental disaster in U.S. history and the second-largest oil spill in the world.

The effect of that oil would reach far beyond the *Deepwater Horizon*, to all whose lives are dependent on the Gulf as well as to many who did not yet realize how large an impact the Gulf of Mexico has on their lives, livelihoods, and political fortunes.

3

Body Count

Oil, Gas, and Dispersant Attack the People, Wildlife, and Wild Places of the Gulf

Having ignored the pending disaster looming below and on the *Deepwater Horizon*, BP tried to hide the amount of oil coming out of the well, the oil itself, and, finally, the oil's devastating effects. Each decision progressively intensified the disaster. Tragically, the Obama administration aided and abetted BP at virtually every step along the way.

Jamie Billiot felt the impact of the Macondo blowout long before the oil came to shore. When she saw the black oil, dyed red and orange from chemical dispersants, streaking across the Gulf, the feeling was overwhelming. "Our waters are bleeding. It feels like we are bleeding," she says. Billiot is thirty years old and has lived on the shores of the Gulf coast her entire life, just as her father, her father's father, and a hundred years of her tribe's people, the United Houma Nation, before her.

Billiot describes where she lives as "in" the Gulf rather than "beside" it. If you hop into a boat on the canal that runs alongside her parents' home in Dulac, Louisiana, you'll hit open water in minutes. When hurricanes strike, you do not even have to travel; the Gulf waters come right to your yard. If your house is not on stilts, elevated off the ground, the water most likely comes right into your living room (oftentimes, even when the house is elevated).

Billiot and her family live deep in the bayou on one of what are most commonly referred to as Louisiana's "fingers." When her grandfather lived here, it was all one mass of land. Today, thanks in no small part to some 10,000 miles of canals cut through the coastal marshes to service off-shore drilling sites, the state is drowning: every thirty minutes a tract of marsh the size of a football field is swallowed by encroaching Gulf waters.[1] As the canals create avenues for the salt water to cut in, the salt kills the vegetation, then the soil, until the land itself disappears. The bits of land that are left behind reach out like fingers into the water. Even this characterization is becoming outdated, however, as the fingers now look more like they are suffering from leprosy.

The effect is stunning. The ocean and sky meet everywhere around you as roads cut through land hardly deserving of the name. The rich blues are in stark contrast to the cypress trees—once lush and green, now gray carcasses stripped of life by salt, standing like scarecrows guarding what life is left until they too are ripped away by the storms.

Jamie Billiot and her father, James, members of the United Houma Nation, in Dulac, Louisiana, in July 2010.

Before the Macondo well blew, Billiot already blamed the oil industry (among others) for her fear that just as she is unable to live on land that her grandfather called home, so too her daughter, Camille, age two and a half, will be unable to live in Dulac when she is Jamie's age. Now Jamie is all but certain of it.

"When the next storm brings in not just mud and water, but oil and dispersant, who's going to stay?" she asks. "How far will our people be spread this time? When the fish die, who will come back?" The questions are rhetorical, but the pain, exhaustion, and anger in her voice demand a response.

Billiot is the executive director of the Dulac Community Center and has devoted her life to protecting her people and their heritage. This means protecting their land, water, and environment.

As Brenda Dardar Robichaux, principal chief of the United Houma Nation, explained to Congress when testifying about the BP oil spill, "The relationship between the Houma People and these lands is fundamental to our existence as an Indian nation. The medicines we use to prevent illnesses and heal our sick, the places our ancestors are laid to rest, the fish, shrimp, crabs and oysters our people harvest, our traditional stories and the language we speak are all tied to these lands inextricably. Without these lands, our culture and way of life that has been passed down generation to generation will be gone." She called the BP oil disaster "perhaps the greatest challenge in our history, as we are at risk of losing the heart of our culture—our homeland" and said that the spill "looms as a death threat to our culture as we know it."[2]

To Billiot, the whole oil disaster "reflects the lack of respect for our place here. The people who run these [oil] companies—not the workers, many of whom are our friends, family, and neighbors, but the CEOs, the presidents, they don't lie down here. This is not their space. People who are making the big money off of this place could really honestly care less about it," she tells me.

Life on the Gulf coast is dependent on the water, and the water has been poisoned. Five million barrels of oil and some 500,000 tons of gas were released from the Macondo well over three long months, from April 20 to July 15. In a case of the cure being, in many ways, worse than the disease, more than 1.8 million gallons of toxic chemical dispersants were simultaneously mixed into the water and sprayed from the air, and at least 410 fires were ignited on the water's surface to burn the oil away. The impact on those living in, on, and from the water is profound, ongoing, and deeply interrelated.

The oil, gas, and chemical monster swept through the ocean leaving a path of death and destruction. The smallest shrimp eggs to the largest sperm whales were harmed, and so were all those creatures that feed on and make their living from them. The monster was carried below water by soot from the burns, mucus from critters that tried to eat away at it, plumes that carried it like clouds, and the monster's sheer weight.

At the water's surface, birds suffocated to death, and the combination of oil and chemicals let off a toxic aerosol that sickened people, whether they were on the water or just living, working, and playing along its shores. The oil-chemical cocktail ate at the roots of wetlands and marshes and leached through the white sandy beaches.

This chapter focuses largely on the effects of the BP oil monster on human health, the environment, and wildlife. But in a place where fish and oil, both of which reside in the water, are so essential to all aspects of life, it is all but futile to separate these effects from those on people's livelihoods and social lives, the economy, or politics. All of the devastating harm converges to create the "worst oil-related disaster in U.S. history"—a label that signifies much more than the amount of oil spilled.

Oil Can Kill You

Crude oil is toxic to humans, plants, and wildlife, capable of causing serious debilitation and even death to any who come in contact with it.

Crude oil contains high levels of volatile organic compounds (VOCs) such as benzene, toluene, and xylene. Benzene is a known human carcinogen. Toluene and xylene affect the central nervous system; exposure can cause headaches, nausea, vomiting, fatigue, impaired speech, tremors, depression, cerebral atrophy resulting in a decrease of the functions that the brain controls, liver and kidney damage, cardiac arrhythmia, and death.[3] VOCs are well-known chemical hazards that can cause acute toxicity as well as longer-term health effects such as cancer, birth defects, and neurological effects.[4]

Dr. Samantha Joye presented testimony to Congress in June on the effects of the oil spill on the Gulf ecosystem. "Everything from the base of the food web—microorganisms—to the higher order consumers—invertebrates, zooplankton, jellyfish, fish, birds, sea turtles, marine mammals—will suffer direct consequences of the BP blowout as long as there is oil in the system, due to the inherent toxicity of crude oil components," she told the committee. "This is why it is essential to recover as much of the spilled oil as possible and to remove it from the environment."[5]

Crude oil also contains polycyclic aromatic hydrocarbons (PAHs) and heavy metals, such as nickel and lead. PAHs are a group of more than a hundred different chemicals that are highly toxic and "tend to persist in the environment for long periods of time, especially if the spilled oil penetrates into the substrate on beaches or shorelines," according to NOAA.[6]

Exposure to certain compounds of PAHs can also lead to cancers in mammals, including human beings. Nickel is a known human carcinogen, exposure to which can affect cardiovascular, dermal, immunological, and respiratory functions.[7] Lead can harm the development of human organs, reproductive capacity, and urinary, kidney, cardiovascular, gastrointestinal, and neurological functions, among others.[8]

VOCs are more likely to be present near the source of the oil leak. As the oil has been weathered by exposure to air and water for some time, the amount of VOCs may decrease, but it still contains PAHs. Whether crude oil sits on water or on shore, the harmful elements in it can be picked up by the wind and other weather conditions and be carried through the air as a toxic aerosol. The elements can also enter the water and be ingested by marine life, which humans then ingest.

Oil is deadly to animals in many ways. When oil coats an animal, it can limit the creature's ability to swim, fly, navigate, maintain body temperature, feed properly, and even reproduce. Oil can harm the eyes, mouth, and nasal tissue as well as the immune system, red blood cells, and organs like the liver, lungs, and stomach.[9]

Oil can kill the root systems of the grassy marshlands that line the Gulf and act as natural storm barriers—a last line of defense against the Gulf's deadly hurricanes. Oil on the beaches harms tourism and the enormous network of industries built up to cater to tourists in the Gulf. Oil in ground water carries the toxic effects from shore inland.

Finally, methane gas released so abruptly and in such massive quantities creates a vehicle to suspend oil in water. Its presence leads to the depletion of oxygen from the water and acidification, both of which kill marine life and all that depend on it.

Oil and Water

Oil and gas are, of course, naturally occurring substances, and when left to their own devices abide in harmony with their surroundings. Dr. Joye has spent her career studying this natural cohabitation. In her congressional testimony, she described how the Gulf community has adapted to the small amounts of oil and gas that regularly leak, or seep, into its waters. The so-called seeps do cause harm, but it is limited.

The Gulf of Mexico, she explains, is "accustomed to slow, somewhat diffuse inputs of oil and gas, and the biological communities have adapted to endure and in some cases metabolize these materials such that negative impacts of such inputs are localized as opposed to widespread." Although oil does enter the Gulf from about 1,000 naturally occurring seeps, the cumulative daily release is just 1,000 to 2,000 barrels—the equivalent of about thirty minutes of one day of the BP blowout. These daily seeps, moreover, are spread out across the entire Gulf of Mexico— a 600,000-square-mile basin stretching from Florida across to Mexico and down to Cuba—the ninth largest body of water in the world.

"While natural seepage varies extensively in space and time," Dr. Joye told Congress, "the BP blowout is an intense, localized input" and "an unprecedented" disruption to the Gulf of Mexico system "that has no natural equivalent."[10]

Among some of the stranger heroes of the Gulf oil disaster are what I call the "mighty Macondo microbes," which are credited by BP-funded scientists with eating up virtually all the oil from the BP oil spill. Although a microbial community does exist that digests some of the small amounts of hydrocarbons that naturally seep in the Gulf, they have never been confronted with anything close to the size of the Macondo well blowout. They have done their darnedest to try to fix the mess the oil industry created by eating away at the oil and gas as it barrels through the ocean, but they cannot possibly come close to eating more than a small portion of this monster—nor do we want them to.

Dr. Joye told Congress, "The direct injection of large quantities of oil and gas into the system has upset the delicate balance of oxygen in the offshore system."[11] When the microbes eat the oil and gas, they consume oxygen. The more oil and gas they consume, the more oxygen they need. The more oxygen they take in, the less oxygen that is available for everybody else; this leads to *dead zones*: areas of water with little or no oxygen in which little or nothing can live.

Dr. Joye personally witnessed and recorded many areas of the ocean in her travels below the Gulf where life once flourished prior to the blowout but where now virtually no life—other than the microscopic kind—could be found.

Not all of the oil that escaped from the Macondo well was released into the Gulf. BP and the Obama administration succeeded in capturing some oil by piping it directly from the wellhead into surface ships that then carted the oil away. These efforts (discussed in chapter 6), however, captured just about 17 percent of the total oil gusher.[12] The remainder of the oil and gas escaped directly into the Gulf of Mexico.

Fortunately, microbes are not our only hope for capturing oil and gas in water. In fighting oil spills, nature has given us one minor advantage: oil and water do not mix. Unless disturbed, oil will largely stay congealed in water, making it possible, under the right conditions, to corral it and suck at least some of it up. Once gas leaves the well, however, it is impossible to recover.

There are three primary responses to fighting an oil spill: stop the source of the spill, contain the spill at the source, and clean up as much of the oil as possible that is released into the environment. From the first day of the Macondo well blowout, it was painfully evident that although BP had asserted its ability to deal with a spill of 300,000 barrels of oil a day in the case of a blowout in the Gulf (at its height and at the highest estimates, less than a third of the size of this disaster), BP, the rest of the oil industry, and the federal government had no idea how to repair a blown-out oil well 5,000 feet below the water's surface. They did not know how to contain the oil and gas that escaped from the wellhead or how to clean them up following their escape. All were caught unprepared and instead engaged in a massive science experiment, the effects of which will be with us for years to come.

Boom and Skimmers

Although the last twenty years have witnessed astronomical advances in the technology of offshore oil drilling, virtually nothing has improved in the methods applied to containing and cleaning up oil spills because the industry has failed to invest in either one.[13] It turns out that the oil industry is not very good at performing either function. The primary devices used for cleaning up oil from water are boom and skimmers.

A boom is typically an inflatable orange or yellow floating barrier filled with air or Styrofoam and pulled by a boat. Boom looks like giant worms. A skimmer looks like a large industrial vacuum; it can either be a vessel itself or sit on another vessel. The boom is used to corral the oil, which the skimmer then sucks or scoops up and salvages for later use. Boom was also used across the Gulf as an artificial barrier stretched out along the surface of the water to try to stop the oil from hitting land.

After the *Exxon Valdez* disaster, the Oil Pollution Act of 1990 required the coast guard and oil companies to maintain both items in case of emergencies. The boom and skimmers deployed in response to the Macondo well, however, were essentially the same as those utilized twenty years ago in the *Valdez* crisis. In that case, only 14 percent of the spilled oil was ever recovered. More than 26,600 gallons of oil remain in Prince William Sound today, readily found oozing up on beaches.

Moreover, the coast guard and BP were caught woefully unprepared to protect the coastlines of five states against the approaching BP oil monster. Nor, for that matter, did any of the other oil companies— Exxon, Chevron, Shell, etc.—that were supposed to be equally prepared in case of their own Gulf oil disasters have adequate supplies.

Boats all across the Gulf were enlisted in these efforts—transformed from vessels designed to capture fish, crab, and oysters into boats hunting for oil. Green fishing nets were discarded and replaced with bright orange and yellow boom.

People tried to forecast the movement of the oil monster, but their efforts were often futile as the monster was wily. As Commander Thad Allen said in early May of the roving oil, "The maddening thing is it's indeterminate . . . it could go any direction. . . . The entire Gulf pretty much has to be on guard."[14] Local TV news programs aired daily "oil casts" during the weather report in an attempt to predict the oil's whereabouts and intensity. The chaos that ensued has been dubbed "the boom wars" as the governors of five states fought both BP and one another over inadequate supplies. All but Texas (thanks to weather, not boom) ultimately failed to keep the oil from reaching their shores.

At its peak on August 14, nearly 700 miles of Gulf coast shoreline was hit with Macondo oil: approximately 386 miles in Louisiana, 117 miles in Mississippi, 73 miles in Alabama, and 108 miles in Florida.[15]

No matter how many times I personally encountered oil from the

Macondo well, it was always a shock. Walking onto a beach or boating along the coast or to an island, I would be immediately captivated by the glorious blue sky, puffy white clouds, purple flowers, deep blue water, rich green grasses, and white sandy beaches. I would forget, just for a moment, why I was there and get lost in the stunning beauty all around. Then, looking out on the beach, I would see what would first appear to be a pool of wet dark sand or a row of scattered bark. When I remembered why I was there, I'd say a little prayer to myself, "Maybe this time it will be different. Maybe this time it will just be a gorgeous beach simply to be enjoyed." I'd walk a little closer, and the pools were oil and the bark rows of oily tar balls stretched as far as the eye could see. On one beach what I thought was a tar ball turned out to be the top of a child's plastic sippy cup caked in thick BP Macondo well oil.

Both boom and skimmers work best when an oil spill is small and isolated and when the oil is thick and on the surface of the water so that it can be seen, held in place, and separated from the water. Both devices also work best in calm waters, without waves, wind, or inclement

Boats once used to haul fish, now hunting oil, their green fish nets replaced by yellow oil boom in July 2010.

The author with two handfuls of Macondo oil well tar balls that washed up on
Dauphin Island, Alabama, in July 2010.

weather. Virtually none of these conditions were met while oil gushed
from the Macondo well. Only the weather, however, was beyond BP and
the coast guard's control.

   In the end, booming and skimming captured a grand total of about
4 percent of the Macondo oil that was released into the Gulf, or some
8.2 million gallons.

   Compared to the other methods used to try to stop the BP oil monster from flowing on the surface and from coming on shore, booming
and skimming were largely innocuous. The other tactics—dispersing
and burning the oil—were both deadly and largely at odds with the goal
of containing the spill.

Carpet Bombing the Gulf

On April 22, the *Deepwater Horizon* sank to the bottom of the Gulf
of Mexico, and a five-mile-long oil sheen spread out across the water's

surface. The coast guard launched its first flight to spray chemicals on the slick shortly thereafter, in an attempt to disperse the oil. BP chose the type of dispersants and made the decision to use them. The coast guard and the EPA authorized its use in the face of great opposition, including by the EPA's own scientists.[16]

The coast guard, followed by the air force, flew the planes that sprayed the dispersant on the slick. The idea was to break up the slick, keep the oil off the water's surface, and keep the oil from coming ashore. Rear Admiral Landry told the *New Orleans Times-Picayune* newspaper, "Our goal is to fight this oil spill as far away from the coastline as possible."[17] By April 23, the coast guard had already flown five sorties and applied 1,500 gallons of dispersant to the oil sheens.[18] The dispersants were applied continuously until July 19, a few days after the well was capped.[19]

The application of chemical dispersants is a standard response to an oil spill—but this was no ordinary spill. It was far deeper than any spill in history. It lasted longer than all but the Ixtoc blowout that poured 3.3 million barrels of oil into the Gulf of Mexico from June 1979 to March 1980. It was also many magnitudes larger than any other spill that anyone had tried to stop or meaningfully clean up, as Saddam Hussein made no such efforts to clean up the oil he purposely dumped into the ocean during the Iraq war and coalition forces had to wait until the end of the war to mount a very limited cleanup.[20]

As each element of the unique nature of this spill became apparent, the method of response should have changed accordingly. Instead, BP and the federal government tried to match dispersant-for-oil. Congressman Markey, who launched an exhaustive investigation into the use of dispersants in the Gulf, stated, "BP carpet-bombed the ocean with these chemicals, and the Coast Guard allowed them to do it."[21]

The use of dispersants against the Macondo oil gusher was unprecedented in history. For the first time anywhere, and without any testing as to its effect, dispersant was applied subsurface at the wellhead in what amounted to be a massive science experiment on the people and ecosystems of the Gulf.

Moreover, the sheer quantity of dispersant was record setting. Nearly 2 million gallons were applied—the equivalent of three Olympic-size swimming pools—and second only to the 2.5 million gallons used against the Ixtoc spill, although the Macondo blowout lasted just one third of the time.

Of the 1,843,786 gallons of dispersants reportedly used, 771,272 gallons, or nearly 40 percent, were applied below the water's surface.[22]

Ultimately, more harm than good was done, and few of the goals set out for using the dispersant were achieved. The oil stained the surface and reached the shore. Moreover, because the oil was so immediately and thoroughly mixed with dispersant, it is impossible to distinguish the effects of one from the other.

Dispersant functions like detergent, breaking up oil into small droplets that will mix easily with water. The result is an oil and water mix composed of droplets of dispersant-covered oil. Because of their small size, the droplets can remain suspended underwater rather than rise to the surface.[23] The dispersant does not *remove* the oil. Rather, as the name states quite clearly, it simply *disperses* it throughout the water.

One week after the explosion of the *Deepwater Horizon*, BP was losing. Its efforts to cap the well, a task that it said would take just twenty-four to thirty-six hours, were unsuccessful.[24] The visible oil slick was an incredible 600 miles in circumference—two and half times the size of Lake Erie—and within days of making landfall. BP and the federal government had been "outed" for the first time for lowballing the size of the spill by the release of John Amos's 5,000 to 20,000 barrels a day flow estimate.

On April 28, BP and the government jointly announced that the spill was five times larger than initially estimated. A reporter present at the announcement wrote, "Eight days after the first explosion on the rig, the tenor of the response team's briefings changed abruptly Wednesday night with a hastily called news conference to announce that the rate of the spill was estimated to be 5,000 barrels a day, or more than 200,000 gallons—five times the previous estimate."[25]

In response to BP's inability to cap the well and the admission of the size of the ongoing spill, two significant changes were made in the strategy for attacking the oil. On the water's surface, the Joint Unified Command (BP, MMS, and the coast guard) began burning the oil. Underwater, the massive application of dispersants began.

On April 30, George Stephanopoulos brought the news to the public, reporting on ABC's *Good Morning America*, "Overnight BP said it would make another attempt to stop the flow of oil by pumping chemical dispersants down to the Gulf floor to break up the oil at the well, a method that has never been used a mile underwater." Coast guard rear admiral Sally Brice-O'Hara appeared on the show, reporting, "That is a technology that is in new stages, and we are working closely with our scientific support from NOAA to analyze what the impact of that dispersant technology will be."

It turns out that it was BP, not NOAA, that was performing the tests. From May 1 to 17, BP applied 45,000 gallons of dispersants underwater and performed three tests for "effectiveness and toxicity." The EPA and the coast guard allowed the use of the dispersant as long as BP continued to monitor both attributes. EPA administrator Lisa Jackson later told writer Jeff Goodell that she made the approval decision on behalf of the EPA and that it was one of the hardest decisions she had ever made.[26]

There was limited scientific information available on which to base her decision, and BP assured her that applying the dispersant underwater would "reduce the overall volume of chemicals discharged into the marine ecosystem," she explained. They believed that underwater application would minimize human contact with the dispersants because the workers would not be applying the dispersants themselves, and, in contrast to aerial spraying, the application could occur in foul weather.[27]

Professor Oliver Houck of Louisiana's Tulane University Law School, one of the original pioneers of the field of environmental law, told me in response to this decision by the EPA, "There is no safe dispersant out there; it doesn't exist. Every dispersant has its own biological, reproductive effect, lethal effects, sublethal effects—they all have these effects. It's just a matter of balancing them. Now, the EPA under President Obama is a strong agency for the first time in memory. [Lisa Jackson] who runs the EPA is one of the strongest leaders in American government today. But she couldn't stop Corexit, not because she didn't have the courage to do it, she didn't have the science to do it."

The science was lacking for several reasons. Most important, as noted, subsea application had never occurred before and had never been tested. BP neither mentioned nor prepared for its use in plans for a 300,000 barrels of oil a day spill. It is not alone. The oil industry had put essentially no money, time, or interest into studying the effects of using dispersants after oil spills whether above or below water or the potential effects of their mass use. For example, the use of dispersants against the Ixtoc blowout has been largely ignored, as neither the oil company nor the Mexican government has investigated the subsequent effects.

There is enough scientific knowledge, however, about the human and environmental harms of Corexit to lead many experts, including those within the EPA, to conclude that Corexit should never have been used in the Gulf.

BP chose Corexit from a list of dispersants preapproved by the EPA for surface application. No such approval existed for subsea use. Although the name of the chemical was publicly released, a complete list of its contents was designated as the proprietary information of the chemical's owner, the Nalco Company of Naperville, Illinois. It would take more than five weeks, hundreds of thousands of gallons of the chemical's application, and enormous pressure from Congress and environmental and public health organizations for the public to gain access to what remains to this day just a partial list of the product's components.

BP first used Corexit 9527A, but when intense public pressure led the EPA to require a less toxic formula, BP switched to Corexit 9500. Three ingredients of the two Corexit formulas were publicly known from the outset. This information was enough for experts to determine that Corexit was a deadly choice. For example, Corexit 9527A contains 2-BTE (2-butoxyethanol), a toxic solvent that can "cause injury to red blood cells (hemolysis), [the] kidney[s,] or the liver."[28] It has been designated a "chronic and acute health hazard" by the EPA and identified as the cause of chronic health problems and even several deaths among the cleanup workers after the 1989 *Exxon Valdez* oil spill.[29]

Corexit 9500 contains propylene glycol, which is also toxic to humans and a known animal carcinogen. Like 9527A, 9500 bioaccumulates up the food chain.[30] In congressional testimony, Wilma Subra, a microbiologist and toxicologist who has studied the petrochemical industry of the Gulf coast for decades, described Corexit 9500 as "slightly less toxic" than 9527A.[31]

There are three concerns about the application of dispersants: the toxic effect of the chemicals in the dispersant; the combined toxicity of the dispersant and the oil, which is higher than the toxicity of either alone; and the saturation effect—the fact that applying the dispersants spreads out the oil-and-dispersant toxic goo, multiplying the reach and the effect of both by an incalculable amount.

"Dispersants increase the time period in which aquatic life is exposed as well as the area extent of exposure in the water column," Jacqueline Savitz, the senior campaign director of Oceana, explained. "Because toxicity is a function of dose and time period of exposure, this increases the number of aquatic animals that are subjected to toxic conditions as well as the degree of toxicity."[32]

Casi Callaway was the first person to warn the public of the negative impact of dispersants on national television when she appeared on CNN on May 1. Callaway is the executive director of Mobile Baykeeper, one of the largest environmental organizations in the Gulf coast. It works for "clean water, clean air and healthy people along with responsible government and a healthy economy," according to its Web site. Callaway is one of the Gulf's most powerful environmentalists and one of the most recognizable people in Mobile, Alabama.

Walking around Mobile with Callaway feels like living an episode of the TV show *Entourage*. More personally, it reminds me of when I worked as a legislative assistant in the Washington, D.C., office of Congressman John Conyers (D-MI) and we would take trips to his district. On any street corner in Detroit, in any church or restaurant, everyone recognizes J.C. They not only know him, they are truly excited to see him, shake his hand, and tell him exactly what they need him to do.

It is the same with Callaway. But instead of being an eighty-one-year-old black liberal lion from the roughest and poorest inner city in the country, Callaway is a forty-one-year-old white, blond, southern Republican member of the Junior League from the more genteel side of Mobile.

Even as one of the most influential people in her region, however, before the disaster, Callaway rarely tangled with the offshore oil industry. "Our attitude," Callaway told me, "was that the industry is here, it's always been here, it's always going to be here, there's nothing you can do about it, and it's basically futile to try." In other words, she said, "I pick battles that I can win."

Callaway is also the first to admit that she was initially slow to discern the seriousness of the BP Macondo blowout. That's because she was listening to the federal government.

"I heard about the explosion through news reports about the eleven missing men. I also heard that several of the guys had been life-flighted to University of South Alabama's hospital. My reaction was, 'How awful, oil drilling is [so] dangerous.'"

She had no indication, however, that the explosion would affect her life or work. When the rig sank, Rear Admiral Landry said that there was no oil leaking from the well, and Callaway believed her. "Nothing to worry about," Callaway assured a friend. "It's fine."

When it became clear that there was a leak, the local TV news called Callaway for a comment. "Would the oil hit Alabama's shores?" "No way," Callaway assured. The spill was too small (just 1,000 barrels a day according to the federal government, remember?) to affect Alabama's water, and the Gulf Stream would "push the oil to Louisiana."

On April 28, for the first time, Callaway got her information not from federal authorities but from personal experience. The air was unbreathable, and a code orange ozone alert was called. BP was burning the oil from the Macondo well on the water's surface. She immediately wrote an e-mail to the listserv of the Waterkeeper Alliance, a network of about two hundred similar organizations around the nation and the world. "I'm writing in desperation," she typed. "It appears that the solution to our oil 'spill' is to BURN IT UP!!!!! Because of Exxon Valdez, they want to avoid this spill landing on shore, so this is their only other option."

Callaway got many responses, offers of help, and plenty more information. One of the loudest responses came back from Alaska, the site of the nation's previously largest oil spill and the only other state in the country to have been bombarded with Corexit. The use of Corexit was hardly comparable in Alaska, however, because less than 4,000 gallons were used there in response to the *Valdez* spill.[33]

When he got her e-mail, Bob Shavelson, the executive director of Cook Inletkeeper in Cook Inlet, Alaska, called Callaway right away. "This is what you will need to fight," he told her of the dispersants. "They're going to do it. They're going to do it in this one like crazy and you need to stop them from doing it."

Shavelson is a self-described "reformed attorney" who, having spent his life working on environmental issues in state and federal agencies, took off for Alaska and never looked back. He has been the head of Cook

Inletkeeper for nearly fifteen years, protecting his watershed from the daily onslaught of offshore oil drilling.

As Shavelson later told me, "Exxon wrote the book on how to cover up a massive environmental disaster; BP took these basic tools, polished them up, and perfected a misinformation campaign that shielded the company from liability while heaping long-term misery on countless Gulf residents."

On April 30, Mobile Baykeeper put out its first press statement on the disaster, calling on the government to "demand transparency, monitoring and caution in implementing crisis clean-up solutions that could make the situation worse, such as controlled burns and chemical dispersants. BP should have been required to have a plan to contain a catastrophic spill like the one we've witnessed."

In the statement, Callaway was one of the first and only people who from the start hit the nail on the head for how long the disaster would last, stating, "This ongoing hemorrhage of oil could continue over the next several months. Clearly, whatever containment plan BP had in place, if they had one at all, has failed the Gulf of Mexico and all those who benefit from its pristine waters and wildlife."[34]

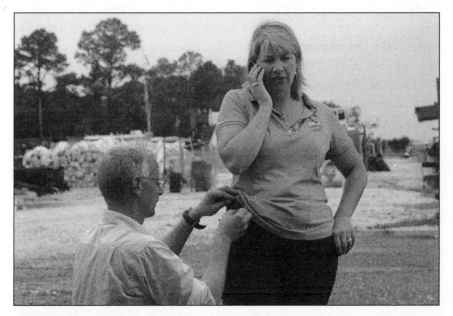

Casi Callaway, the executive director of Mobile Baykeeper, prepares for an interview on Dauphin Island, Alabama, in July 2010.

On May 1, Callaway appeared on CNN wearing her trademark Mobile Baykeeper's T-shirt. Callaway often has a windswept look on camera, partly because she is usually coming straight from monitoring the oil hitting her beaches and partly because of how much she is literally on the run. I've had more than one anxious car ride sitting in the passenger seat while Callaway simultaneously applies makeup, drives, gives instructions to a staffperson in the back, and mentally prepares for a public appearance.

Callaway told CNN host T. J. Holmes that nearly 70 percent of the oysters and 80 percent of the shrimp generated in the United States come from the Gulf coast and that both were at risk not just from the oil but also from the dispersants. "Gulf shrimp come from the bottom, and the bottom is where we are pushing the oil with the dispersant they are using, so it's laying down on top of the shrimp beds" and on the oysters, Callaway warned.[35]

"The impacts could last for twenty years, maybe even longer," Callaway reported. "It just depends on how bad we decimate them. There are ways to protect our grasses, and I assure you everyone on our coast, everyone on Louisiana's coast, Mississippi's, and Florida's coast, is doing everything they can with or without permission right now to try to protect our shorelines and our grass beds. They are just so important. It's not just the environment. That's what I do, sure, but it is the economy of our community."[36]

Like Bob Shavelson, those with the greatest experience with dispersants, the Alaskans and the EPA scientists, raised the loudest concerns about their use. Marine biologist Rick Steiner, a professor at the University of Alaska who had also worked on the *Exxon Valdez* oil spill, said of dispersants, "It's about PR. It's about keeping the oil out of sight, and out of the public mind, so fewer people really understand what is happening in the Gulf and get outraged by it." During the *Valdez* response, he added, "Corexit earned the nickname: 'Hides-it.'"[37]

Dr. Riki Ott is a commercial fisher turned fisheries professor from Alaska. Her Ph.D. from the University of Washington School of Fisheries includes an emphasis on the effects of heavy metals on benthic (deepwater) invertebrates. "I basically have a Ph.D. in oil," she tells me. Her life and her work were transformed by the *Valdez* disaster. She immediately brought her knowledge to the Gulf coast when the Macondo blew.

Dr. Ott is sitting in Barataria, Louisiana, in the backyard of Tracy Kuhns (another in a small family of U.S. commercial fisher*women* and a member of the Louisiana Bayoukeeper). Kuhns has given Ott a home away from home from which to join in the fight against the BP oil monster. Kuhns's fishing boat sits in the water directly behind Dr. Ott, the nets removed and stacked up in the yard, as Dr. Ott speaks with me about dispersants.

"The dispersant is a delivery mechanism for the oil," Dr. Ott tells me. "It breaks the oil up faster and puts it into bite-sized particles. That's not a good thing, because we're talking bite-sized for crab, for fish, for zooplankton. So all it does is enhance bioaccumulation and pickup through the whole food web. I mean, it's insane. It's totally insane."

In addition, Corexit is "an industrial solvent. It's a degreaser. It's chewing up boat engines off-shore. It's chewing up dive gear on-shore. Of course it's chewing up people's skin," Dr. Ott explained.[38]

Many scientists in the EPA shared these concerns and opposed the decision to use dispersants, particularly given BP's approach, which is best described as intense overkill. All but one, however, were afraid to go public for fear of retaliation.

Public Employees for Environmental Responsibility (PEER) is a nonprofit organization that represents government whistleblowers. Jeff Ruch, PEER's executive director, has reported that nearly a dozen EPA staffers, including several toxicologists, came to PEER to raise concerns about the use of dispersants in the Gulf. "[The EPA] appears to be making decisions at the behest of BP and not exercising much, if any, independent judgment," said Ruch.[39]

The scientists raised their concerns to the EPA but were excluded from the decision-making process, which was limited to just a handful of EPA representatives working with the Joint Unified Command. "The concern was the agency appeared to be flying blind and not consulting its own specialists and even the literature that was available," Ruch said.[40]

There was one EPA employee who did go public. Hugh Kaufman is a senior policy analyst at the EPA. After serving in Vietnam as a captain in the air force, he went to work protecting the environment at the EPA in 1971. He has dedicated his professional career to the agency and has never shied away from a fight if it has meant holding the agency to its own highest professional standards, regardless of the administration in power. Under the Bush administration, he successfully challenged the EPA for downplaying the health impacts of the World Trade Center attack on cleanup workers. He is now challenging the Obama administration and BP for downplaying the harmful impact of dispersants.

Kaufman has exposed not only the harmful health effects of the dispersants for workers, wildlife, and the environment but also the tremendous economic incentives for BP. "The EPA has all the information on what [the] ingredients are," he reported. "And when the ingredients are mixed with oil, the combination of Corexit, or any dispersant, and oil is more toxic than the oil itself. We know enough to know that it's very dangerous."[41]

Kaufman explained that the impact of the use of dispersants in the Gulf included internal hemorrhaging in dolphins and humans as well as making the cleanup far more difficult. He concluded, therefore, that "the sole purpose in the Gulf for dispersants is to keep a cover-up going for BP to try to hide the volume of oil that has been released and save them hundreds of millions, if not billions, of dollars . . . not to protect the public health or environment. Quite the opposite."[42]

BP is not only financially liable for each barrel of oil and gas spilled, it is also liable for all of the harm caused by their release. Dispersants applied at the wellhead can make it more difficult to measure exactly how much oil is escaping from a well. It is also clearly much more difficult to measure the harmful effects of oil and gas in the water and on marine species than on the shore, because the latter effects are far more visible and tangible. Thus, breaking up the oil has the potential to vastly limit BP's damages.

Dr. Joye warned Congress in June (when official public estimates of the spill were still just 20 percent of the actual size), that "the application of dispersants at the riser makes it impossible to estimate the size of the leak solely from surface observations (e.g., using satellite imagery). . . . Dispersed oil cannot be cleaned up, rather it moves with the water and the oil and dispersants are likely to influence oceanic ecosystems for years to come."[43]

Months later, Dr. Joye concluded that the use of dispersants "does one thing really well. It masks the magnitude of the spill, and it potentially does many, many things badly."[44]

"Dispersants do not remove oil from the ocean, and therefore it is important that we not adopt an 'out of sight, out of mind' attitude," explained David C. Smith, professor and associate dean of the Graduate School of Oceanography, University of Rhode Island.[45]

The application of the dispersants subsea was also far from an exact science. Remote-controlled vehicles at the wellhead sprayed the dispersant into the gusher with what I would describe as a glorified squirt gun. Dr. Joye has seen footage of the chemicals simply spraying out directly into the ocean rather into the gusher due to the force of the water and the gusher itself. She told me that her colleague, Dr. Ira Leifer, who has studied this application, is convinced that dispersant was in the gusher only 5 to 10 percent of the time—meaning that 90 to 95 percent of the dispersant was sprayed not on the oil but directly into the ocean. Dr. Joye has subsequently seen pockets of orange and pink goo that she suspects (she has not yet performed an analysis) to be dispersant lying on the sea floor miles from the site of the wellhead.

Public concern about the use of dispersants mounted steadily during the course of the disaster. By May 20, the EPA, concerned about the dispersants' use in "unprecedented volumes and because much is unknown about the underwater use of dispersants," asked BP to identify, within twenty-four hours, a less toxic and more effective dispersant than Corexit that could be used.[46]

BP responded on May 21 that Corexit 9500A was the only dispersant that met the criteria. The EPA called this reply "insufficient" and "focused more on defending [BP's] initial decisions than on analyzing possible better actions." The EPA then decided to conduct its own tests.[47]

As a result, the EPA ordered BP to "eliminate surface application of dispersant except 'in rare cases' approved by the Coast Guard, to limit subsurface application to 15,000 gallons a day and to reduce overall dispersant use by 75 percent from its maximum daily levels." Any dispersant application beyond this would require prior approval from the U.S. Coast Guard.[48]

After this directive was issued on May 26, people throughout the Gulf experienced one of what would come to be countless instances of feeling as though they were living in an alternate universe from the rest of the nation. They saw oil that they were told did not exist, they were made ill by chemicals that they were told were harmless, they saw oiled birds that they were assured had disappeared, and they watched planes flying overhead and spraying dispersants that they were assured were no longer in the air. Proof that the planes were real came after two long months.

On August 1, Congressman Markey revealed that after the EPA's directive was issued, BP and the coast guard essentially ignored it. More than seventy-four daily exemption requests were sent to the coast guard by BP, and all of them were approved, usually on the same day. "After we discovered how toxic these chemicals really are, they had no business being spread across the Gulf in this manner," Congressman Markey concluded.

He also found discrepancies between what BP reported to Congress and what BP reported to the coast guard as to the amounts of dispersants being applied. According to what was publicly disclosed, more than 1.8 million gallons of toxic dispersants were applied. The validity of that number, however, is in question. "BP was lying [either] to Congress or to the Coast Guard about how much dispersants they were shooting onto the ocean," said Congressman Markey. His investigation remains ongoing.[49]

In whatever amount it was used, dispersing the oil created many ill effects, including hindering the goal of recovering the spilled oil.

The primary tools for blockading and capturing oil, as noted earlier, are boom and skimmers, and both work best when the oil is as solid as possible. The dispersant had the opposite effect, permitting the oil to mix with the water and become much more liquid.

Dispersed oil in water is impossible to miss. The water has a chemical-looking tinge, usually red or orange. Depending on its concentration, it looks like black, brown, red, or orange slime or slop. It is a concoction that readily slips and flows over and under boom and skimmers. It was all too common to see orange boom sitting in the middle of a lake of Macondo well oil-and-dispersant slop throughout Gulf waters and along its shores.

"Normally, the oleophilic skimmers should have been the backbone of our operation," Mark Ploen, BP's offshore operations section chief, told author Jeff Goodell. "But with all the dispersants being used, we found that less oil was sticking to the skimmers, and they were far less effective."[50]

The EPA and the Joint Unified Command stated repeatedly that the trade-off for using the dispersant was that less oil on the surface of the water would mean fewer harmed animals, less oil would come ashore to damage the coastline, and spreading the oil would increase the ability of microbes to eat the oil.

The dispersant did keep some oil from hitting the shore that otherwise would have done so. On the other two points, however, the dispersants not only failed, they did more harm than good.

The National Oil Spill Commission reported its preliminary findings on these points. "Less oil on the surface means more in the water column, increasing exposure for subsurface marine life. And, while the smaller droplets may accelerate biodegradation, their smaller size increases the dissolution of potentially toxic compounds and exposure to aquatic organisms." The commission also stressed that "the assumption of increased biodegradation may not always be accurate," citing studies that have found that dispersants had either no effect on biodegradation or may even inhibit it. Moreover, there is no reason to suppose that "all dispersants act in the same manner." And finally, that all of this is awash in unknowns, as "it is also only largely in the aftermath of the Macondo well explosion that scientists have begun to research the extent to which oil-eating bacteria are present at the low temperatures of deepwater."[51]

One problem is that Corexit 9527 and 9500 have not been made available to researchers to determine their effect on Gulf of Mexico microbes. Dr. Joye, for example, has sought access and has thus far been denied. The only scientists who have had access are those who have signed three-year confidentiality agreements with BP.

Dr. Joye and other researchers have found vast pools of oil on the ocean floor and dispersed throughout the water, contradicting the notion that microbes have eaten large amounts of the oil.

There was one study, however, that claimed to find special microbes that were not only eating so much Macondo oil that the plumes were gone, but they were also doing so without removing oxygen from the water. The study, released in late August, received a great deal of media attention. Unfortunately, the study was not independent but rather was partially funded by BP. Moreover, the science behind it was lacking.

On August 24, Dr. Terry Hazen, a microbial ecologist with the

Lawrence Berkeley National Laboratory (LBL), a U.S. Department of Energy national laboratory, released the article, "Deep-Sea Oil Plume Enriches Indigenous Oil-Degrading Bacteria," in *Science* magazine. The research for the study was paid for by an existing grant Dr. Hazen held with the Energy Biosciences Institute (EBI).[52] Dr. Hazen serves on the executive committee and the executive board and has conducted other research funded from EBI. The EBI is supported by a ten-year, $500 million grant from BP.[53]

Dr. Hazen reported to the press that he was "convinced that bacteria have already eliminated the hazard posed by the plume" and that "we no longer see any deep plumes that can be attributed to the leak."[54] The LBL press release announcing the findings stated that "this degradation appears to take place without a significant level of oxygen depletion."[55]

Dr. Hazen's findings were, however, very isolated. They were not supported by the work of other scientists, including Dr. Joye's, below, as well as, for example, the documentation of two plumes discussed in the previous chapter by the Woods Hole Institute that were released at the same time as Dr. Hazen's article. Moreover, the conclusion that microbial activity did not lead to an oxygen drawdown is not demonstrated in the article, which does not show rates of oxygen consumption nor does it show oil consumption coupled to oxygen.

The Microbial Snot Highway

In September 2010, I walked onto the *Oceanus* research vessel docked in Gulfport, Alabama. Chief scientist Dr. Joseph Montoya, professor of biological oceanography at the Georgia Institute of Technology, greeted me briefly as he hurried back to a meeting under way around the ship's kitchen table. Among those seated staring at laptop computers and a raised video projection screen were Dr. Joye and Dr. Tracy Villareal, a biological oceanographer at the University of Texas at Austin Marine Science Institute. Dr. Villareal was chief scientist on the *Oceanus*'s partner research vessel, the *Cape Hatteras*. The scientists had hoped for one large ship but instead got two small ones.

For a newcomer, it is shocking to see the amount of equipment packed into these vessels. Every available inch of space is transformed to provide room for computers, flow cytometers, the MOCNESS (pronounced like "Loch Ness")—the Multiple Opening/Closing Net and Environmental Sampling System—and more, with just a few inches left over for bodies to squeeze in. The sleeping bunks are tight, but fortunately they are not used much. The vessels and equipment are too expensive and rare to waste a minute on such luxuries as sleep, so work goes on around the clock. As Dr. Joye explains, "You don't want to miss anything so you better stay awake."

The team, which included microbiologists, isotope geochemists, chemical ecologists, physical oceanographers, geologists, and biogeochemists, had already been out for two weeks and was docked briefly before heading out for another two-week voyage. As I joined them, they were working out some of the technical problems associated with communication between the two boats. Out at sea, cell phones did not work, nor did e-mail or Facebook. Dr. Villareal had a suggestion. He had found the best luck communicating with his wife via Second Life. What did they think, he asked jokingly, about communicating through avatars?

Getting back to business, the scientists began comparing maps. Where would they go in their hunt for the BP Macondo oil monster this time? This was now the third research cruise for many involved. After the *Pelican*, Dr. Joye, this time no longer benched by her back, had joined Dr. Vernon Asper and others on the *Walton Smith* research cruise from May 25 to June 6. The *Walton Smith* hunted for plumes, not an easy thing to do in a huge ocean given that deep waters are constantly on the move. They chased the plumes for days by following the signature of the microbes that eat away at them: the absence, or marked reduction, of oxygen in the water. Dr. Joye describes it as "chasing shadows of the real thing." Eventually, the game of cat and mouse ended and they found plumes once again lurking deep at the bottom of the ocean.

Back in the Gulf again, the team was looking for oil on the ocean bottom and once again playing plume raider.

Dr. Montoya showed me their two prized research devices: a huge CTD (conductivity, temperature, depth) rosette used to take water samples and a giant steel multicorer used to take sediment samples. Both are lowered by a large crane off the back of the boat that sends them thousands of feet below the ocean's surface. Using the devices is an acquired skill. The rosette, for example, operates using sensors. Deciding when to open the capsules to capture the water sample requires patience, precise timing, and knowledge of what the sensors are telling you. Dropping the multicorer is an entirely different skill. Lower too fast and too far and you'll crash on the bottom of the ocean. Fail to go far enough and you will miss the oil.

The team was experienced, and except for one significant mishap that resulted in a broken finger for Dr. Joye, they got both excellent and startling results. They found the plumes' hallmark signature, the destruction they leave in their wake: areas of depleted oxygen. They also discovered oil spread out across miles of the bottom of the ocean from the wellhead, in some places as much as two inches thick. They found dead sea creatures, including recently dead shrimp, worms, and other invertebrates. And in areas where they should have seen abundant life, only microbes survived.

As Dr. Joye told Diane Sawyer on ABC's *World News*, "We're finding it [the oil] everywhere we've looked."[56] And in an article published and widely reported by the Associated Press, Joye said in a ship-to-shore telephone interview, "'I expected to find oil on the sea floor. I did not expect to find this much. I didn't expect to find layers two inches thick.' Joye said 10 of her 14 samples showed visible oil, including all the ones taken north of the busted well. She found oil on the sea floor as far as 80 miles away from the site of the spill."[57]

I spent a good amount of time on the *Oceanus* and the *Cape Hatteras*, and even more talking to Drs. Montoya, Villareal, and Joye. Let's just say, excluding the discussion of avatars, a lot of what was said I did not understand. Lose track of the trail of discussion for one minute and before you know it the trail has led right up to life on Mars (or lack thereof), and trying to figure out how the discussion got there is virtually impossible for a layperson. I did catch on very quickly, however, when Dr. Joye pulled out her photographs of core samples demonstrating how the oil got to the bottom of the ocean.

The route was what Dr. Joye then called the "microbial snot highway." She later cleaned up her language for popular consumption. As *Time* magazine quoted her in its verbatim section, "It's kind of like a slime highway from the surface to the bottom. . . . Eventually the slime gets heavy, and it sinks."[58]

What Dr. Joye and her colleagues found was that microbes were able to eat minuscule portions of the oil and gas. As they chomped, they not only used up oxygen, they also excreted mucus. That mucus created a drippy slide by which oil moved down until it settled at the bottom of the ocean floor. With so many microbes at work, there was a great deal of mucus and a great deal of pathways to carry the oil ever downward.

Once the oil settled, there arose many concurrent problems. The oil is toxic and can kill a good deal of what it comes in contact with. But the mere fact of the additional layer of oil creates its own hazards. Dr. Vernon Asper was on the first leg of the *Oceanus* voyage. As he later explained, "Let's suppose you're a worm down here, and you make a living sticking your head up through the mud and feeding. Now you've got two inches of goo on top of you. You're going to have a real hard time surviving. And not only that: the worms help to oxygenate this sediment. With this stuff, whatever it is [oil, oil and mucus, or oil and dispersant], on top, it's going to be really hard to get oxygen into the sediments. So this could be potentially a large area of very serious impact." That worm's fate could have ripple effects up the food chain, too, starving "the deep-diving things that feed on deep organisms. Sperm whales dive really, really deep, and they're eating squid and whatnot. All kinds of fish dive way down there and graze. Any time you affect part of the ecosystem, it's definitely going to have an effect throughout the ecosystem."[59]

Moreover, hurricanes can pick up the oil off of the bottom and carry it once again to shore at a later date.

In response to these and similar findings, an outraged Casi Callaway asked in a radio interview, "And then frankly, we've got to know the plan for sucking oil off the bottom of the Gulf of Mexico. Is that even possible? How can that be done?" So far, she has not received an answer.[60]

Back in the cramped quarters of the laboratory of the *Oceanus*, I waited while Dr. Joye "had dinner" via a Skype hookup with her husband, Christof, and Sophie, their two-and-a-half-year-old daughter. Sophie is very blond and very bright, and very much does not want to sit and finish her dinner. Dr. Joye coaxes her to stay at the table by starting a song. She then fights to hide the tears while Sophie sings "Itsy Bitsy Spider" but lets them fall freely when the call ends. This was only her second research voyage since the birth of her daughter. "I wanted Sophie to know I was going out here to do something," Dr. Joye explains. "I didn't want her to think I just left her for no reason. So, I told her that the ocean is broken and I'm going to fix it." Whenever she gets the chance, Sophie now asks, "Did you fix the ocean yet, Mommy?"

As Dr. Joye would later discover, oil was not only traveling to the bottom of the ocean via mucus, it was also traveling via soot left in the water following the burning of the Macondo well oil.

Armageddon

Across the Gulf, media access to oiled beaches was tightly controlled and often fully restricted. Nonetheless, daring reporters and nimble authors could slip along rocks and the sides of cliffs to see oil that the local sheriff and BP security company officials did not want us to see. Access to the air was another story, particularly access to the spill site itself, which was all but completely denied. Planes could not fly overhead and boats could not approach the site without prior permission, which was rarely granted.

The operators of Southern Seaplane in Belle Chasse, Louisiana, for example, called the local coast guard and Federal Aviation Administration (FAA) command center for permission to fly over restricted airspace in the Gulf of Mexico with a photographer from the *Times-Picayune*. They got a BP contractor who questioned them extensively. "Who was on the aircraft? Who did they work for?" Rhonda Panepinto, who owns the company with her husband, Lyle, recalled. "The minute we mentioned media, the answer was: 'Not allowed.'"[61]

This was not an isolated incident, and the Panepintos wrote their senator, David Vitter (R-LA), in protest. "We are not at liberty to fly media, journalists, photographers, or scientists," they wrote. "We strongly feel that the reason for this massive [temporary flight restriction] is that BP wants to control their exposure to the press."[62]

When I was arranging a flight to cover the spill the pilot, Lance Ryberg, asked where I wanted to go. "The spill site," I said. He just laughed in response. "Sure. Seriously, though, where do you want to go?" It was a spectacular flight. It was arranged by Caroline Douglas of South-Wings, a unique nonprofit organization that provides skilled pilots and aerial education to "enhance conservation efforts across the Southeast."

We went up in Ryberg's tiny two-seater 1976 Citabria and flew so low that I could open the window to take photographs (and get air so I would not get sick). I even steered from the backseat. While breathtaking, our flight path was strictly limited to hugging Louisiana's coast.

Thus, it was both in rare and uniquely horrifying moments of television that the public was allowed to view the burning of the Macondo well oil on the ocean's surface. These opportunities were only ever granted from the air or at a great distance from the water. In fact, when Jeffrey Kofman of ABC's *Nightline* was granted one of the few opportunities to film from onboard a vessel at the spill site, there were no burns taking place even though he was there in June, a month in which dozens of burns were otherwise taking place daily.[63]

Even from a distance, the scene looked like Armageddon. The Gulf itself was on fire. Flames licked the surface, and dark black smoke billowed far into the horizon. As the Gulf burned, the coast guard, the air force, the National Guard, and every other force gathered at the scene not only allowed it to blaze, they were making it happen. The ships brought out to capture the oil simultaneously burned off the natural gas and part of the oil they captured, such that flames sprayed out of their sides as if they were fire-breathing dragons. The surrounding waters, meanwhile, were filled with oil and dispersant as far as the eye could see while dozens of cold steel ships stood by, as though gathered for battle.

The "controlled burns" of the oil on the Gulf surface began on April 28 and continued through July 19.[64] The Joint Unified Command reported that 411 controlled burns were conducted, but it does not indicate whether this number accounts for each fire (there were several fires at a time) or each series of burns. The command reports that the burns removed more than 11.14 million gallons of oil from the open water "in an effort to protect shoreline and wildlife."[65] The burnings ultimately removed approximately 6 percent of the oil released into the Gulf.[66]

Burning the oil was a desperate response to a desperate situation, to be sure. The oil was gushing from the wellhead, and no one wanted to see the oil on shore. Not everyone was convinced that it was necessary, but all agreed that it came at an enormous cost.

Microbiologist Wilma Subra told me, "If you can burn oil, you can siphon it up." Before oil can be burned, it is encircled in flame-retardant

Hundreds of "controlled burns" of Macondo well oil took place on the surface of the Gulf from April 29 through July 19, 2010.

boom. The oil stays enclosed in the boom while it burns. Given that the oil is encircled and held in place, it would be just as simple to siphon off the oil as burn it, Subra argues. The reason it was not siphoned appears to be straightforward: BP did not have enough vessels to hold all the oil it was gathering until months into the disaster. Moreover, the use of dispersant made it impossible to utilize even available vessels.

The EPA's Hugh Kaufman reported, "One of the things that happened is they brought this big boat, *The Whale*, in from Japan to get rid of the oil, and it didn't work because the majority of the oil is spread throughout the water column over thousands of square miles in the Gulf. And so—and there's been a lot of work to show [that] the dispersants, which is true, make it more difficult to clean up the mess than if you didn't use them."[67]

Bob Grantham of TMT, the company that owns *The Whale*, said in a statement: "When dispersants are used in high volume virtually from the point that oil leaves the well, it presents real challenges for high-volume skimming."[68] In the end, *The Whale* was sent home, its services unusable.

At her home in Dulac, Jamie Billiot smelled the burning long before she saw it. The smell was "unnatural," she recalls. "It was terrible. It was a harbinger of something awful."

Not only was the crude oil burning, so too was the dispersant. Both released deadly toxins into the air that sickened workers at the spill site and people all along the coast. People used to breathing the fresh, crisp ocean air were suddenly exposed for three long months to air pollutants in amounts "typical of urban areas in U.S. cities," according to NOAA, which noted that "15 to 70 kilometers downwind from the oil spill, concentrations of certain hydrocarbons were much higher than found in typical polluted air."[69]

Jamie's father, James, describes that when the burning took place, "Everyone knew it. It was like being in New York or Chicago, like standing next to a motor car running all day. You can't see it, but you can feel it in the air." He tells me that when you walked outside, "you could pass your hand on a car and your fingers would come up dirty" from the soot. "Just wait and see," he predicts; in twenty years "we're gonna have the highest cancer rates you ever seen."

"God Help Us All"

A hand-painted wooden sign posted on the highway leading to Grand Isle, Louisiana, declares GOD HELP US ALL. This same plea ends one of the thousands of amateur videos taken by Gulf coast residents and visitors horrified by what has become of their waters and beaches. Greg Hall posted his video on YouTube in June; he was just one of several people to capture an event reported in several locations across the Gulf.[70]

Hall's video shows the Pensacola, Florida, surf, the camera pointing down and out to sea. The waves pouring in are not normal. They are literally bubbling, as if the water were boiling. The bubbles are rising from circles of white-orangish goo, the telltale mark of chemical dispersant. "Look how the water's bubbling. There's definitely something in it that's causing it to do this," Hall says. "God only knows what's that in the water that's making it bubble like it's got acid in it." Footsteps splash loudly in the surf as the camera turns around toward the beach. The all-too-familiar sight does not cease to shock: thick wet gooey black oil smeared in globs, stretching out along the coast. "It's unbelievable to think these used to be called the world's whitest beaches, and look at them now," he tells us. "God help us all."[71]

I went to Pensacola Beach for the first time in my life in July. It was a surreal experience. Pensacola is timeless. Everything about the beach screams 1950s Americana—from the giant PENSACOLA BEACH sign greeting visitors in classic '50s yellow, red, and turquoise, with a seahorse painted blue, orange, green, yellow, and white looping over on top, to the rows of red-and-white peppermint candy–striped umbrellas dotting the white sandy beaches.

Children ran along the beach, and adults sunbathed, in images that could be from any decade in the last fifty years. Once I got a little closer, however, I could see that something was off. First came the realization that few people were actually in the water. The closest most came was walking along the water's edge. Next came the reason. When I walked up to the water, visible oil and dispersant washed in on waves, and tar balls littered the beach as far as the eye could see. As I walked into a more populated area of the beach, the effect became much more bizarre.

In a scene that looked straight out of the movie *Jaws*, tourists sat under umbrellas shielding them from the sun's ultraviolet rays, while side by side BP cleanup crew workstations and workers in hazmat suits scooped up the oily mess.

Early on, restrictions were placed on BP's cleanup crews to keep them from speaking to the public or the media. By July, these restrictions had been eased, although the crew members remained reluctant to speak.

I found crew leader Brian Jones almost eager to speak, however, as though his desire to tell his story had been bottled up for weeks. He had come to Florida from the coalfields of West Virginia so that when his six kids, ages three through fourteen, asked, "'What were you doing when the *Deepwater Horizon* exploded, Dad?'" he could say he was doing something to help. It was also good money. He describes watching teenage boys the day before playing in the ocean, diving directly into the oil and rising, laughing at the oily sheen on their skin. They played with the oil tar balls as if they were footballs. "I would never let my kids play on this beach," Jones protests. "It's awful."

The only thing missing to make this scene complete was *Jaws'* Mayor Vaughn of Amity Island, coaxing his wife and son into the water to prove that all was safe. It turns out that I had missed that part of the story just two weeks earlier, when, right on cue, the governor of Alabama announced that he would spend Father's Day weekend with his wife, Patsy, two of their daughters, and four of their grandchildren in Orange Beach. A press release announced that the governor, "who has been encouraging people to visit Alabama's beaches," would be spending the weekend "with his family on the state's coast."

In the most amazing use of spin I think I have ever heard, the governor told the gathered reporters who followed him there, "You're going to see people cleaning up tarballs on the beach. They started at seven this morning, but what it does is make the beaches as clear as they've ever been since I've been coming down here."[72] We do not know what was in the water the day the governor and his family showed up, but one month later, a local TV news team conducted water and sand samples that severely contradicted the governor's assurances.

A sunbather walks past workers as they clean up oil from the sand in Orange Beach, Alabama, in June 2010.

News anchor Jessica Taloney, reporting for local WKRG-TV news, took her crew to five separate beaches in Alabama—including Orange Beach, Dauphin Island, and Gulf Shores—to test the water and sand. Taloney reported that it had been more than a week since Alabama's beaches had seen significant oil. Swimmers were "taking their chances" and going into the water, beachgoers were walking along the shore, and children were playing.

Wearing a blue baseball cap, a T-shirt, and white shorts, Taloney looked like any tourist as she walked along the beach, occasionally stopping to dunk jars into the ocean where people swam, the sand where they walked, and the water off the edge of a boat next to an orange boom, "where people must be working." On Governor Riley's Orange Beach, a child was digging in the sand at least twenty feet away from the water's edge. Taloney took a sample from the ocean water that collected between the child's legs. To the naked eye, the samples looked normal.

Taloney took her samples to a lab, where Bob Naman, an analytical chemist with nearly thirty years of experience, tested them. The amount of oil and petroleum on Alabama's beaches should be undetectable. At most, Naman said, he would expect a sample to contain 5 parts per million (ppm) of oil. Instead, the samples tested five to forty-five times higher—ranging from 16 ppm to an incredible 221 ppm. The water where the child was playing tested at 51 ppm—ten times higher than should have been allowed. In Gulf Shores, where people had been swimming, the water tested at 66 ppm. The sand where beachgoers were walking had the highest concentration, 211 ppm. The same concentration was found at Orange Beach—where children were playing and the governor had been visiting.

Even more disturbing was the sample that could not be tested because it exploded in the lab. This sample came from Dauphin Island Marina, near the orange boom. "I've never seen anything like it in thirty years," Naman said, looking perplexed at the blown-out beaker in his hand. "It is either due to the methanol or methane gas or the presence of the dispersant Corexit."[73]

Oil and Corexit are both toxic. People across the Gulf coast reported illnesses associated with exposure to oil and dispersant in the air, in the waters, and on their beaches that continue to this day.

Wilma Subra received the prestigious MacArthur Fellowship (aka "Genius Award") in 1999 "for helping ordinary citizens understand, cope with and combat environmental issues in their communities." Subra, now age sixty-seven, has lived in Louisiana her entire life, and in New Iberia, population thirty-three thousand, since 1965. New Iberia is situated about forty miles from the Gulf coast on the Bayou Teche; the land is lush and green and far more likely to be encountered on the road to somewhere else than as a final destination. Being a woman, a microbiologist, and a chemist in the 1960s deep in the back roads of Louisiana was a challenge. It gave Subra both a strong backbone and a willingness to take on fights others shy away from. She has dedicated her life to exposing and redressing the harmful impacts of the omnipresent and uniquely powerful petrochemical industry on Gulf coast residents. Louisiana is the second largest producer of petrochemicals in the nation, and the Mississippi River corridor alone, from Baton Rouge to New Orleans, is home to more than two hundred chemical and petrochemical facilities.

Subra works for herself out of her home, right next door to her daughter Emily and her grandchildren Kaitlyn and Troy Jr., whom she takes care of while her daughter is at work. Subra worked for fourteen years for Gulf South Institute and contracted with, among others, the EPA, conducting cancer and environmental impact studies and developing mechanisms and protocol for national programs. Then she went out on her own, and today she is a consultant for state and federal governments, as well as nongovernmental organizations such as the Louisiana Environmental Action Network (LEAN) and the Lower Mississippi Riverkeeper (LMRK).

Since the explosion of the *Deepwater Horizon*, Subra has traversed the Gulf coast holding public briefings, meeting with government officials, testifying before Congress, talking to reporters, listening to every concern from every community member she meets, and sharing people's concerns with everyone who will listen. The concerns have been many, constant, and consistent. People all along the coast have reported suffering from headaches; nausea; respiratory problems, irritated eyes, nose, throat, and lungs; and asthma attacks. Those who came into physical contact with oil and dispersant have reported rashes, peeling, and even burned skin.

Dr. Ott has conducted similar briefings all along the coast and has heard identical reports. Between the two of them, Subra and Ott have drawn thousands of attendees to their events.

Illness has been caused by several sources: sprayed dispersant, controlled burns, oil and dispersant being carried as an aerosol along the coast, direct contact with oil and dispersant (such as on beaches), and the extreme exposure of workers involved in all aspects of the cleanup and recovery-well drilling.

Subra told me that based on the chemical composition of the oil and dispersant, those most at risk are children, pregnant women (and their unborn children), the elderly, and the already sick.

I have spent much of my career studying the impact of oil operations across the United States and around the world. As I met and spoke with people along the Gulf coast, one very strong comparison was constantly brought to mind: it is as though these communities, instead of living and working in and on the ocean, were picked up and plopped down next to rows of massive dirty oil refineries.

Along the coast, emissions were created from several sources, including the burning of oil and the burning off of natural gas and oil on the containment ships. The combination created particulate matter and a visible haze. In many of the fishing communities Subra visited, BP cleanup workers were reluctant to speak about the impact on their health, even though they suffered some of the most extreme effects. She was told by many that they feared for their jobs if they spoke out. Moreover, their contracts with BP stipulated that workers could not speak with the media and in the case of an accident during the cleanup efforts, BP would not be liable.[74] For a time, the "wives were speaking out for their husbands, but then word came back that if you don't keep quiet, they will lose their jobs," Subra tells me. "So BP succeeded in shutting down the fishing communities as well as the wives."

Not all of the wives were silenced, however. Kindra Arnesen's father was a commercial fisherman, her husband is a commercial fisherman, and, in her words, "every man that I've ever known, loved or respected is on the water. They're good men."[75] Arnesen is young and angry. She described herself as "an uneducated housewife" who has turned her worry about her husband's, children's, and community's health into action.

Arnesen lives off Highway 23 in Plaquemines Parish, "right out in the middle of nowhere in the boondocks," almost as far out into the Gulf as homes are able to stand and in the area hardest hit by the oil gusher. Thick oil washed up daily on Arnesen's shores. It is a place where people are far more likely to encounter one another while boating than while walking or driving. I had the unique experience of being in a boat off of Plaquemines Parish and encountering a person I was set to interview in an oncoming boat. Rather than go ashore, we sat out in the middle of the water and conducted the interview there, boat-to-boat.

Arnesen's husband, David, was shrimping downwind from the disaster site on April 29 when he called to say that not only was he sick, so too were men on the seven other shrimping boats working near his boat. Kindra told CNN, "I received several calls from him saying, 'This one's hanging over the boat throwing up. This one says he's dizzy, and he's feeling faint. Everybody's loading up their stuff, tying up their rigs and going back to the docks.'"[76]

Arnesen knew that something was seriously wrong, because "shrimpers work through illness," and cutting a trip short can cost $1,000 per boat. The shrimpers' symptoms became all too common in the Gulf: vomiting, dizziness, headaches, shortness of breath. After three weeks of coughing and feeling weak, her husband finally agreed to see a doctor. He was diagnosed with respiratory problems and prescribed medicines, including an antibiotic and cough medicines.

Arnesen knew that it was the vapors from the oil and the dispersants that made her husband and the others sick. The smell was "so strong they could almost taste it."

When asked at a news conference about people getting sick while out on the Gulf, BP CEO Tony Hayward said, "I am sure they were genuinely ill but whether it has anything to do with dispersants and oil, whether it was food poisoning or some other reason for them being ill. You know food poisoning is surely a big issue when you have a concentration of this number of people in temporary camps, temporary accommodations, it's something we have to be very, very mindful of."[77]

This quote was repeated with incredulous anger more times around the Gulf than even Hayward's now infamous stated desire to get his life back. Many people could understand a man wanting to skip out on this trauma, if he could. They could not, however, tolerate his dismissal of their debilitating health crises caused by *his* oil as mere "food poisoning."

In Alaska these symptoms are called the "*Valdez* crud." It is "crud in their lungs that's been there for twenty years," Dr. Ott tells me. After the *Exxon Valdez*, people went to health-care professionals complaining of coughs, sore throats, and respiratory problems. They were understandably misdiagnosed as having a common cold. These were not cold symptoms, however, but rather symptoms of intense toxic chemical exposure, which became chronic, lasting for decades.

Like so many other women Subra encountered, Arnesen said that she was afraid to go public with her husband's illness for weeks because he had signed a contract to work with BP. But when she heard Dr. Ott speak at a public forum, Arnesen decided to speak out and "to organize other wives to ask questions."

Now speaking out at public forums herself, Arnesen told an audience in June what her life had become since the disaster. "We're out in the middle of it," she explained. "This is on all three sides of my home. I walk outside and there is a haze. They're called 'bad air days.'" On

Kindra Arnesen, who lives at ground zero of the Macondo blowout, speaks out in November 2010 about the impacts on her family and community.

these days people are told to stay inside, shut their windows and doors, and turn on their air conditioners. It's an event all too familiar to people who live near oil refineries. Their response is the same as Arnesen's. She looks incredulously at the audience and asks, "Where do you think the air comes from that's inside the house? Outside. These people—they never cease to amaze me!"

The effects on her children, five-year-old David Jr. and eight-year-old Aleena, have been severe. They have repeatedly broken out in rashes that "magically" clear up when Arnesen takes them away from their home. When she returns, so do the rashes. "Today she is broke out again," Arnesen said of her daughter. "Not to mention that my beautiful, healthy, straight-A student, gorgeous daughter has a double ear infection and respiratory problems."

Arnesen then asked for help. "I know that my parish only makes up 2 percent of Louisiana's population, but does that make my people expendable? They are slowly poisoning every person that I have ever been close to in my entire life. My people are more important to me than their bottom line, and that is my bottom line."[78]

In June, Subra reported to Congress that in order to capture and document the human health problems resulting from the various aspects of the spill, she had developed a human health survey instrument and, through LEAN, provided the instrument to community members in the affected areas and the areas to be affected. "The results of the human health impacts documented in the surveys," she told Congress, "will be used to assess the severity and magnitude of the human health impacts associated with the *Deepwater Horizon* disaster."[79]

I asked Subra for the results. In December, she wrote to tell me that there were none, because the community was too scared to complete the survey, fearing that it would hinder their ability to receive claims compensation from BP or through the federal process set up by Kenneth Feinberg. Unable to get the information from the people, Subra prepared a list of health symptoms associated with the crude oil and the dispersants, and "the community used the health symptoms to educate their doctors and receive appropriate health treatments." The list was critical, because a constant problem throughout the Gulf region has been getting the appropriate information to health-care providers.

WOIA-TV in San Antonio, Texas, for example, reported in June that local hospitals were tracking "patients with suspicious symptoms coming in from the Gulf Coast," and doctors were having trouble "distinguishing it from the flu." Dr. Claudia Miller from the University of Texas Health Science Center told WOIA that many patients were suffering from toxicant-induced loss of tolerance. Chemical exposure from burning oil, toxic fumes, or dispersants from the spill caused a loss of tolerance to household products, medications, or even food.

"Things like diesel fuel, exposure to fragrances, cleaning agents that never bothered them before suddenly bother them," Dr. Miller reported. She began educating primary care doctors in the area on how to diagnose the illness. "What makes it challenging is that patients show up with nonspecific symptoms—headaches, fatigue, problems with memory and concentration, upset stomach," she explained. She warned the TV audience to wear protective equipment and stay out of areas with smells. "If they feel sick, the smells are usually chemicals that can make them ill," she said.[80]

Another frequently reported problem was that doctors and the public were categorizing symptoms as heat-related that could instead be the far more serious results of chemical exposure, requiring very different remedies.

The Louisiana Bucket Brigade (LABB) is a nonprofit organization founded in 1999 that works with communities affected by the petrochemical industry to, among other things, train them on how to monitor the quality of their air and water. Much of its work has been done in neighborhoods across the street from refineries.

LABB was founded in response to inadequate regulation, oversight, and enforcement of the industry by state EPAs and other government agencies, a particularly problematic issue in Louisiana. LABB launched an online oil spill crisis map to track the oil spill, including the health impacts reported by communities themselves. LABB also went to affected areas to listen to and learn from those most affected.

Callie Casstevens of LABB reported in a blog posting in July, "Time and again I have heard fears of chemical exposure categorized as effects from the heat. . . . It is very hot in Louisiana at this time of year," she concedes, but the health assessments are being made based not on examinations of the patients but instead on misinformation provided by BP and the state of Louisiana.[81]

Casstevens attended a public forum in Thibodaux, Louisiana, organized by BP and several government agencies. She reports that every single flyer on the health and safety table described the symptoms of heat stress but "nothing about the dangers of being exposed to the oil, dispersant 9527, or Corexit 9500. Nothing."[82] Many people who are suffering symptoms, particularly those exposed to prolonged exposure, may never report their symptoms, believing them to be heat-related. If they do report them, they are likely to have the cause misdiagnosed.

Fortunately, people were willing to use LABB's Web site (www.oil spill.labucketbrigade.org) to post anonymous reports on health issues. A few examples are indicative of the commonly reported concerns:

7/10/10 *Burning feet after sand gets in flip-flops, Long Beach, MS*
My feet burned after sand from toxic beach in Long Beach, MS, got in my shoes.

7/2/10 *Health problems for my three-year-old son, Pass Christian, MS*
My 3-year-old son was diagnosed with pneumonia on Monday morning. He was admitted to the hospital Monday afternoon and finally discharged Wednesday afternoon. He was a perfectly healthy and happy 3-year-old boy until this incident. I read that children have been susceptible to dispersant-related pneumonia. If this is true, I have a feeling that this was his problem, as he has had no significant health problems up to this point. He was in the hospital for three days, with the fourth day at home. I had to miss nearly a week of work.

6/30/10 *Respirators for workers, Port Sulphur, LA*
Two local tankermen in BP hazmat classes were concerned that respiration of toxic chemicals was not addressed during the course. When the course's director was questioned regarding respiration of chemicals, the question "was basically ignored."

6/14/10 *Foot Burned, Grand Isle, LA*
I was walking and had flip-flops on. I stepped in what appeared to be sludge, it was green and smelled toxic. My foot felt like it was on fire, like someone took a match and was holding it

underneath my foot. It had actually given me what appears to be a second-degree burn.

6/12/10 *Oil spill clean-up worker with open sores on his hands and arms, Grand Isle, LA*
Supervisor for BP subcontractor reported to first-aid tent that he had a worker referral from a physician to the nurses for open sores on his hands and forearms. The sores contained blood and pus. Reported that this worker is "known for safety violations," like not wearing protective gear.

People working for BP near the site of the explosion, laying boom around the Gulf, or cleaning oiled beaches suffered the greatest exposure and many of the worst health problems. When operations began, BP initially not only refused to provide adequate health protection to its workers but also threatened their jobs if they protected themselves.

In May, LEAN and LMRK purchased and distributed protective equipment, including respirators, goggles, gloves, and sleeve protectors.

Workers attempt to clean up a beach drenched in oil in Port Fourchon, Louisiana, in May 2010.

In response, BP threatened to fire the workers if they wore the equipment. LEAN and LMRK took BP to court and won. In spite of orders from the court and from the Occupational Safety and Health Administration (OSHA), BP continually failed to readily provide adequate protective gear to response workers, further exacerbating the impact on workers' health.

Commercial fisherman John Wunstell Jr., for example, was part of a crew that was burning oil. He spent a night on a vessel near the source of the spill and was hospitalized after experiencing a severe headache, upset stomach, and nosebleed. "I was completely unable to function at this point and feared that I was seriously ill," Wunstell reported. He is now, with others suffering from illnesses related to the disaster, part of a class-action lawsuit against BP and Transocean.[83]

The Louisiana Department of Health and Hospitals (LDHH) noted in June that dozens of complaints, most from spill workers, had been reported that were related to oil exposure. In just one week, eleven workers involved in cleanup operations out on the water visited the emergency room at West Jefferson Medical Center, just outside of New Orleans. They complained of "respiratory problems, headaches and nausea."

Dr. Damon Dietrich, who treated them, called their concerns "a pattern of symptoms" that could have been caused by the burning of crude oil, noxious fumes from the oil, or the dispersants dumped in the Gulf to break the oil up. "One person comes in, it could be multiple things," he said. "Eleven people come in with these symptoms, it makes it incredibly suspicious."[84]

In July, the LDHH reports related to exposure to pollutants from the oil spill skyrocketed to 227 cases (cumulative since April).[85] Again, the vast majority were by workers involved in the spill operations and cleanup. Seventeen people had been hospitalized as a result of their illnesses. The most frequently reported symptoms were headache, dizziness, nausea, vomiting, and upper respiratory irritation.

The LDHH made its final report on illness related to the oil disaster in September. The symptoms remained consistent and totaled 415 cumulative cases, including 329 workers and 86 people from the general population. Eighteen people had been hospitalized.[86]

There is a concern, however, that just as the general population is not reporting health problems, so too are workers' illnesses going unreported or undocumented in publicly available data. There are several categories of workers involved in spill operations: those on BP vessels, at the spill site; those in BP work crews, primarily on the beaches; and those employed through BP's Vessels of Opportunity Program, who work on their own boats and who, though being paid by BP, rarely come into contact with BP while working. The third category appears to have sought out and received the most formal health-care services, whereas the first two categories, those most directly under BP's guard, have not only been discouraged from reporting health-related problems, they also have reportedly been forced to receive only private care under BP's oversight when they do.

In June, Anna Hrybyk and Shannon Dosemagan of LABB visited a so-called BP zone—the area off limits to the public or the press—in Grand Isle, Louisiana. They spoke to a nurse who was staffing the medical tent within the BP zone. This nurse was part of an official local parish government response that was advertised as the place where workers and others involved in the spill efforts should go if they experienced health problems.

"The nurse was incredibly frustrated," Anne Rolfes, the founding director of LABB, recounted in congressional testimony. "She had arrived on the scene to treat medical emergencies . . . but was told she could only offer aspirin and Band-Aids. 'BP is running its own Emergency Medical Service,' the nurse reported. 'The sickest people are being taken there and avoiding the parish emergency center.'"

Two weeks later, Hrybyk returned to the site to get more information on what types of health problems the workers were being treated for. She found that all oil cleanup workers under contract with BP were required to receive care from the privately contracted firm Care EMS. While she was at a West Jefferson Medical Center (WJMC) tent, Hrybyk reported, a BP cleanup contractor came in to ask about a worker who had open sores and blisters on his hands and forearms after having come into contact with the water. The doctor wanted the worker to be treated by the WJMC staff. However, "much to the nurse's discontent, she was bound by the protocol to refer the worker to the BP EMS even though his doctor referred him to the WJMC." According to the nurse, "Contractors who know and trust the work of the WJMC are livid about this BP imposed protocol."

Hrybyk reported that even though BP's Care EMS area was heavily guarded, she managed to speak with members of the emergency medical team on duty. She was told that they were creating detailed incident reports for every worker they saw and that those reports were sent to the Joint Unified Command Center in Houma, Louisiana. "I have been chasing the head nurse at the Houma Command Center for weeks trying to get those reports," Hrybyk said. "I am now going to submit a Freedom of Information Act request to the U.S. Coast Guard for their reports on worker health incidents."[87]

Fumes from the oil and dispersant were common. I, and everyone else I spent time with in the Gulf, suffered headaches regularly when spending even short periods on soiled beaches and near the coastline. However, the most common health problem that seems to be emerging from BP and the LDHH is heat-related illness. I spoke with several BP crew leaders over the months and was repeatedly told that the most common serious health problem reported by their crews was related to heat exhaustion.

Hospital records confirm this, with the highest percentage of problems documented by the LDHH being heat related. Even before I learned that information on chemical exposure was not being provided to workers, the public, or health-care professionals, this reporting seemed odd to me. All the people doing the work were from the area, very accustomed to being in and working in the Louisiana sun and capable of determining when they should take shelter from it.

When I was conducting these interviews, the crew leaders also seemed very aware of the need for breaks, shelter from the sun, and access to cold water and sports drinks. Provided that these were actually given to the workers, it seems far more likely that their illnesses were instead, at best, a combination of the sun and chemical exposure.

What appears to be similar misreporting occurred in the *Exxon Valdez* disaster. Court records show that more than sixty-seven hundred workers involved in the *Exxon Valdez* cleanup suffered respiratory problems, which the company attributed to a viral illness, not chemical poisoning.

The Associated Press reported that Dennis Mestas represented the only known worker to successfully settle with Exxon over health issues. "According to the terms of that confidential settlement, Exxon did not admit fault. His client, Gary Stubblefield, spent four months lifting workers in a crane for 18 hours a day as they sprayed the oil-slicked beaches with hot water, which created an oily mist. Even though he had to wipe clean his windshield twice a day, Stubblefield said it never occurred to him that the mixture might be harming his lungs. Within weeks, he and others, who wore little to no protective gear, were cough- ing and experiencing other symptoms that were eventually nicknamed *Valdez* crud. Now 60, Stubblefield cannot get through a short conversa- tion without coughing and gasping for breath like a drowning man. He sometimes needs the help of a breathing machine and inhalers, and has to be careful not to choke when he drinks and eats."[88]

Is the Air Safe to Breathe?

Soon after the disaster began, the federal EPA began an air monitoring program to determine the health effects of the disaster on Gulf coast communities. LABB, which has been following existing programs in Louisiana and across the United States for more than ten years, reported that the EPA's efforts appeared to be "the most comprehensive air mon- itoring program ever conducted in this part of the United States. The agency is to be commended for its effort." Nonetheless, LABB identified many deficiencies, noting that its findings held true not only for the air monitoring after the disaster but also for water and soil monitoring.

Under normal conditions, air, water, and soil monitoring is a state- level responsibility. Historically, however, Gulf state governments have not given these issues high priority. Thus, there was little existing infra- structure and information when the disaster struck.

Nonetheless, in its analyses in the months following the blowout, the EPA repeatedly stated in public what is also written on its Web site: "EPA's air monitoring, to date, has found that air quality levels for ozone and particulates are normal on the Gulf coastline for this time of year."[89] However, the EPA has no data from past years to back up this claim. With no comparative data, there is no way to assert that the levels are "comparatively normal."[90]

To conduct these analyses, the EPA utilized just thirteen stationary air monitoring stations to cover the full 1,600 miles of shoreline of the five affected Gulf states. Though a historical high mark, this number was utterly incapable of adequately measuring air impact on such a vast region. LABB wrote, "The EPA continues to use a limited number of monitoring sites to extrapolate to a broad region. Given the relatively limited scope of the sampling, data should not be used for general characterizations. If the EPA does not have the data then they should simply state that fact." Moreover, the EPA provided no scientific or public health rationale for the location of its stationary monitors. "Availability of power appears to have been an important consideration," LABB reports, rather than the location of the hardest hit areas or those reporting the most impacts.[91]

The EPA did start using mobile laboratories, the Trace Atmospheric Gas Analyst (TAGA) buses, which performed much of the EPA's benzene monitoring. Rather than monitoring along the coastline and where people live, however, the buses monitored almost exclusively along major highways, which both limited and corrupted the data.

Finally, the EPA's decision to report the measurements as daily averages significantly reduced the utility of the data. Pollutant and chemical exposure typically came in concentrated bursts, most often carried by the wind. Thus, if a thick concentration occurred in the morning but the rest of the day was clear, the data would convey a relatively low exposure for that area. Anything or anyone exposed to the "pollutant burst," however, would suffer a high impact.

The Centers for Disease Control (CDC) conducted a preliminary review of the EPA's data, including the results for air, water, soil, sediment, and waste oil samples, and concluded that while the pollutants from the crude oil could have caused temporary illnesses, "the samples collected in places where non-response workers would spend time showed none of those substances at levels high enough to cause long-term health effects."[92] The CDC did not consider the impacts of the chemical dispersants in its analysis.

Federal government officials concluded that exposure along the coasts following the disaster was "typical of urban areas in U.S. cities." The phrase was frequently repeated by officials as though it were a good thing. Typical American cities are in constant violation of 1970's Clean Air Act regulations, daily exposing their inhabitants to harmful, cancerous, and even deadly pollutants. There is no reason to take solace from such an effect.

Moreover, there remain reasons to worry that the government data are not capturing the full potential harm of the disaster. For example, the EPA "was not looking for all of the potential chemicals" released by the oil and dispersant, Wilma Subra told me in December. "The Asphaltines were later identified by the agency as the cause of some of the health impacts. However, these chemicals were not being analyzed for in the air. In addition, the EPA was not sampling for polycyclic aromatic hydrocarbons [PAHs] until the very end of the ongoing spill" even though experts such as Subra and Dr. Ott identified PAHs as potential hazards at the outset of the blowout.

The EPA has stated that these findings are not conclusive and do not represent a complete or final analysis.

For now, anecdotal reports from the Gulf of Mexico are all that we have, for comprehensive studies have only begun. For example, the National Institutes of Health announced a $20 million study on workers' exposure to oil and dispersants, but half of it is funded by BP.[93] Universities and nonprofit organizations have launched studies as well. All of these are limited, however, by what people feel comfortable reporting, how doctors report and treat the symptoms, and how much access the public has to the data. Understanding the information is made all the more difficult by the fact that no one has ever seen anything like the Gulf oil disaster before.

Thus, the most common comparison—to the *Exxon Valdez*—is insufficient, because the magnitude of the Gulf oil disaster is unprecedented, in terms of the oil spilled and the dispersants used as well as in the fact that the dispersants were used underwater. Past studies have confirmed that there are serious health effects from exposure, but these studies have most likely severely underestimated the impact of the BP oil spill.

In discussing the effects on workers, for example, John Howard, the director of the National Institute for Occupational Safety and Health, told Congress, "It is important to note that in recent years several studies of previous oil spill response efforts have reported acute and chronic health effects in response workers. These studies may underestimate the health effects associated with oil response work since the magnitude and duration of the *Deepwater Horizon* response is unprecedented."[94] Also unprecedented is the level of toxic exposure to people living on the coast.

The Restricted

As mentioned earlier, media access was restricted within days of the oil making landfall. Throughout the Gulf, reports poured in of media waved off of beaches and restricted from air and boat travel. I was personally restricted from beaches all across the Gulf by private security officers hired by BP, sheriffs, and police officers. In May, a CBS news crew on a boat was attempting to film oil-covered beaches when they were threatened with arrest and told, "This is BP's rules, not ours."[95]

The policy was formalized when the Joint Unified Command issued written regulations on June 30 restricting anyone from approaching within twenty meters (about sixty-five feet) of "booming operations, boom, or oil spill response operations." Violation of "the safety zone could result in a $40,000 fine and a Class D felony."[96] The restrictions were imposed by the Joint Unified Command and enforced by local sheriff and police departments, private BP security guards, and even military personnel.

CNN's Anderson Cooper became a local hero throughout the Gulf for his extensive and personal coverage of the disaster. In a scathing opening statement on his show *AC 360* in July, he noted, "Transparency is apparently not a high priority with [coast guard commandant] Thad Allen these days." The new rules to "keep prying eyes out of marshes, away from booms, off the beaches, is now government policy. . . . What this means is that oil-soaked birds on islands surrounded by boom— can't get close enough to take that picture."[97]

Although the restrictions were placed on the general public and certainly served to protect people from oil and chemical exposure, they seemed most often enforced when reporters were present. For example, coast guard personnel turned away Jean-Michel Cousteau, the son of ocean explorer Jacques Cousteau, from a wildlife sanctuary after they discovered that an Associated Press photographer was onboard his vessel.[98]

Senator Bill Nelson (D-FL) tried to bring a small group of journalists with him on a trip through the Gulf on a coast guard vessel. His office told the *New York Times* that the coast guard had agreed to accommodate the reporters and the camera operators, but at about 10 p.m. on the night before the trip, the Department of Homeland Security's legislative affairs office called to inform the senator that no journalists would be allowed.[99]

I found myself stranded on many occasions by private boat captains who, when they learned that I was a writer rather than a mere tourist, refused to take me out in the water, much less near boom, for fear of the fine and the felony charge. Most often, it was out-of-work fishermen who were willing to take me out. When I did find myself in the water within a few feet of a boom, I had to think briefly about whether it was worth going to jail for. My answer was always a reluctant yes.

In order to cover the disaster by air, journalists and authors had to receive FAA permission. Even when it was given, a 3,000-foot restriction was often enforced. "Although there's a tremendous amount of oil, finding out exactly where it's washing ashore or where booming is going on is very difficult," said John McCusker, a photographer with the *Times-Picayune*. "At 3,000 feet you're shooting through clouds, and it's difficult to tell the difference between an oil slick and a shadow from a cloud."[100]

In response to the restrictions, on June 28 the American Civil Liberties Union (ACLU) of Louisiana sent an open letter concerning media and public access to the BP oil spill to the sheriffs of all Louisiana coastal parishes, reminding them of their obligation to respect the First Amendment rights of a free press and public assembly. The letter chronicles several incidents of deputies of various parishes denying access to journalists who were attempting to report on or document the spill.[101]

Similarly, the Society of Environmental Journalists sent a letter to Admiral Thad Allen reporting denial of access to the media and stating that the group is "deeply disturbed at the growing number of reports we have received that journalists are being prevented from doing their Constitutionally protected jobs: to provide information to the public about the mammoth oil disaster playing out on the Gulf Coast."[102]

The Dead

As oil hit her beloved beaches, Casi Callaway told a reporter in May, "We have white sugar sand beaches that stretch for 55 miles in Alabama, and a total of nearly 600 miles of tidal shoreline at stake. For them to now have oil deposits is just the beginning of what is going to be a 24-hour-a-day fight to protect one of the most valuable natural resources in America."[103]

One of the first tasks that Callaway took on in response to the disaster was organizing citizens as volunteers to monitor the beaches. There was no shortage of volunteers. Within ten days of the initial explosion, Mobile Baykeepers, the Alabama Coastal Foundation, and the Mobile Bay National Estuary Program received calls from seven thousand people volunteering to help. "I want to help! I want to do something! Give me a job!" Callaway recalled them virtually yelling into the phones.

The groups created the Volunteer Field Observer Program to monitor the shoreline, see whether oil was coming ashore, where oil was found and what the impact was, and to contact BP and the Joint Unified Command for help.

Now, when Callaway talks about her life, be it an argument with her mother-in-law or a major environmental victory, all events fall into two distinct categories: "pre-oil" and "post-oil." As in, "We were organizing for Earth Day—this was pre-oil, and we contacted the governor's office." Or "Post-oil, Coleman started day care, and it was far more terrifying to me than to him." She uses the terms so frequently and so offhandedly that I am fairly certain she is unaware that she does it at all. When I bring it to her attention, Callaway freezes in midsentence, stares, and the tears begin to fall. This is not the life she asked for or wanted.

As the oil, workload, stress, and worry increased, Callaway came home later and later at the end of each day, feeling more and more overwhelmed. Eventually Coleman, her two-year-old son, asked, "Is the beach over, Mommy?"

The oil and dispersant did more than soil beaches. As these toxic substances made their way through the ocean to the shore, they seeped into rich and vital wetlands, national wildlife preserves, and the homes and habitats of thousands of species.

In June, the U.S. Fish and Wildlife Service announced that as many as thirty-eight federally listed species protected under the Endangered Species Act of 1973 could be affected by the oil spill in the Gulf of Mexico. Of those, twenty-nine were already endangered.[104]

The Gulf of Mexico is a rich, diverse, and extremely fragile ecosystem that has provided the inspiration for great literature, art, and music for hundreds of years. It contains both beautiful sandy beaches and half of the country's wetlands—more than five million acres. Wetland ecosystems, such as marshes, tidal flats, bays, sea grass beds, and mangrove forests, serve as vital bird nesting and feeding sites and as spawning and nursery areas for many commercially important fish species. They trap nutrients and sediment from rivers as well as carbon from the air to reduce global warming pollution. They protect inland areas from wind and storm surges, acting as a vital last line of defense against hurricanes.

The area is home to some of the most lush and endangered coral reefs in the United States. Pulley Ridge, located about 150 miles west of Cape Sable, Florida, is the deepest known coral reef in the continental United States and extends for more than 60 miles. Located about 110 miles south of the Texas-Louisiana border, the Flower Garden Banks National Marine Sanctuary includes the northernmost warm-water, shallow coral reefs in the continental United States. Amounting to more than 300 acres of protected reef, this bank is home to more than 23 species of coral, 250 species of invertebrates, 175 species of fish, and 80 species of algae.

Among the endangered species that make their home in the Gulf are the sperm, sei, fin, blue, humpback, and North Atlantic right whales and the Florida manatee. Recently, killer whales have been found in the Gulf as well. Several species of fish are in trouble, including the endangered smalltooth sawfish, the threatened Gulf sturgeon, and certain sharks and grouper, listed as "species of concern." Many highly endangered birds, such as the whooping crane, piping plover, brown pelican, peregrine falcon, and even the bald eagle all have extensive habitat here. A 125,000-square-mile area in the Gulf has been identified as a critical spawning habitat for bluefin tuna, whose population has decreased by 75 percent in the last thirty years.[105]

The disaster struck at the peak of the breeding or spawning periods of a large number of species. Sea turtles, many local bird species, including brown pelicans and least terns, fish, shrimp, and smaller invertebrates that are critical species at the base of the ecosystem were just in the process of reproducing.

Millions of migratory birds that range across the Western Hemisphere—but ultimately spend the winter in or migrate through the estuaries, marshes, and other coastal areas of the Gulf—could all still be affected. Many are currently north and are expected to fly in later, "including hundreds of millions of neotropical migratory songbirds that rest and feed in these habitats during both their spring and fall migrations," explains Jane Lyder, the deputy assistant secretary for the Fish and Wildlife and Parks divisions of the U.S. Department of the Interior. These migratory birds are likely to be "exposed to oil as they forage, or possibly face starvation as a result of depleted insect, marine, and plant life due to oil incursion," according to Lyder.[106]

It will take decades to measure the impact of oil on this system. Alaska's deputy attorney general, Craig Tillery, stated that "one of the most stunning revelations" in the last ten years is that "*Exxon Valdez* oil persists in the environment and, in places, is nearly as toxic as it was the first few weeks after the spill." It has been widely estimated that "the ecosystems affected by the 1989 *Exxon Valdez* spill may take thirty years to fully recover."[107]

Simply accounting for all the dead and injured wildlife is a challenge in itself, given the vast area affected by the oil spill and, consequently, the large numbers of wildlife, including aquatic life, affected. There is also no precedent for measuring or monitoring the impact of this quantity of dispersant on such a large number and great diversity of affected habitats and species.

Seventeen federal staging areas were set up from Grand Isle, Louisiana, to Pensacola, Florida. Using daily wind and current projections from NOAA, the U.S. Fish and Wildlife Service, along with private companies hired by BP, tried to identify where the oil was moving and dispatch wildlife rescuers to those locations. But even with a coordinated multiagency effort, finding wildlife was a considerable challenge, given the area to be covered, the density of the wetlands, and the maze of islands and bays. Retrieving dead and oiled animals is also often far easier said than done.[108]

In September, a preliminary study of water off the coast of Louisiana showed that levels of deadly PAHs increased by approximately forty times between May and June.[109]

Dead marine life, of course, can be identified only when bodies wash ashore. Scientists and fishermen reported total dead zones in the areas where the oil and/or the dispersant passed—nothing living, no fish, no mammals, nothing. I visited barrier islands off Louisiana that are accessible only by boat. The cleanup workers there told me that if they kept all the dead fish and birds, they would have mountains of both, and their wives would never let them come home at night for the stench. Instead, they put them back into the ocean. Both birds and marine life that died at sea became food for other organisms. For larger mammals that died at sea, Dr. Joye explains that if they were not eaten by other species, their bodies were more likely to sink to the bottom than float to shore.

Dr. Ott noted that in the *Exxon Valdez* disaster, "only one percent of the dead seabirds made landfall in the Gulf of Alaska, for example. That means [that] for every one bird that was found, another 99 were carried out to sea by currents."[110]

Timothy J. Ragen, the executive director of the U.S. Marine Mammal Commission, concluded, "Observers will undoubtedly fail to encounter all of the affected marine mammals, as some are likely to die and sink—their loss being neither detected nor documented."[111]

Early in the spill, TV cameras caught images of dolphins stuck or moving through thick oiled waters dyed orange with dispersant. These images came to an abrupt end when the media's aerial movements became much more restricted. John Wathen of Hurricane Creekkeeper in Tuscaloosa, Alabama, flew flights in defiance of these restrictions, and his videos of dolphins and one of a sperm whale caught in the deadly toxic muck went viral. Sharks, whales, and other mammals and marine species were reported to have changed their migration patterns in order to avoid the soiled waters. The results of these changes will not be known for years.

Of course, the effects on birds were the first and most powerful images of the disaster. In the first days of the oil making landfall, the photographs and film footage of birds soaked in oil permeated the media. There is something uniquely powerful and gripping about these images. They were immediately reminiscent of the *Exxon Valdez* disaster, in response to which the public was galvanized nationally to act. As the photographs in the Gulf gripped the hearts of the public in this way once again, restrictions on both professional media and private citizens were put in place, so that as the number of oiled birds and dead wildlife increased throughout the months of the disaster, the access of the press to oiled beaches declined and the number of heartbreaking photos dwindled.

A pelican drenched in Macondo well oil swims in oil-soaked water in Barataria Bay, Louisiana, in June 2010.

Adult brown pelicans wait in a holding pen to be cleaned by volunteers at the Fort Jackson International Bird Rescue Research Center in Buras, Louisiana, in June 2010.

The Count

The Joint Unified Command issued a daily "Consolidated Fish and Wildlife Report" documenting the impact of the disaster. Among the dead documented in these reports were an endangered sperm whale and 2 melon-headed whales. Nearly 100 dolphins were killed. More than 600 endangered sea turtles died, and an additional 535 were found oiled, their fate yet to be determined. Of the turtles found dead, 801 were Kemp's ridley, 201 were green, and 87 were loggerheads. Two hundred seventy-eight turtle nests had to be removed from their natural habitats and transported to new locations. Fifteen thousand of their hatchlings were subsequently successfully released. More than 6,000 birds died, and more than 2,000 birds were found oiled but alive; these included more than 900 brown pelicans, nearly 3,200 laughing gulls, 627 northern gannets, and more than 340 royal terns.[112]

Death came in several forms, including the burning alive of endangered sea turtles.

Sea turtles are among the earth's oldest living inhabitants. They have lived on earth for 150 million years, before the time of the dinosaurs. Yet today all eight species of sea turtles are endangered. Five of these species—the leatherback, green, loggerhead, hawksbill, and Kemp's ridley—live in the Gulf.

Of these, the Kemp's ridley is the most endangered, making it one of the rarest sea turtles on earth. It also relies the most extensively on the Gulf coast for its survival. There are just twenty-five hundred nesting females left in the world today, whereas only fifty years ago, forty thousand were filmed nesting in just one day on a single beach. The Kemp's ridley is the smallest of all the sea turtles, weighing about ninety pounds.[113] Its wide mouth and its big eyes give it the appearance of having a constant smile, and its shell is often heart-shaped.

Disaster struck as the Kemp's ridley began nesting in the Gulf. Boat captain Michael Ellis had been rescuing sea turtles from oiled water, beaches, and wetlands when he came across a controlled burn. The crew, working for BP, blocked his team from the areas that had been set afire, where Ellis believed the animals were trapped, effectively shutting down his rescue operation and condemning the turtles to being burned alive. Wildlife officials at the National Marine Fisheries Service subsequently confirmed that turtles were being caught in the fires and burned alive.

Miyoko Sakashita, the oceans director for the Center on Biological Diversity, said, "Newly hatched sea turtles are swimming out to sea and finding themselves in a mucky, oily mess. News that BP has blocked efforts to rescue trapped sea turtles before they're burned alive in controlled burns is unacceptable. Hundreds of species in the Gulf are being killed or harmed by the toxic oil, but the plight of the Kemp's ridley is particularly heartbreaking since it had been poised to become an endangered species success story. Now, once again, the species is moving toward extinction."

Protests of the burning went unheeded by BP, including 150,000 people who signed a petition organized by the center and CREDO Action.[114] The center and the Turtle Island Restoration Network sought a temporary restraining order against BP and filed a lawsuit. On July 2, the lawsuit was settled, with BP and the coast guard agreeing to clear burn boxes of sea turtles and place wildlife observers onboard every vessel.[115]

This was a unique and limited success story, however, as the burns themselves continued and they were just one mechanism through which death came to the wildlife and wild places of the Gulf.

It is difficult to track the trajectory of deaths. The best source of information, the Joint Unified Command, did not begin releasing data to the public until May 28, and it is not necessarily provided in real time—that is, as the dead were found. However, the data do support anecdotal reports that the number of dead animals did not decrease over the life of the gusher. The only thing that decreased was the media's ability to report accurate numbers.

The reports document that June 5–30 was the deadliest period for turtles. About 204 turtles were reported dead during this period, much higher than any other period between May 28 and November 2.

The number of birds reported dead grew steadily, week by week, with 103 birds reported dead from May 28 to June 5 and 622 birds reported dead July 21–29, just after the well was capped. Large numbers of dead birds continued to be reported from July 14 to August 29, with a record 725 birds reported dead August 21–29. The number of birds found alive yet oiled followed the same trajectory, although far fewer birds were found alive than dead.

Finally, August 21 to September 22 was the deadliest period for marine mammals, with 53 (out of a total of 100 since the disaster began) being reported dead during this time.[116]

There is also concern that underreporting is taking place. BP is financially liable for the restoration of the wildlife and the environment destroyed or harmed by the disaster. Under the Endangered Species Act, BP is liable for up to $50,000 for each dead animal on the endangered species list, such as each dead Kemp's ridley turtle.[117] BP, contractors for BP, and scientists on its payroll have participated in the formal assessment process with the federal government. Anyone it hires is required to sign a three-year confidentiality clause. In order to gain access to government data related to the disaster, the National Wildlife Federation (NWF) was forced to file a Freedom of Information Act request for documents related to the disaster.[118] Along with scientists from several other organizations, the NWF has also written to Attorney General Eric Holder and BP CEO Robert Dudley, expressing its concerns about the confidentiality clause and asking for "full and prompt" release of information assessing the damages, particularly "key data relating to wildlife mortality and injury."[119]

Research has only just begun on the effects of the disaster on coral reefs and the diversity of animal and plant life that call them home. There are two primary concerns: hypoxia, which would deprive life at the coral reefs of oxygen needed to survive, and the toxic effect of the dispersant. Dr. Gretchen Goodbody-Gringley of the Mote Marine Laboratory's Tropical Research Lab in the Florida Keys has been studying the effects of Corexit on coral larvae. She could not gain access to the formulas of Corexit that were actually used by BP and was instead testing with Corexit 7500. The preliminary results show that dispersants can kill coral.

"In both the oil-plus-dispersant and the dispersant-alone samples, survival and settlement were extremely low," Dr. Goodbody-Gringley reported. "In fact, in the high concentrations that we used [to mimic the early days of an oil spill] we had 100% mortality."[120]

A similar study by the University of Miami's Rosenstiel School of Marine and Atmospheric Science found that the coral tissue dissolved as a result of dispersants. "When the corals were treated for 12 hours with mixtures including dispersant or oil plus dispersant, there was no coral to measure," the study noted. With oil alone, there was "no effect," but "with the dispersant, and the oil plus dispersant in a 12-hour exposure, the [coral] tissue mostly just dissolved."[121]

Oiled grass and marshes were a common sight in the Gulf throughout the months of the spill. In Louisiana alone, the wetlands and estuaries in eight of the nine coastal parishes, St. Bernard, Plaquemines, Jefferson, Lafourche, Terrebonne, St. Mary, Iberia, and Vermilion, received "floating crude oil, dispersed crude oil plus dispersants and tar balls from the BP Deepwater Horizon Drilling Disaster," reported Wilma Subra in congressional testimony. "The wetlands and estuaries in these parishes have been severely impacted and continue to be impacted as the crude oil continues to come on shore."[122]

Marsh and other vegetative death was reported throughout the region, but comprehensive data are not yet available. Thus, the long-term impacts have only begun to be understood. If the contamination causes a mass dying off of smooth cord grass and other vegetation, for example, the marsh soils will no longer be held in place, and more land will simply dissolve into the Gulf.

"It's like taking the air out of a balloon," Irv Mendelssohn, a plant ecologist with Louisiana State University in Baton Rouge, said. "If the plants die, not only are the roots no longer able to hold the soil, and therefore it's more subject to erosion, the land actually sinks. That's enough to stress any regrowth of new plants after the oil spill. They can't get re-established."[123]

"Oil both kills the marshland vegetation directly and poisons the complex mixture of algae, microbes, and detritus—known as periphyton—that coats the individual leaves," the World Wildlife Fund explains. "From there, it moves on up the food chain—from young shrimp and other animals that graze on marsh vegetation, up through birds that consume crustaceans and small bait fish, to carnivores such as tuna, brown pelicans, and dolphins."[124]

The Long-Term Effects

The effects of this disaster on the environment of the Gulf coast are expected to continue for decades. Jane Lubchenco, a marine ecologist and the NOAA administrator, warned at an August White House press briefing that "oil that was released and has already impacted wildlife at the surface, young juvenile stages and eggs beneath the surface, will likely have very considerable impacts for years and possibly decades to come." The lack of data made available to the public, including to scientists and researchers, means that it will be a fight to even gain access to information to answer these questions.

Emphasizing the long-term effects and uncertainties, Audubon chief scientist Thomas Bancroft stressed, "We can't begin to fathom what the long-term effects on the marine food chain will be. This remains a giant, uncontrolled science experiment, with birds and all the communities that depend on the Gulf as the unwitting subjects."[125]

For people like Jamie Billiot and her family, the science experiment is wreaking havoc on their entire way of life. Like so many in the Gulf, the Houma live off the marine food chain not only for their economic livelihood but also for their own sustenance. It is deeply integrated into their history as a people. The United Houma Nation has characterized the oil disaster as placing the Houma at high risk of "cultural extermination" and has suggested that perhaps it was time to introduce an Endangered Native American Cultures Act to protect Native Americans in the same way that the Endangered Species Act protects animals.[126]

The Gulf oil disaster now threatens centuries-old ways of life not only for the Houma but also for people throughout the Gulf whose lives depend on the water's rich bounty. So too does it threaten all those who enjoy the fruits of this rich harvest.

4

When the Oil Kills the Fish, Can the Fishers Survive?

The Hydra is one of the fiercest beasts of Greek mythology. The multi-headed monster has poison for breath and its appearance is so hideous that people who encounter it are frightened to death. The Hydra has been known to guard both the entrance to Hell and the Golden Fleece. Jason and the Argonauts were able to slay the beast, only to have the "children of the Hydra"—skeletons of people who had been killed by the monster—spring from the dead Hydra's teeth, the latter many times more deadly and terrible than the original monster alone. When Hercules tries to slay the Hydra, his sword cuts off its head, only to have dozens more heads emerge in its place.

A similar story is told in the 1980s film *Gremlins*. A single gremlin is nothing to fear, but give it water, and thousands of tiny ferocious gremlins spring from its fur, destroying all in their wake.

As if 120 million gallons of oil and 30,000 tons of gas were not enough to terrorize the fisheries and fishers of the Gulf, BP and the

147

Obama administration attacked the monster with toxic chemical dispersants.

The dispersants slowed the monster's assault, only to have it break up into hundreds of millions of smaller more terrible parts, continuing the deadly attack.

The trade-off BP and the administration made was to sacrifice the ocean for the shore. But those who live on the shore live off the ocean. And when it comes to seafood, we are all Gulf coast locals. In Los Angeles, La Paz, or London, we might not feel the sting of poisoned water or inhale the burning air, but we eat from the same table as the people of the Gulf.

When the oil and the chemicals kill the fish, the question remains: can the fishers survive?

Bayou La Batre, population twenty-five hundred, is the seafood capital of Alabama. Locals like to tell visitors two facts about the town. First, that it's got "four seasons: shrimp, oyster, crab, and fish." Second, that it's the place where Forrest Gump and his friend Bubba went to "get in the shrimpin' bidness."

Vinh Tran was nearly born in a crab-processing plant in Bayou La Batre. His mother, Rot Thi Lam, was standing at the table, digging meat out of a crab shell, when her water broke. She went home briefly, took a shower to clean up, went to the hospital to give birth, and returned to work shortly thereafter.

Tran began accompanying his mother to the plant at age four and was picking crab by age six. His hands were too small for the gloves that crabbers wear to protect themselves while they work, but because he was young, he "worked more slowly than everyone else, so I was okay," he assures me. Twenty-one years later, Tran was still working at his mother's side when they heard about the *Deepwater Horizon* from another picker. Three days later, all of them were out of work, forced to stop when Alabama's waters were closed to fishing for fear of the spreading oil.

Federal and state governments began closing waters to fishing within days of the explosion, and by early May, the closures affected the entire Gulf coast. At the height of the closures, on June 2, more than 88,500 square miles, or nearly 40 percent, of the entire Gulf of Mexico's federal waters were closed to all fishing.[1]

Depending on the state, federal waters begin anywhere from three to seventeen nautical miles from shore. State governments own the land and resources within their jurisdictions. Each state government except Texas closed its own waters to fishing, and at their respective heights, all the state waters of Alabama and Mississippi and 30 percent of Louisiana's state waters were closed. Data for Florida were unavailable.[2]

The closures continued throughout the summer, fall, and at the time of this writing. For example, the day before Thanksgiving, the National Oceanographic and Atmospheric Administration (NOAA) closed an additional 4,213 square miles of federal waters off Louisiana, Mississippi, and Alabama after a commercial fisher hauled in not just shrimp but tar balls as well.[3] As of January 2011, some 32,000 miles of federal waters remain closed as oil, dispersant, and tar balls persist.

The closures and reopenings have been sporadic, often changing daily, and vary by state, the type of catch, and the weather. They have also been highly controversial, with many people arguing that the federal and state governments were opening waters prematurely in an attempt to rush an "end" of the crisis.

Without access to waters, fishers cannot fish. Without fish, fish markets have nothing to sell, fresh seafood restaurants nothing to cook, and seafood processors nothing to process. The BP oil and gas monster, now an oil-gas-and-dispersant beast, spread its harm from the owners of the largest shrimp companies to the smallest individual fisher who uses his catch to feed his family. As well as to deckhands who work on boats that head out to the Gulf for three months at a time, dockworkers who clean the decks when the boats return, drivers who bring the catch to processing plants, and those who pop the heads off of shrimp with their thumbs and cut the meat out of crabs with their knives.

For towns like Bayou La Batre, a major processing center for seafood where an estimated 80 percent of the workforce makes its living from the commercial seafood industry, the spill was immediately devastating.

But, as the oil mixes with dispersant and both settle on the ocean floor, and as plumes of oil and gas spread through the ocean, leaving dead zones in their wake, there are many more concerns. What will happen to the habitats that the Gulf's fish call home? What will be the effect on next season's harvest, and the seasons thereafter? Will the next storm season bring the oil and dispersant onto the land and into the marshes?

Will the toxic mixture permeate the beaches and the wetlands? Will it enter the groundwater?

The many unknowns about the future are almost as difficult as the unemployment and the illnesses faced today. All of this means that the economic, social, physical, emotional, and political tolls of this crisis have only just begun to play themselves out.

For the fishers of the Gulf coast, BP's oil has meant far more than the end of a season. For many, it may mean the end of their entire way of life.

With more than 70 percent of all U.S. fish and shellfish produced in the Gulf of Mexico, including 70 percent of the nation's oysters, 80 percent of the shrimp, and 30 percent of the blue crab, for America and all those who it sells its products to, it is from places like Bayou La Batre, Dulac, Grand Isle, and so many other fishing communities, that BP's monster threatens to pour across the world.[4]

Pickin' Crab

Alabama, which is the smallest of the Gulf states' seafood providers, produces 17 million pounds of shrimp, 1.8 million pounds of crab, and 73,000 pounds of oysters annually. The total Gulf seafood trade was worth nearly $37 million in 2009 alone.[5]

Vinh Tran grew up picking crab, but it wasn't supposed to be where he ended up. Tran joined the navy shortly after 9/11. After the navy, his plan was to attend college and have a career outside Bayou La Batre. Had he grown up in Los Angeles or New York, I could easily imagine that career to be modeling. He has classic good looks. He's tall, with dark olive skin, a strong square jaw, and an easy charm.

When Hurricane Katrina hit, "I came home to help out my parents," Tran tells me. "They're old. They don't speak English, and times were tough. It was only going to be temporary." Tran thought that after six years in the navy, he might be able to find work in Bayou La Batre's only other industry, shipbuilding. But without a college degree and very few job opportunities, his options were limited, so he started crabbing again. After a few years, things were looking up, and his thoughts returned to college, but "then the BP oil spill hit."

Vinh Tran of Bayou La Batre, Alabama, July 2010.

His mother, Rot Thi Lam, is his inspiration in the workplace and out. She's been crabbing for twenty-five years. The only time she stops is when something or someone makes her. In fact, there have been just two significant instances: Hurricane Katrina and the BP oil spill. According to Tran, Katrina was easier.

"Katrina only lasted like three months and then she was back to work again," he tells me. "This thing is over six months already, and people are still out of work and have no idea what the future is going to bring. It's the unknown that's the real killer."

At age fifty-six, Rot Thi Lam could pass for a far younger woman. Her energy is constant; she is like a piston charged and ready. She has to be. Crabbing is brutal work, and Tran's awe at his mother's ability (and that of all the women he has worked with) to withstand the work for a lifetime is palpable.

Crab pickers can work all year, but the bulk of their income comes in just a few months, from April through September, so they have to make those months count. These are the same months of the oil's most frontal attack.

A typical workday is twelve hours, from 4 a.m. to 4 p.m. At the height of the season, in August and September, the hours can be even longer, and Tran's mother always takes them whenever they are offered.

"If there's work, she'll go," Tran explains. "That's the mind-set. Not a single time when there's work are they are not working. That's everyone down here." More often than not, this means working seven days a week, fifteen or more hours a day. Tran is a veteran of two tours in Iraq, but "the navy is a lot easier than crabbing," he assures me.

Tran takes me to a processing facility in July to show me how the work is done. In most ways it is a typical wholesale processing plant, including the fact that it is closed down at the height of the season because of the spill. We are joined by the shop owner and manager, Ta Taponpanh, and a former employee, Chanty Prak.

Taponpanh left her home country of Laos at age seventeen. After spending two years in a refugee camp, she arrived in the United States in 1980. She and her husband worked their way up from, in Taponpanh's words, "nothing," first to renting the facility, and then, in 2006, to owning it. Things had been going very well. She finally was even able to bring her eighty-three-year-old father and fifty-four-year-old sister from Laos a few years ago to live with her. It was the culmination of a lifetime of hoping and saving money. But by July, in just three months without work, Taponpanh could no longer support them, and they had to return to Laos. When she tells me this, it is hard to tell whether heartbreak or shame most dominates her feelings. Before I have a chance to ask, she moves on, quickly shifting the discussion back to business.

When the processing plant, Chan-Ta Seafood (her husband's last name combined with her first name), is operating, it employs thirty to forty people but supports many more. For most families in the area, the men generally work out in the water catching seafood while the women work on shore and in the processing facilities. Taponpanh buys her crab from the local small fishermen. They set traps out in the water and wait all day and night, then come back in the following morning with their catch, which is loaded up in boxes and brought to the facility, where the

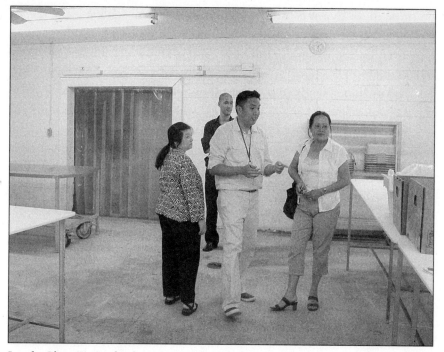

Inside Chan-Ta Seafood processing facility, once bustling but now idle due to the oil spill, in July 2010. From right to left: Shop owner Ta Taponpanh, David Pham, Vinh Tran, and Chanty Prak.

women then get to work. It is common for many members of the same family to work together in the same facility, and that's how it was at Chan-Ta.

Chanty Prak is from Cambodia. She worked at Chan-Ta with her aunt, mother-in-law, father-in-law, and sister-in law. "It is a community here," she explains. "We come here, work together, talk together, get all the news. That's now gone."

Chan-Ta Seafood is essentially a concrete box located deep in the bayou. Thick trees and brush surround the facility. Its neighbors are the Laotian Buddhist Temple across the street and a few ramshackle houses hidden in the trees down a dirt road.

The building is about the size of a relatively small one-story American home. It has low ceilings and is both gray and pristine inside. Though industrial, it somehow maintains a homey feel. Perhaps it's the touches of color, such as bright red mounted giant shrimp from Vietnam on the wall, or that it is broken up into three rooms, giving it the feel of having a living room, a dining room, and a bedroom. Or maybe it's just how friendly Taponpanh is and how comfortable everyone seems to feel in the space.

The entrance leads to a small office with a bed where Taponpanh often sleeps. There's a large workroom and a break room with a microwave oven and a refrigerator. Tran is impressed. The break room alone is a rare luxury. Most workers do not take breaks or lunch, but if they do, they typically do so sitting in cars or outside on the ground.

Tran points around the barren workroom as he describes what it would look like if it were alive and buzzing with dozens of workers. When the crab arrives, there are a variety of jobs that must take place; and Tran has done them all. At the height of the season, the work takes place around the clock; they can process twenty thousand pounds of crab in twenty-four hours.

The first step is cooking the crab in a giant stove. This happens at night. After it's cooked, the next line of workers rips the top shell off and puts the crab into the freezer. Early the next morning, the pickers get to work. Standing around a large gray concrete table in the middle of the room, the pickers remove the meat from the shell with knives.

Even though they wear gloves, all crabbers, as well as everyone else I met in the fishing industry, bear its telltale scars: deep red notches lining hands and arms. The size depends on what the specific line of work is. Crabbers have thick scars the width of a knife. Tran has them, as does his mother.

Tran stands at the table to demonstrate the process. He looks down at the table, pretending to hold a knife in one hand and a crab in the other. His eyes move as the imaginary knife cuts into the imaginary crab, his body shifting into a position that is as familiar as walking.

"You stand at the table, you must stand there all day. Your body hardly moves. Your hands hardly move. You hold the knife in the same position all day, just moving your arms to scrape out the meat. Crab after crab after crab. When you let go of the knife, your hand gets stuck like this." Tran's hand freezes, locked around the invisible knife. "Your shoulders ache. Your back aches. Your hand cramps. It's cold."

The work can be relentless. "You can take as much of a break as you want," he says. "But you get paid either by the pound or the cup, not by the hour. So it's your choice. Most people would rather starve than take thirty minutes to eat."

Tran first went to work helping to clean his father's shrimp boat when he was four. When he wasn't there, he was at his mother's side at the processing facility. Even though he started working at age six, he had to stop when he was eight because "child labor laws were enforced," he explains. When he was fifteen he got a work permit. He worked at the plant in the morning, after school, and sometimes on the weekend. In between he squeezed in homework.

When he recalls this time, it is with tangible shame. "I didn't hold the weight that my parents did. I took breaks. I didn't know responsibility. I really regret that. Some people enjoy themselves. I got distracted and slowed down. But you don't want to get distracted. You want to do your best; if you're not careful, you lose your worth," he tells me.

When Tran returned to the plant after his time in the navy, he was older and wiser. "I worked long hours with my parents. Drank Gatorade. Would eat a few things, then be back at work in a few minutes."

The work is essentially done in a freezer, but sweat builds up from the effort nonetheless. When the shift is over, the workers exit into the hundred-plus-degree heat of the bayou, now smelling of fish, plus a cold and a hot sweat. "You wouldn't want to stop by a store or a restaurant," Tran explains. "It would inconvenience the other people [because of the smell]. So you go home and start up again the next day."

Shuckin' Oysters

A visit to Bayou La Batre is not complete without a stop at what I have dubbed the "giant oyster mountains." These are two massive piles, about two stories high each, of giant gray, purple, and white oyster shells. The remnants of the daily oyster shucking efforts of hundreds of the town's inhabitants, the mountains are not static but are constantly being rebuilt; older shells returned to the sea and fresh replenishments dropped on top by cranes.

Tran has worked all four of Bayou La Batre's seasons, and he assures me that while crab picking is tough, oyster shucking is far more difficult.

The oyster shuckers' job is to crack open the oyster shell and remove the meat. This may sound easy, but the shell is very hard and does not want to be opened. American workers, Tran explains, use a knife, because "they are strong. Asians use a hammer. Beat it once, crack open the shell, and then use a knife to pop it open."

Local oystermen bring the catch in from the water in sacks of fifty to seventy-five pounds. It is brought in to the processing plant by the ton, or two thousand pounds worth. This goes into the freezer until first thing in the morning, when the shuckers arrive.

One man lifts the sack of oysters and pours it out for the shuckers down the line. Tran admires these men: "He's built strong, just look at their backs." The shuckers wear big thick gloves, but "a lot of them, their hands are cut up. They are swinging a hammer. They are always tired already and sleepy, and they miss and hit their hands."

The fastest shuckers are the older women, who can do an oyster a minute. "I tried that; the seventy-year-olds always outdid me. My whole body aches. I could never do it as fast as them." After the oysters are shucked, another line of people cleans them, while yet another packs them for shipping. Oyster shuckers get paid by the gallon. The fastest shuckers can do a gallon an hour, earning as much as $30 an hour.

There are likely two reasons that the older women are the fastest: one is their work ethic, so admired by Tran, and the other is that they have been doing the work their entire lives. Most, like Tran's family, are immigrants from Vietnam, and others are from Laos and Cambodia, having come to the United States in the 1970s and 1980s. More than 15 percent of Bayou La Batre's residents were born in East Asia, and a third of the population is of Asian descent.[6]

Across the Gulf coast, approximately a third of those who work in the seafood industry are Southeast Asian.[7] The effects on this community have been severe.

In Biloxi, Mississippi, for example, the daughter of a Vietnamese American fisherman galvanized with many of her peers a coalition of community groups to organize bilingual community forums for Vietnamese American fishers. "Our families are falling apart," she reported in June. "Our lives as we know it are gone. We will no longer get to eat the seafood our father and brother catch. We won't have the opportunity to come help with unloading the shrimp when their boats come in after two weeks out at sea. We won't have financial support from them because they can't do the work they have done for the past 20 to 30 years—catching shrimp, fish, crab, oysters. It is very sad to see our family members' careers as fishermen ending because of this BP oil spill disaster."[8]

Many families throughout the Gulf coast have stories very similar to Tran's. To share their story with me, Tran translates for his parents. Much of it is new to Tran, including that his mother nearly gave birth to him in the crabbing facility.

Tran's parents fished and farmed on the small island of Hon Nghe, off the southeast coast of Vietnam. After the U.S. military withdrawal in 1975, "the Vietcong took over the north side," Tran explains, "and they were forced to sell what they produced" to the new government for about a quarter of what they had formerly earned. Food was also rationed, and "you could only get a handful to feed your family." Unable to make a living or support his family, Vinh's father, Dom Van Tran, escaped. In 1980, in the dead of night, Vinh's mother, his father, his five older siblings, a cousin, and eleven others loaded into Dom's small boat and sailed across the South China Sea to the Philippines. Had they been caught, Dom insists, they would have been put in jail.

They came to the United States because Dom had good memories of the U.S. Navy officers stationed in South Vietnam who had treated his father in their medical facility for a burn. Vinh has never heard this story and is clearly moved to learn of the familial connection to the U.S. Navy. The family first went to Atlanta, where Dom was a dishwasher at the airport's Marriott Hotel.

Eventually, they heard of Bayou La Batre and the opportunities available there. When they arrived, it felt like a home away from home. The entire family lived in a small two-bedroom apartment with another family from Vietnam. Dom worked as a deckhand and a crab picker, eventually earning enough to buy his own shrimp boat and a house for his family. Health problems have now forced him into early retirement—early for the bayou—at age sixty-five, making Vinh and his mother the only income earners in the home. In his heart, however, Dom is still a farmer. The backyard of their home has been transformed into an Asian jungle, with giant banana trees, peanuts, papayas, and other vegetation.

The people of Bayou La Batre got by before the disaster, but just barely. It was a long and difficult struggle to recover from Hurricane Katrina, which demolished the town, leaving more than two thousand of the city's residents temporarily without homes, city officials said, "after a tremendous tidal surge covered two-thirds of the city and left about 800 of its 1,600 structures uninhabitable," reported the *Mobile Register.* "All of the boatyards and seafood processing houses along Shell Belt Road were damaged, if not destroyed."[9]

The devastation was some of the worst in Alabama. Secretary of State Condoleezza Rice visited Bayou La Batre in September 2005 and held a press conference at the town's community center.[10] Everyone in the town, like the rest of the Gulf coast, has a Katrina destruction story. But they also did recover. The townspeople remained or returned. They rebuilt homes and infrastructure, and the seafood industry was reborn.

Nonetheless, the estimated median income is less than $30,000 a year today and nearly 30 percent of the population lives below the federal poverty line.[11] In other words, there was no cushion when the BP oil spill happened. No one had savings, family members with savings, or friends with savings. Any "extra resources" had been tapped out five years ago. This story is repeated throughout the Gulf coast. People, businesses, communities, and an entire economy had just fought their way back from extinction by the storm only to be thrust back into crisis by the spill.

Headin' Shrimp

I met Misty Ellis in the waiting area at Boat People SOS, a Vietnamese American direct service and advocacy nonprofit organization. Ellis is not Vietnamese; she is white and blond. Boat People SOS does not discriminate; its services are available to anyone.

Ellis is here for food. She grew up in New Orleans but has lived in the Bayou La Batre area since age sixteen.

She and her husband had been getting by. He unloaded shrimp from incoming boats, and Misty "headed" the shrimp at a processing facility. Between them they earned $350 to $450 a week. "And then the oil spill hit," Misty says. "It just wiped everything out. No boats were coming in for the shrimp. He couldn't unload, and I couldn't head shrimp because there was no shrimp coming in."

The disaster struck just as the shrimp season was about to open, killing the season that normally runs from mid-May through December.

Just like Tran, Ellis describes her work in pantomime, re-creating the minute gestures that quickly put her in a near trance of concentration. The process depends on the type of shrimp. "You stand all day on your feet at the table. You put three or four shrimp in each hand, you take your thumb, and you just pop the heads off. Well, if you get an easy shrimp, you can just do it like that with your thumb, and then you can dig in it and get the guts out and everything. I do that with brownies paint shrimp, mud shrimp, and so on."

The heads go into a bucket, where they are weighed to see how many pounds have been produced. The rest of the shrimp is then "cleaned, processed, packaged, and shipped." The work is hard for many reasons, including the fact that shrimp have sharp stubs on their faces. Hit these the wrong way, it's like getting stung by a strong bee. "Trust me, it kills, but I know I got to do it. I can't complain."

When they both lost their jobs, there was no financial cushion, and, Ellis explains, the stress gave her husband a heart attack. "He had it right there on the front porch." Like almost everyone else in the fishing industry, they do not have health insurance. His care has been provided by a state version of Medicaid, but Ellis can't afford the copay on his medication, which means he has been going without it.

Now she's worried that if the worst happens, she won't have money to pay for the funeral. "BP didn't think about that, did they? Are they going to pay?" So she's trying to shift what money she does get from BP's claims process to his medication, and she's asking for help for food. Her daughter lives with Misty's mother in New Orleans. "It's embarrassing for somebody like me to ask your parents for money that's helping you raise your kid because you can't because of an oil spill."

Boat People SOS

Boat People SOS opened a small office in Bayou La Batre in 2005 in the wake of Hurricane Katrina. Since the oil spill, the small staff has become the central service provider for the town, helping with food drives and distribution, access to state and federal social services, critical translation services for the Southeast Asian community, and with the BP and Kenneth Feinberg claims processes. The need for basic social services has utterly overwhelmed Boat People SOS and its service providers.

David Pham is the program associate at Boat People SOS. He tells me that his parents "were boat people from Vietnam who emigrated to the United States in 1983," after spending two years in refugee camps in Hong Kong. His father had been imprisoned by the Vietnamese government for being in the South Vietnamese Army during the war and was being "reeducated." His father worked at a dry cleaners and his mother at a sewing shop in Pomona, California, but then they moved the family to Mobile when his uncle opened a nail salon and asked his mother for help.

During his first year of college, Hurricane Katrina hit, and David Pham began volunteering at Boat People SOS. He graduated from college, and although he had a job, he returned to Boat People SOS to volunteer in his free time. It was only going to be temporary, but one month before the oil spill, the only Bayou La Batre office staffperson left, and twenty-five-year-old Pham took over.

Before the spill, Pham would get perhaps twenty walk-ins at the office per month. After the spill, on one day in late April, Pham alone saw two hundred people. In May and June combined, twelve hundred people came through Pham's door—almost half of the town's total population.

Then he got help. Pham had met Tran when they were in high school. When Pham got some money from the State Department of Human Services to pay for someone to help people get welfare and other social service payments, he hired Tran, who also worked for free as a translator. With additional grants, Pham was able to hire a few more people.

The human need was hardly confined to Bayou La Batre. Statistics are not yet readily available, as studies have only begun to be conducted on the effects of the oil spill on Gulf coast residents.

One study of twelve hundred Gulf coast residents, conducted in July by Columbia University's National Center for Disaster Preparedness and the Children's Health Fund, offers some preliminary findings: more than 40 percent of the population living within ten miles of the coast experienced direct exposure to the oil spill; one in five households has seen its income decrease as a result of the oil spill; more than one-third of parents reported that their children had experienced either physical symptoms or mental health distress as a consequence of the oil spill; and more than one-quarter of the coastal residents thought they might have to move from the area because of the oil spill.[12]

Emotional Stress on the Rise

The need that Pham and Tran filled was not only economic, it was also emotional. This too is not confined to Bayou La Batre. As early as June, a Louisiana State University survey of nine hundred Gulf coast residents found that "effects of widespread psychosocial stress are substantial," with nearly 60 percent of the people reporting "feeling worried almost constantly . . . because of the oil spill."[13]

Similarly, the Louisiana Department of Health and Hospitals reported in June that crisis counseling teams had already "engaged and counseled" almost two thousand individuals in the affected areas and had found "palpable increases in anxiety, depression, stress, grief, excessive drinking, earlier [in the day] drinking, and suicide ideation." Community-based organizations, the department reported, have recorded similar findings. "These are early warning signs of developing substance abuse and dependence, mental illness, suicide, and familial breakdown, including divorce, spouse abuse, and child abuse and neglect."[14]

There are the expected stressors, such as economic problems. But there are others that one might not expect that were common throughout the Gulf. One stressor came from the simple fact of being forced to take government and/or BP's assistance. "These people want to work. They are ready to work. They do not want to take handouts," Pham tells me with obvious aggravation. "They were ready for the season. Now it's like a lost hope."

There is also a distinctive lifestyle in fishing communities that was disrupted by the spill. "It's a unique family dynamic, you know," Pham explains. "Here in the Gulf coast it's normal for the father to be gone most of the summer, out at sea fishing. The mother works from 6 a.m. to sometimes 6 p.m., at two processing shops. Now you have the situation where the dad's at home all day out of work. The mom's home all day out of work, the children are at home 'cause it's summer, so the stress level rises."

The mental stress is not just on the community; it is also on the service providers. Like Dr. Joye and Casi Callaway, Pham has had his life consumed by the oil spill: "I've pretty much devoted all of my time to the oil spill," he explains.

I visited Pham several times over the course of the disaster, and by October he was clearly showing signs of stress. He was unable to sleep, he was having problems with his relationships, and he was feeling depressed by his "inability to meet the need" he saw all around him. He says that it was especially hard when he had particularly bad cases, like a "mother of five, no job, BP is not giving her enough money—it stays in your mind until you can't sleep at night."

He sought therapeutic help and was told to take time off. "Take a vacation?" he says, incredulous. "Then I'd just have an additional week's worth of work to take on when I got back." He looks at me, demanding that I share his incomprehension at the idea of a break.

Casi Callaway has had a similar experience. "It plagues every waking moment that I have," she tells me. "I have two things I need to think about: how many things I have to do. Then the second thing is, what do I do?" She adds, "It is so unbelievably daunting. When you're facing millions upon millions of gallons of oil, you're not big, you're tiny. What can you do?" Callaway also points out that even though the federal government and BP officials get to take time off, "we don't have replacements. We have to work every day, nonstop, without a break. It definitely wears you down."

I was talking with Casi and her husband, Jarrett, in their home in Mobile in October. Casi isn't really able to describe her breaking point, so Jarrett does it for her, and even he has a difficult time articulating it. In the end, her breaking point is the same as Pham's: she reaches it when she feels ineffective, when she thinks she's not accomplishing the job that BP and the federal government have laid at her feet. It is, of course, an impossible standard.

The Heart of Seafood

Alabama was not alone. Louisiana, the heart of U.S. seafood production, was the epicenter of the spill. Nearly 400 miles of shoreline was oiled.

Try to imagine oil smeared from Jacksonville to Miami, or from San Francisco to Los Angeles, and you'll get the idea.

The effects on Louisiana's economy and its people are incalculable. Louisiana is the largest shrimp and oyster producer in the nation and the second largest producer of crabs. Louisiana alone produced nearly 13 million pounds of oysters in 2008, 40 percent of the total U.S. production, leading the state government to proclaim, "Each and every day 1.3 million Louisiana-produced oysters are consumed somewhere in the U.S."[15] The same year, the state produced 89 million pounds of shrimp and 41.6 million pounds (approximately 26 percent) of all blue crabs landed for the nation.[16]

The Commercial Fishermen of America estimated that the cost to the Louisiana fishing industry from the disaster could total $2.5 billion.[17]

Grand Isle

Grand Isle, Louisiana, just forty miles from the site of the explosion, was one of the earliest and hardest hit areas. It is a tiny strip of land, just seven miles long and less than a mile wide. Today, offshore oil and natural gas platforms have replaced the barrier islands that once dotted Grand Isle's horizon. Just as the barrier islands once provided a natural blockade from storms, so too could they have provided a barrier to the oil.

Instead, there was nothing to stand in the way of the thick waves of oil and dispersant that washed up from the Macondo well week after week, month after month, soaking Grand Isle's beaches and devastating the economy. The waters were promptly closed to fishing after the explosion, and the vacation and tourism industry that drove the rest of the island's income evaporated overnight.

Some of this income, however, was recovered from the many journalists who went to Grand Isle. Photographs taken on Grand Isle became some of the most famous of the disaster as the community there quickly responded to the visits of President Obama and other government and BP officials—and the large media contingent that joined and then followed them—by expressing their anger and anxiety in signs and artwork. The large wood and cardboard signs, handwritten in red and green paint, sprouted up all along the island, reading BP, WE WANT OUR

BEACH BACK!, SHAME ON YOU, BP. A painting of the cartoon characters Spongebob Squarepants and his sidekick Patrick was done on wood and hung on a street post. In it, Patrick asks, "What's an oil plume?" There was even a makeshift graveyard dedicated to "the memory of all that is lost courtesy of BP and our federal government." The field of white wooden crosses erected in the grass alongside the beach bid farewell to redfish, sand castles, marlin, seafood gumbo, and more.

The most famous display was on the road to Grand Isle, in Larose, just before reaching Highway One. There, tattoo artist Bobby Pitre turned his parlor, Southern Sting Tattoo, into a gateway of shame for BP and a challenge to President Obama. His art stopped traffic not only because it was so good but also because it portrayed so well the feelings of those devastated by the spill.

I drove down to Grand Isle with microbiologist Wilma Subra and a crew of German documentary filmmakers in July. When we stopped to look at the art, Pitre came out to greet us. With his hair in a classic black

Microbiologist Wilma Subra in front of Southern Sting Tattoo in Larose, Louisiana, in July 2010. Its signs depict the anger and anguish of many in the Gulf.

Artist Bobby Pitre of Southern Sting Tattoo in Larose, Louisiana, standing beside his work in July 2010: a skeleton in a black BP cape.

gelled pompadour, long sideburns, and black sunglasses, Pitre describes his motivations for his art. One is a deep concern about the toxins his young son is being forced to breathe. Another is his inability to provide for his son as his business, like everyone else's, suffers the loss of both tourist and visitor traffic along the highway and the lack of locals with enough income to afford a luxury like a fine tattoo. Finally, Pitre grew up in the Gulf, and he worries that his son will be unable to enjoy the beauties and pleasures that, until the BP oil disaster, he had taken for granted.

Pitre's artwork includes a two-foot-tall wooden child standing in the parking lot shielding his face with his tiny arms while cowering behind a sign held by an adult skeleton wearing a gas mask. He holds a dead fish and a drinking glass filled with black oil in one hand, and in the other he holds a sign reading GOD HELP US ALL. A painting of a skeleton dressed in a black BP cape, hands dripping with blood, reaches out across the blue waters toward the Louisiana coast under the words YOU KILLED OUR GULF, OUR WAY OF LIFE! Inside the shop, there are two works in progress depicting then BP CEO Tony Hayward. In one, oil is smeared on his face while he holds the book *Deepwater Drilling 101*. In the other, Hayward has sprouted donkey ears and teeth, and the words BURRO OF BAD NEWS hang above his head.

As for most people in the state, BP is not the only source of condemnation, and Obama is not the hero. A bloody severed white torso is the centerpiece of a yellow sign that reads WHILE TRYING TO GET AHEAD . . . BP TOOK OUR ARMS. THE GOVERNMENT IS TAKING OUR LEGS. HOW WILL WE STAND? Below this is Shepard Fairey's quintessential red, white, and blue image of President Obama from the 2008 campaign with WHAT NOW written across the president's forehead and question marks surrounding his face. In the upper right-hand corner of the display sits a sign that reads BOBBY JINDAL FOR PRESIDENT. Louisiana governor Jindal emerged as a champion among the state's residents in the wake of the disaster.

During this trip to Grand Isle, Clint Guidry and Dean Blanchard took us out in a boat to see the waters both have fished for three generations but which are now closed to them. Guidry is acting president of the Louisiana Shrimp Association and served on the Louisiana State Shrimp Task Force. Blanchard is owner of one of the largest shrimping businesses in the Gulf. He tells us that his company alone accounts for about 11 percent of the U.S. shrimp supply and employed some ninety people before being shut down by the disaster.

Blanchard stresses two things about his roots: he's a Cajun and he's a fisher. His brother and nephew, however, work in the oil industry. Blanchard's anger at the oil industry since the disaster, and his brother and nephew's anger at the Obama administration for implementing a moratorium on offshore drilling, has become so great that they can no longer speak to one another.

It is not a unique conflict. Two of the primary employers in the Gulf, particularly for those without college educations, are oil and fishing. When the oil killed the fish, families and communities were cut in two, forced to choose sides in defending their industry over the other.

As Blanchard shows us around, his pride in his business is palpable, and so is his heartbreak at its loss. He worked long and hard to rebuild after Hurricane Katrina, spending, he told us, about $5 million to do so just in time for what was expected to be the banner year of 2010.

All across the Gulf I heard the same angry refrain. After the double whammy of Hurricanes Katrina and Rita, followed close behind by Gustave and Ike, both the seafood and the fishers had finally recovered. So too had the global seafood industry, whose prices had been depressed for two years straight. Like Blanchard, fishers across the Gulf, large and small, had invested heavily in the past year to ensure that they would profit from long-awaited bounty. They hired new staff, spent money fixing old boats or buying new ones, and stocked up on fuel and ice. Now, not only was the seafood lost, so too was this money—and the hope that went with both.

When we talked in July, Blanchard described how he was still showing up to work every day but has nothing to do. "I walk in circles around the office. I'm a man who likes to work. I don't know what to do without work. I fight with my wife. I fight with my family. I have nothing to do," he tells us. As he speaks, we can hear what sounds like near-suicidal desperation in his voice.

There was reason for worry. Just days earlier, William Allen Kruse, age fifty-five, a charter boat captain from Gulf Shores, Alabama, had been found dead as a result of a self-inflicted gunshot wound to his head. His suicide was considered the first fatality as a result of the oil spill (as opposed to the explosion of the *Deepwater Horizon*). Although recently hired by BP as part of the Vessels of Opportunity (VOO) program, Kruse was "quite despondent about the oil crisis," Stan Vinson, the coroner, told the *Los Angeles Times*. "All the waters are closed. There's no charter business anymore. You go out on some of the beaches now, with the oil, you can't even get in the water. It's really crippled the tourism and fishing industry here."[18]

Faith

I attended services at Holy Family Catholic Church in Dulac, Louisiana, on Sunday, July 25. It was nearly over by the time I got there, but I had a good excuse for being late. It was only my second time in Dulac, and the town is so small that I blinked, missed the church, and found myself at the end of the bayou with nowhere else to go but out into the Gulf.

Once I turned around, I found the church and saw Jamie Billiot, her father, James, and her two-year-old daughter, Camille, seated quietly in their pew. At the end of the service, the priest announced that the following Friday, there would be a Gulfwide interfaith prayer service for the BP oil spill. "We've been praying about this spill, praying for those harmed, praying for it to end. Now we can all do it together," he said.

Sharon Gauthe, the director of Bayou Interfaith Shared Community Organizing (BISCO), and her husband, David, organized the interfaith service, which was a national event. BISCO, which is based in Thibodaux, Louisiana, uses "faith-based community organizing to empower residents to effect positive change related to typical social justice issues such as poverty, illiteracy and racism, and to Katrina-Rita-Gustav-Ike recovery issues," according to its Web site.

That David and Sharon have lived and worked in Louisiana their entire lives is evident from their thick Louisiana accents and their predilection for classic Louisiana jokes.

Here's a taste. David tells this joke over a crab dinner: "Boudreaux's friends come home one day. 'Boudreaux, we got bad news for you, Marie is dead! She drowned. But maybe there is some good news in this. When we found her, she had ten huge crabs stuck to her. So we're taking her out again tomorrow!'"

Sharon Gauthe, the director of Bayou Interfaith Shared Community Organizing, speaking at the BP Oil Spill Interfaith Prayer Service in Houma, Louisiana, in August 2010.

The idea for the interfaith prayer service came up at a pastors' breakfast hosted by BISCO. "It was noted that with all the events happening around the oil disaster, the people were going through so much stress and so much uncertainty and fear that they, as pastors, should do what they do best, which is to pray for the families and the community," Sharon later tells me. They asked not only for prayers on behalf of "the communities affected by the disaster" and "the ecosystems stressed and devastated" but also for "just and comprehensive action from our policy makers." The call and the response were nationwide.

"Prayer is necessary for all of us in this very sacred part of the country," Sharon tells me. "Faith is not taken lightly, and no matter what the faith, there is respect for it and we know how important it is for us all."

Much of BISCO's work prior to the disaster focused on stemming land loss and coastal erosion in Louisiana, particularly around Grand Isle.

Sharon and David tried to help address the pain of this community from the outset in many ways, including staffing a food-voucher distribution center on Grand Isle. The distribution center had been operating for a month, however, before people began to come in. "The people are very used to living on the land and making money last and stretch and doing what they can do on their own," Sharon tells me. "They are proud. When BP told one fisherman, 'You can apply for unemployment,' he said, 'I'm not unemployed, I'm a shrimper and I want to shrimp.'"

Eventually, just as the need for social services increased in Bayou La Batre, so too the people on Grand Isle started coming for help in droves. "With a hurricane, you know it's coming," Sharon explains. "You get ready, you prepare, it hits, you take care of it, then it's over." With the oil spill it was different. "Now they can't go out because oil's in the fishing area. All their bills are starting to come due. There was anger before. Now it's just despair."

Dulac

Jamie Billiot faced similar problems at the Dulac Community Center. "We are generally impoverished down here anyway," she tells me. "But we have always been able to eat." Not only do people know how to stretch a dollar, they catch their own food. "We eat what we fish. Shrimp, oyster, fish—that has always just been in the freezer. Now the freezers are empty and the waters are closed."

Billiot describes Dulac "as a small fishing community" where people live "the fisherman's life." The town of Dulac has about twenty-five hundred inhabitants. Approximately 40 percent are Native American, 90 percent of whom are, like Billiot and her family, members of the United Houma Nation. About 50 percent of the people in Dulac and the surrounding communities, which Billiot's community center services, make a living from fishing. "From the get-go, every last fisherman was out of work," Billiot explains, causing a devastating impact on the economic and social fabric of the small community.

The Dulac Community Center serves an area of six to seven thousand residents, providing social, emergency, and recreational services. With the center not even fully rebuilt from Hurricanes Katrina and Rita, the needs that Billiot and her staff of five were already struggling to meet suddenly doubled overnight. The center has a food bank and provides school supplies, clothing, shoes, and other essentials, as well as recreational activities for children.

"We always have a pretty standard list, and that has doubled. Everything doubled," Billiot explains. "People would call and say, 'Hey, I really need this,' because all of a sudden they had no job and they might have gotten a little bit of money from their BP claims, it was never enough, or the claim check was late, as it always seemed to be, whatever the circumstances."

The center has tried to expand its services, offer free meals and other extra goods, but it has been extremely difficult to match the growing need. It has also become a central hub for helping people to traverse the difficult landscape of water openings and closings as well as BP and government claims.

Dulac used to have many seafood processing facilities, but only a few remain today. The rest closed after being unable to rebuild following the hurricanes. As they have for generations, most men in Dulac make their living directly from fishing for shrimp and, to a lesser degree, for oysters and crab. They take short trips out just as far as Lake Boudreaux, which is directly adjacent to Dulac, or longer trips out into the Gulf that range from overnight to several months.

That came to an end when the waters were shut down because of the spill. An early casualty was one of the surviving processing plants, a small oyster factory, forced to close in the wake of the spill. "They haven't worked at all since it happened," Billiot explains. "They haven't had anyone who has been able to go out and get any of their oysters from their beds. So no one's processing, either."

As in the rest of the Gulf, the toll here is not only financial but emotional as well. Jamie describes a set of problems that is virtually identical to what David Pham has struggled to address in Bayou La Batre. The financial and emotional stresses combine with a change in lifestyle that is, for some, impossible to deal with.

"Family dynamics completely changed," she says, "because the men were used to being away either a week at a time or every night, and then all of a sudden you've got a whole bunch of men who can't work, which already makes them feel 'less than.' And then they're thrown at home with their families."

The result is that a community that already struggles with alcohol and substance abuse has seen both—along with domestic violence—increase. It is a painful but increasingly common sight, for example, to see "men show up at the grocery store to start drinking beer first thing in the morning," Billiot explains. Both the economic and the emotional strain are then felt by the children.

The center provides activities for children, including traditional Native American dance, at which Billiot excels. These activities have greatly increased in use and in importance. "We try to keep them busy, you know," Billiot says of the children, "to sort of keep them out of that [troubled home] space."

The center has some successes, but it can only do so much, and the children are showing the signs of stress, worry, and depression that have permeated the region.

Are the Fish Safe?

The problems that plague the fisher communities of the Gulf were not only caused by the BP oil monster. They are also the result of the toxic chemical dispersants used to attack the monster, which poisoned the air and water and spread the dispersant-covered-oil throughout the water.

The Gulf coast was home to a massive science experiment: untested and unprecedented chemical assault. The combined effect of the oil and the chemicals is a series of potentially deadly unknowns, which fishers, scientists, government officials, and advocates have sought to untangle.

Within the knot are oftentimes competing motivations. The protection of the health of consumers could yield a slower, more cautious approach to determining whether the fish are safe. Keeping fishers employed could motivate some to be slow and cautious, thinking of long-term effects, while others may respond by demanding greater speed and risk taking so that fishers get back to work right away. For others, the desire to get the BP oil spill behind them as quickly as possible could also yield both speed and risk.

NOAA's LaDon Swann was ready to allay everyone's fears. "The fish are safe," he declared on August 19 at a community forum in Bayou La Batre. This should have been good news to the 150 people filling the Alma Bryant High School auditorium, virtually all of whom made their living in one form or another in the fishing industry.

Having spent four months with a severely reduced or nonexistent income, all were now desperate. Even the lucky few who had been able to participate in BP's Vessels of Opportunity (VOO) program were now largely out of work because BP had all but shut down the program in the area. Many could no longer afford rents or mortgages, medical bills, or even food for their families.

Vinh Tran and others from Boat People SOS were translating for the very large contingent of Vietnamese, Cambodian, and Laotian fishers and seafood processors who attended the meeting. Dr. Riki Ott and Casi Callaway were waiting for their turns to speak. None of them, however, appeared relieved by Swann's reassurances. They simply did not agree with him, because many continued to see far too much evidence that both oil and dispersant remained in the water.

As will be discussed in chapter 7, Dr. Joye and her colleagues at the University of Georgia had concluded just days before that almost 80 percent of the estimated 4.1 million barrels of oil released directly into the Gulf still remained there.[19]

Less than twenty-four hours after Swann's declaration, I ran into some of that oil in Alabama waters that had been opened to shrimping less than two weeks earlier. I was in a boat off Dauphin Island, about seventeen miles from Bayou La Batre. Originally named Massacre Island, the area was renamed in 1707 in honor of Louis XIV's great-grandson and heir, the Dauphin. It is a barrier island three miles off the edge of Alabama's southmost point. All fourteen-by-one miles of the island are designated as a bird sanctuary. Pat Carrigan, the boat's captain, is among its thirteen hundred permanent human residents.

An Alabama native, Carrigan has lived on the island for more than twenty years and has fished its waters for far longer than that. He was a commercial fisherman and then, when the waters were closed to fishing, a VOO worker, cleaning up BP's oil. Before we set out in Carrigan's boat, my companions, Rocky Kistner of the Natural Resources Defense Council and Zack Carter of Mobile's South Bay Community Alliance, and I each purchased a commemorative "1st Annual Offshore Tar Ball Fishing Rodeo 2010" T-shirt from the local convenience store. The tongue-in-cheek shirt featured a large oil rig in the background, and in the foreground fishers pull an oil-soaked brown pelican and swordfish out of oil-filled black waters. The "rodeo" is "sponsored by British Petroleum."

We hit an oil slick within fifteen minutes of setting out in the small boat parked in Carrigan's backyard. We were in the Mississippi Sound, heading toward the Katrina Cut, a shortcut to the Gulf of Mexico that had opened up when the hurricane split one portion of Dauphin Island off from the rest of the island.

"That's dispersed oil," Carrigan said, as we passed through a slick of light brown foamy goo that stretched out into the visible distance. Our photographer, Greg Westhoff, snapped photos while I dunked a clean white oil-absorbent rag into the mess. The rag emerged brown, orange, and slimy. Clint Guidry of the Louisiana Shrimp Association looked at the photographs the next day and confirmed Carrigan's assessment: "That's oil, oil with dispersant."

The oil was not limited to the water. After passing through the sheen, we landed at a western strip of Dauphin Island accessible only by boat. We trekked through a completely untouched and unpopulated strip of wild brush, green grass, and blue flowers. As we emerged on the other side of the island, the beach opened to clear blue sky, white clouds, and a stunning white sandy shore.

Kistner looked ecstatic as he gazed at the beach—that is, until he looked down. Tar balls, some as large and as thick as an open hand, stretched in a line where the waves had left them, as far as the eye could see down the beach. Carter, using a piece of bark he found on the beach, bent down and started scooping the tar balls into a white bucket. He looked like the little Dutch boy who put his thumb in the dike to stop an erupting dam.

"It's not only on the beach, it's in the water," Carrigan said. He stood in the ocean, bent down, and gathered more tar balls in his hands as they washed up.

On our trip back to Carrigan's house, we made another disturbing finding. Discovering the oil slick had been problematic enough, because it meant that the waters should not have been opened to shrimping. However, we actually did not even need to see the oil to draw that conclusion. The best evidence that the waters were not ready was demonstrated by the fact that no one was shrimping.

"These waters should be filled with shrimpers," Carrigan explained. Instead, there was not a single boat on the water shrimping during the several hours of our trip. "They're just not shrimping."

The fishers knew what the government didn't—that oil, dispersant, and fish don't mix.

As Swann had reminded the audience at Alma Bryant the day before,

Zack Carter of Mobile's South Bay Community Alliance scooping up tar balls on Dauphin Island, Alabama, in August 2010.

NOAA and the FDA, in consultation with the EPA and Gulf state governments, had established a protocol for reopening both commercial and recreational fisheries in federal and state waters. The process of reopening and reclosing these waters was a daily battle. Reporting all of the changes would easily take up fifty pages.

The protocol states that the visual observation of oil or "chemical contaminants" on the surface of the water is cause for the recommendation that the fishery be closed "until free of sheen" for at least thirty days in federal waters and seven days in Alabama state waters.

Specifically, there are four criteria that have to be met before closed waters can be reopened: (1) they have to be oil-free "for some time"— this is determined by visual, including aerial, observations, and "some time" is defined as thirty days for federal waters but varies by state; (2) the risk of recontamination of the water (such as by current movement) has to be low; (3) the fish have to pass sensory tests—no bad smell, and so forth; and (4) the fish have to pass chemical tests, which look for oil and dispersant residues and PAH chemical hazards.

Numbers three and four are the now infamous "smell tests." Most people could barely contain their laughter when the images of men with their noses poked into the stomach of shrimp, twitching like Elizabeth Montgomery as Samantha in *Bewitched*, started to appear in the media. Many people in the Gulf, however, assured me that this is not as crazy as it looks. Fishermen can tell by sight and smell if large amounts of oil are in certain species. They are used to looking for it, and the sight and smell are distinctive. However, they all equally assured me that there is no sensory test for Corexit or for dispersant-soaked oil mixed into fish— a point that the federal government later conceded.

The whistleblower support group, Public Employees for Environmental Responsibility (PEER), filed a legal petition with the FDA in August, asking the agency to test seafood chemically for dispersants before declaring the seafood safe. Noting that the FDA and other agencies did not have any such tests in place for dispersants, PEER cast doubt on the FDA's claims that the seafood was safe for consumption. "FDA cannot say for certain whether the seafood coming out of the oil spill areas in the Gulf is safe," said PEER staff counsel Christine Erickson. "The high levels of uncertainty should dictate a policy of striving to be safe now rather than having to say you're sorry later."[20]

The FDA's Donald Kraemer testified to Congress a week later, conceding PEER's arguments. He testified to the need to conduct additional studies to "reaffirm that dispersants do not accumulate in tissues of fish and shellfish." At the same time, he admitted that the FDA and NOAA did not have an efficient and practical chemical test for dispersants in seafood.

"FDA is refining its ability to test for dispersants by working with NOAA to develop a practical and efficient chemical test for dispersants in edible portions of seafood that can be deployed in federal and state labs to provide rapid yet reliable results," he said. "FDA will continue to study the long-term impacts of chemical dispersants on food safety."[21]

Even with all that has befallen her and her community, Misty Ellis shares concerns I heard repeatedly throughout the Gulf. Even though her livelihood depends on the waters being opened, she does not want that to happen prematurely. "If one person gets sick from that shrimp, it's all over for all of us," she tells me. "The shrimp are at the bottom of the Gulf. That's where those chemicals are, too. They have to make sure those chemicals haven't hurt the shrimp. I think they're just rushing it, hurrying it, to make it look all better before they know if it is."

The Future of Fisheries and Fishers

The deep fear of what next season and the seasons thereafter will bring permeates the Gulf. The oil and the dispersant did not just affect the living seafood, it also affected their living eggs, their future reproductive capacity, their habitats, and their food sources. The long-term effects could therefore last for generations.

Jane Lubchenco, the NOAA administrator and a trained marine ecologist, has explained, "For example, bluefin tuna, who spawn at this time of year, have eggs and young juvenile stages called larvae that would have been in the water column when the oil was present. If those eggs or larvae were exposed to oil, they probably would have died or been significantly impacted. And we won't see the full result of that for a number of years to come."[22]

Jamie Billiot's father, James, began fishing when he was seventeen.

By age twenty-one he was the captain of his own boat, and he continues to fish today, at age sixty-nine. His father fished until he was ninety-two years old, and his grandfather fished before him.

James knows the waters of Dulac. He warns that the real problem lies at the bottom of the ocean, where the oil and dispersant have taken rest. "Just because we're fishermen doesn't mean we're stupid," he tells me. "They think hiding the oil on the bottom will fix the problem. It won't. Water is continuous. What is on the bottom has to come up. The Gulf current is strong. We have to worry about this not just today and tomorrow, but for future years."

Jamie draws the same conclusion, "They must think we're complete idiots," she says. "The idea that the oil just disappears throws me for a loop. Something has to go somewhere. Period. It doesn't disappear. How much is going to rest in with the shrimp and oysters? How much will wash up with the next flood?" It's these uncertainties that plague everyone.

Repeating words said by countless others, fisherwoman Tracy Kuhns said from her home in Barataria, "With a hurricane you know what to expect. It comes, it hits, it's over, you rebuild. With this [the oil disaster], we have no idea what's really happening, and most important, when it's really going to be over. Even when the oil stops flowing, there's still the dispersant to contend with. And we have no idea what the long-term effects of both will be."

Clint Guidry shares these concerns. "Unprecedented use of toxic dispersants during the BP *Deepwater Horizon* Disaster without prior scientific study and evaluation on the effect to Gulf of Mexico marine ecosystems and human health was a horrific mistake that should never have been allowed to happen," Guidry stated in a press release announcing a court action demanding better testing of dispersants. "Potential ecosystem collapse caused by toxic dispersant use during this disaster will have immediate and long-term effects on the Gulf's traditional fishing communities' ability to sustain our culture and heritage."[23]

Jamie and James's long-term concern is also for the long-term survival of their Houma heritage. "Being fishermen is who we are," Jamie stressed, "and this may be lost forever." If more than one season is destroyed because of the spill, so too may be Jamie's community. James has an eighth-grade education, but through fishing he has been able to provide for his family and his community. He has built and rebuilt after each storm a beautiful home where he, Jamie's mom, Jamie, and Camille all live in comfort.

Less than 5 percent of the people in Dulac have more than a high school education. Forty percent of the Native Americans live below the poverty line. They are fishers. They do not have other employment skills, options, or, according to Jamie, "the desire to learn." Nor do they have any financial cushion to withstand season upon season of lost income and food supply.

Jamie also worries that although the younger generation may want to become fishers, they will now feel that there is too much uncertainty in following this course and will instead decide to make their living in another way and, most likely, in another place. "There is a whole culture that is literally being washed away because of this spill. This is not just about making a living," Jamie says. "Our entire way of life is going to disappear."

There is another way of life, but it brings its own deep trade-offs and potential threats to the Houma. Jamie comes from three generations of fishermen, but her sister's husband is an oil field worker; as is about 50 percent of the Dulac community. Although the oil killed the fish, and the oil production is killing their land, the oil may now be the community's only "hope."

The oil industry both giveth and taketh away for the Houma, as for the rest of Louisiana.

In the wake of the disaster, many fishers across the Gulf were forced to turn directly to BP for their survival, signing up to clean BP's mess and to take its money in the form of claims checks. In this, as in so much else related to the catastrophe, the efforts to "make BP pay" have come with decidedly mixed results.

5

Making BP Pay

Cleanup Workers, Vessels of Opportunity, Claims, and Protests

Before dawn on June 2, 2010, fishers from across Alabama blockaded Bayou La Batre's waters out to the Mississippi Sound about 100 yards from BP's operations center. More than half a dozen small boats formed a line across the water while dozens of supporters lined the docks. Holding a sign that read BP LIARS FROM THE BEGINNING, LIARS TO THE END!!, Johnny Bosarge, an oysterman, protested BP's implementation of the Vessels of Opportunity program.

A small gray boat was spray-painted, in bright orange letters, 44 DAYS STILL PUMPING, BP LIED. When he refused to remove the boat, Brent Buchanan, its owner and an out-of-work fisherman, was taken from the docks in handcuffs by local police as his two sons watched from a small pier a few feet away. Buchanan later told the local *Press Register* that he brought his two boats and his family out "to get a point across to the community. We need a job."[1]

· · ·

BP and Transocean were notified by the Department of Homeland Secu- rity and the coast guard of their designation as "responsible parties" in the *Deepwater Horizon* incident on April 28. As such, they were liable for all cleanup costs from the resulting oil spill, as well as other damages, including environmental restoration and economic loss.[2] According to the letters each received, they were required to "publicly advertise this designation and the procedures by which claims may be presented to you" in newspapers and other modes of notification, within fifteen days.[3]

The federal government decided a few days later that in order not to confuse the public, Transocean should coordinate claims processing with BP and not advertise on its own.[4] It is expected that BP and other parties involved with the *Deepwater Horizon*, including Transocean, Hal- liburton, Mitsui Oil Exploration, Cameron International, and Anadarko Petroleum, will arrive at a settlement among themselves once the results of the investigations into the disaster are complete.

BP then launched one of the most massive disaster response oper- ations of any corporation in world history in the Gulf of Mexico. As a result, it implemented cleanup operations employing tens of thousands of workers, including Gulf residents harmed by the disaster. It was also required to compensate victims through a formal claims process. By December 2010, however, only about one-third of the more than 450,000 claim requests filed had received any payment whatsoever.

In total, the company's public response efforts, while both enormous and historic, were ill planned, poorly executed, and insufficient to meet the needs that its oil disaster created.

BP began its response by launching an advertising blitz, both in com- pliance with the law and, of course, in response to the horrific and historic public relations nightmare it suddenly found itself in. BP spent more than $93 million just on advertising between April and the end of July 2010—more than three times what it spent on advertising during the same period in 2009.[5] By comparison, BP spokesman Robert Wine said in June that BP spent just "$29 million over [the last] three years on safer drilling-operations research" and no money on researching oil- spill-cleanup technology.[6]

Part of the media blitz was BP's Gulf of Mexico Response Web site. Designed in the vivid yellow and green colors of BP's logo, the site is bright, clean, and sunny. It is more of a source of PR for BP than information for those in the Gulf, however. Photographs of people actively cleaning, building, working, and helping fill the screen. Most of the headings on the site direct users to pages that detail BP's response to the disaster in a variety of formats, including time lines, pictures, videos, and maps. There is also information on claims, contact numbers, and a link to BP's internal investigation. You can even follow BP on Twitter, Facebook, YouTube, Flickr, and an RSS feed.

In late May, some pretty odd tweets started coming, presumably from BP, by way of the account @BPGlobalPR. They included the following, which was sent shortly after BP surpassed Exxon for the largest oil spill in U.S. history: "Just got the concession call from Exxon Valdez. They were great competitors and remarkably evil about everything." Other tweets said, "Negative people view the ocean as half empty of oil. We are dedicated to making it half full. Stay positive America!" and "Catastrophe is a strong word, let's all agree to call it a whoopsie daisy."

It turns out that these tweets were not coming from BP but from an impersonator. Nonetheless, as sixty thousand people signed up to follow the fake account, the real one had just fifty-seven hundred followers.[7] When BP started asking that the account be closed, the comedian Jimmy Fallon joked, "BP wants Twitter to shut down a fake BP account that is mocking the oil company. In response, Twitter wants BP to shut down the oil leak that's ruining the ocean."[8]

For its part BP tried to direct Internet traffic to its Web site in the old-fashioned way: by buying it. Beginning in June, if you typed "oil spill" into your Internet browser, the first item to pop up was BP's Web site with the tagline: "Learn more about how BP is helping." This is not because BP's site ranks highest among users' choices but because BP bought this and other terms related to the spill from Google and Yahoo![9]

Under the heading "Making It Right," BP's Web site provides details on the company's response. It states that BP's efforts grew from 2,000 people during the initial weeks of the response to more than 45,000 at its peak in July. The height of BP's work came the week of July 13, when, BP states, it mobilized 46,000 people, 6,850 vessels, and 117 aircraft and deployed more than 3.4 million feet of boom. By the end of August, there were approximately 21,000 people, 3,000 vessels, and 61 aircraft mobilized and 1.8 million feet of boom deployed. By December 2, the numbers were just 6,000 people, 445 vessels, and 11 aircraft.

The people and the equipment were engaged in all aspects of BP's response, including sealing the well, dispersing and capturing the oil, cleaning up the water and land, addressing harmed wildlife, and promoting environmental and economic restoration. Although there are a lot of modes of information on the site, it does not provide much detail. It does not break down the numbers by areas of work, nor does it provide much information on what people were actually doing.

In response to several queries to BP for this book, company spokesman Daren Beaudo replied to one. He explained that BP decided to not break down this information "for press" because it was "lots of work without much merit."[10]

BP's response efforts to stop the oil gusher are discussed in the following chapter.

The Web site does provide some general information on the three main response efforts that directly affected and employed the people of the Gulf coast: Vessels of Opportunity (VOO), beach cleanup, and the claims process.

Reporting on these efforts here is not as easy as one would hope because BP limited access to information on its actual—as opposed to its PR—efforts in a variety of ways. BP restricted access, under threat of fine and arrest, to virtually all of the areas where it worked. It required that the people employed in VOO sign confidentiality agreements keeping them from divulging what they did, saw, and collected in their jobs.[11] BP kept those engaged in beach cleanup from speaking to the press or the public for well over a month after their efforts began.

A clause in the claims process states that those who file claims cannot sue the company—thus limiting access to information that would be gained in a discovery process. BP required all those involved in the cleanup to do the same, forcing them to sign a "voluntary waiver of release," in which they gave up their right to sue BP. Wilma Subra and George Barisich, president of the United Commercial Fisherman's Alliance, appealed the waiver (and other issues) in U.S. District Court on May 2. Two days later, the judge ruled that the language in the contract was "over board" and required that the waiver be removed.[12] BP CEO Tony Hayward later apologized for the requirement, calling it a "misstep."[13]

BP coordinated its deployment of people and resources from command centers and staging areas in Louisiana, Mississippi, Alabama, and

Florida. The staging areas were generally fairly remote docks. Although I never gained access to one, I assume from what I could see from afar and learn from others that they were primarily used for vessels and people setting out to lay boom, release dispersant, or clean up oil.

These areas were fully enclosed by security fencing, with access for cars and people restricted by the twenty-four-hour presence of private security guards. Often, when one drove up to an area, a car would come down the road to wave unwanted visitors off from even approaching. I tried to gain access to several of these areas and thought that a better strategy was to park down the road and walk up to the gate in order to engage the guards on a more personal level or at least have a better chance of peeking through the fence and maybe even snapping a photo.

Whenever given the chance, I'd ask very politely if I could come in. The nice guards would laugh then look over their shoulders to make sure that nobody had noticed. The less friendly would wave me away with a grunt. The distinct impression I came away with each time was that I was engaging in criminal activity by the mere act of approaching a BP staging area.

On one special occasion, however, I did enter the National Incident Command Center in Mobile, Alabama. I had gained access as part of a tour organized by the Gulf Coast Fund for Community Renewal and Ecological Health. Founded in the immediate aftermath of Hurricane Katrina, the fund provides sustaining grants and links some 160 small grassroots organizations across the region. As soon as BP's Macondo oil monster was released on the Gulf, the fund began organizing, educating, and supporting its participant groups to better address the disaster's harmful effects. Through connections with state elected officials, LaTosha Brown, the fund's director, had arranged this important tour for about fifteen of the fund's leading community organizers.

Visitors are not allowed to approach the command center by car; rather, we were directed to park at a nearby shopping mall, where a bus picked us up and drove us to and from the center. During the bus ride, our handler, Jon Kreischer of the Department of Homeland Security, warned, "I don't want any outbursts." He also told us that we would not be permitted to interview anyone there. "Please," he urged us, "cooperate," his tone indicating that this was more warning than request.

Casi Callaway of Mobile Baykeeper was with us. As the bus

approached the facility, I could see her emotions vacillate between nerv-
ousness and defiance. Callaway had twice been escorted out of the
facility by armed police guards. At the onset of the disaster, Callaway
had hoped to work with Unified Command in its response efforts. She
came to the building days after its establishment (back when you could
drive up to it) and asked to speak with a public relations representative.
She was told she could not see anyone and that the area was restricted.
At a NOAA representative's suggestion, Callaway returned the next day
to attend a press briefing that was supposedly open to the public. Within
minutes of her arrival, three large uniformed police guards surrounded
her and escorted her out of the building. She tried again, this time
through a coast guard contact (a friend from church) assigned to Unified
Command. He not only got her in, he also had her set up as an environmental liaison. But she made the mistake of speaking to reporters she
knew inside the building. "I wasn't spilling state secrets," Callaway
recalled, "I didn't know any!" Apparently, only those given authorization
could speak to the press. For anyone else, the result was expulsion. Once
again, two police officers, this time joined by a BP representative, made
Callaway leave. When she returned to gather her things, the cold hard
stares of the guards "charged with keeping a 5'4" blond environmental-
ist" out of the building, and out of BP's way, terrified and enraged her.
Now, she was back again.

Even for those of us who had never been thrown out of the building,
it was obvious that the tour was a unique event. As our group moved
through the center, heads turned as the staff took notice, some in shock
and others with outrage. A few started to stand in defiance, until they
saw first, our guide, Kreischer; followed by two BP public relations men,
Dick Brewster and Kenneth Phillips; and finally, the armed officer close
behind us every step of the way (I never caught his name).

BP leased the one-story, largely open office space, now converted to
a command center. It buzzed with the activity of hundreds of people
hard at work grouped at cafeteria-style tables with computers and
phones placed before them. In addition to the BP staff, every branch of
the military was represented, as well as federal and state agencies. The
workers were dressed in a variety of brightly colored vests, indicating
their branch and role; for example, purple for "BP community outreach"
and green for "coast guard administrative function."

A child's crayon drawing taped to the wall in one area depicted a deceased oil rig worker being airlifted out of oily ocean waters.

Kreischer calmly ignored his stunned colleagues and allowed us to (almost freely) snap photographs and ask him questions as he described the work being done by several (but not all) of the groups and answered (some of) our questions. He showed us deployment charts of VOO and beach cleanup workers and community outreach efforts to facilitate the claims process.

We did not gain great insights into what Unified Command was doing, but we did leave by our own volition. As we did, Callaway breathed a sigh of relief in spite of herself. It was clear that the facility and those within it were protecting themselves against an enemy. But, we asked ourselves, wasn't the foe the oil-gas-and-dispersant monster tearing across the ocean, and weren't we all on the same side in that battle? The conclusion we were forced to draw was that in far too many aspects of the war, the answer was no.

Cleaning Up after BP

Vessels of Opportunity

As mass closures of Gulf waters were implemented, BP launched the Vessels of Opportunity (VOO) program "to put Gulf vessel owners to work helping clean up the BP Gulf oil spill."[14] The program began about a week after the explosion and ran through August; it cost BP approximately $450 million.[15]

The program filled two desperate needs for BP. First, BP did not have the resources it needed to boom and clean up even a fraction of the oil it had released. Second, it wanted not only to get money into the hands of desperate fishers but also to keep them busy and stem the urge for any unrest, such as the Bayou La Batre protest described earlier.

The program was very popular in that every person with a boat or able to work on a boat wanted to participate, and the money that was earned was desperately needed and welcomed. VOO workers successfully skimmed oil; laid, maintained, and retrieved boom; removed tar balls; transported supplies; protected shorelines; saved wildlife; and escorted government personnel all around the Gulf.

The program was highly unsuccessful, however, because there was never enough money or boats hired, there were extreme problems with implementation, and the program ended abruptly, before the need for it had been filled.

Dulac's Jamie Billiot and Bayou La Batre's David Pham and Vinh Tran became reluctant experts in the program, helping boat owners and crew members in their communities to wade through contracts and BP Web sites. In the end, only a very small percentage of the people from their communities who needed and wanted jobs got them—just 10 to 15 percent, at best, a pattern reported throughout the Gulf.

Pham and Tran tried to get more of the people from their communities signed up for BP programs, only to hit dead ends. As Pham told

David Pham (third from left) of Boat People SOS surrounded by out-of-work fishermen asking for his help in Bayou La Batre, Alabama, in August 2010.

me in July, "We're trying to get people into BP training for onshore clean-ing and VOO. We called BP. They take your contact info, we gave them all that, they never call back."

That left contacting BP online—an equally frustrating process and one that was not always an option. "Most of my people don't have com-puters," Billiot tells me. "We had to collect the information for them, make sense of it for them, file it for them, and follow up to see what happened." Pham and Tran reported the same process.

To participate in VOO, workers were required to sign a twenty-five-page contract that included twenty-five articles, three exhibits, and a confidentiality clause.[16] As a result, virtually all reports on VOO are sec-ondhand or from BP.

The program was beset with problems. As oil drenched waterways and shores, people could not understand why more boats were not hired to capture oil and lay boom. In an all-too-common refrain, Callaway bemoaned BP's ineptitude "at every stage of its response." Following the explosion, Callaway recalls, "BP had three weeks to prepare for oil to hit ground in Louisiana. Why weren't they ready? Why wasn't boom in place, berms built, and volunteers ready? Why was the shore deci-mated?" Instead, volunteers, including thousands trained by Callaway and her cohorts, were turned away, and boats waiting to carry boom were left sitting idle on shore.

Similarly, Diane Huhn of Bayou Grace Community Services in Houma, Louisiana, says, "We know we can't trust BP. A few days after the rig sank, BP was promising it would mobilize tons of boats. On April 25, we only saw two boats—one with skim, but not skimming. Basically BP had no capacity to respond to this disaster. It makes me furious."

Of the nearly nine thousand boats that registered with the VOO pro-gram at its height, only thirty-two hundred were deployed at any one time. Only the people on the deployed boats got paid; the rest waited by the phone.

"That caused a lot of stress," Tran tells me. "Many of the people only have landlines, not cell phones, so they'd sit at home all day long, afraid to leave because the phone might ring, they'd miss it, and someone else would get the job." Those with cell phones sat on their boats all day, keeping them ready so they could deploy at any time. "Waiting by the phone, waiting by the phone. People could barely sleep at night; they'd watch the phone all the time. In the end, only a handful ever got called."

Crew members earned $200 a day, and boat owners earned $1,200 to $3,000 a day, depending on the size of the boat—for many, this is about half of what they could earn on a really good day fishing.

As the number of people hoping to join the program increased and vastly outnumbered those actually being hired, BP did not increase the amount of money it spent on the program; rather, it shortened the amount of time each vessel was hired—spreading out the benefits to more people, but lessening how much each person could earn.

I was standing on a dock in Bayou La Batre with Pham and Tran when a group of men working on a nearby boat joined us. Speaking in Vietnamese, the men inundated Pham and Tran with questions about VOO and other BP-related concerns. As we were leaving, one of the men, who desperately wanted to get hired by the program, took my pen and notebook and wrote, "Tell BP I'm a good worker, David Yang."

A lack of translators plagued not only VOO but also all of BP's

Out-of-work fishermen in Bayou La Batre, Alabama, in August 2010. David Yang (right) asked me to "tell BP I'm a good worker."

programs in the Gulf. Groups like Boat People SOS contacted BP immediately, alerting the company to the desperate need for translation services. When BP did not respond, they reached out to local elected
officials and the media. Their leading government advocate (Pham
would say the "only" government advocate) was Congressman Anh
"Joseph" Cao (R-LA). Born in Vietnam in 1967, Cao escaped to the
United States at the age of eight. After serving as the in-house legal
counsel for Boat People SOS, Cao became the first Vietnamese American elected to Congress in 2008. Following the oil spill, Congressman
Cao became a constant thorn in BP's side by demanding that the special
needs of the Asian community be met. Unfortunately, he had only limited success.

When translation was provided, it was generally inadequate and
insufficient. For example, Pham and Tran report that when BP finally
did provide a translator in the Bayou La Batre area, there was just one
man, he only spoke Vietnamese, and he spoke a Northern dialect, which
"few of our people could understand." They say that people would regularly wait six to seven hours to see this man only to be frustrated by
their inability to speak with him.

Boat People SOS and other service organizations did their best to
fill the need, but it was not easy. Mary Queen of Vietnam Community
Development Corporation of New Orleans was another group helping
out. Project manager Daniel Nguyen explains that in the Gulf, "There
are roughly about forty thousand south-east Asian people involved in the
fishing industry. The oil spill pretty much affected eighty percent of the
Vietnamese population overnight. It was a huge issue for our organization to provide direct assistance on top of what we're already doing to
help rebuild after Katrina. It's a very difficult struggle. The oil spill is
definitely a huge disaster for us."[17]

Not only were there not enough translators to help those in need,
but the paperwork they were asked to translate was time-consuming and
extremely difficult to understand, even for the best-trained lawyers.

Other problems faced by VOO workers were health concerns
described in the previous chapter. Thanks to Wilma Subra's and Tracy
Kuhns's efforts in court, BP did eventually provide VOO workers
with safety gear and require them to get Hazardous Waste Operations
and Emergency Response (HAZWOPER) training. Nonetheless, health

problems—from headaches to skin rashes to vomiting, sometimes putting people in the hospital or confining them to bed for days—beset
those involved in the work.

Dr. Riki Ott remains concerned that the boats used by VOO may
now be saturated with chemicals that could leach through the wood and
poison their occupants for years to come.

There were also the problems that drew the fishers out in open
protest. Recreational boaters from as far away as Texas were getting jobs
over out-of-work local commercial fishers, and even when the fishers
were hired, they did not get paid for weeks on end.

In Bayou La Batre, a sign reading COMMERCIAL 1ST: WE NEED WORK
NOW hung from one of the vessels idling in the waterway as Ray Foster,
sixty-two, explained his reasons for being at the protest to a local newspaper. "I ain't against nobody making money, but I think BP ought to put
all our commercial fishermen's boats to work." Foster was hired by the
VOO program. He had submitted an invoice for his work and was on
standby for another deployment. "They told me to stand by the phone,"
he noted. More than nine days later, he hadn't been called back to work,
nor had he been paid for his service. "I need to go to work," he added.

Wendy Buchanan—whose husband, Brent, was arrested while
protesting—said that Brent too was working for BP but had yet to get
paid. "We're frustrated and tired of the runaround," she said. "We
had six people employed under us. A lot of these guys, this is all they
know."

Michael Sprinkle, owner of two boats, said that if it weren't for the
oil contaminating the water, he would be out catching shrimp. "It's
ruined us," he said of the spill. "I don't think we'll get to do it again.
Especially if they don't get it stopped anytime soon."[18]

On July 15, the well was capped, and the Obama administration and
BP went on the offensive to declare that the oil had now disappeared.
Not only did the administration's own data and scientists fail to support
this contention, the oil was also still in the water, washing up on beaches
and wetlands and poisoning the Gulf.

Nonetheless, BP started pulling back its workers from all of the
cleanup activities. The result was a virtual halving of employment within
a matter of weeks, far less boom deployed, fewer areas protected, and
less oil cleaned up once it hit the shore.

On July 29, I attended one of the many community forums hosted by BP, the coast guard, and other agencies throughout the Gulf during the crisis. The forums had very mixed results. Although residents were very happy to have real people to talk to, they usually left understandably frustrated by the lack of meaningful information provided in the presentations and by the inadequate answers (or no answers at all) to their questions.

Sharon Gauthe of Bayou Interfaith Shared Community Organizing (BISCO) had worked with BP at first to help organize some forums, which also included community organization representatives. But then BP decided that it could do the forums without her and without certain members of the community present.

At the forum in St. Mary Parish, Louisiana, the people standing at the tables looking helpful nearly outnumbered the community members who sought their help. I spoke briefly to an out-of-work fisherman desperate to be hired by the VOO program (one of many I met that night) as he was leaving. I asked why there were so few people there. "What's the point?" he answered. "They never tell you anything." Another fisherman said that the process was just so aggravating. He asked, "Who wants to come to just get mad?" He had been fishing these waters his entire life. His nineteen-year-old son wanted to follow in his footsteps, but he wasn't so sure. "We're a dying breed," he told me.

When the parish president announced that St. Mary did not, does not, and would not have oil from the spill, he was immediately surrounded by the fishermen, one of whom said loudly, "Then why does Kermit have oil in his bag right now?" At that point the president turned off the microphone, and, in Kermit's words, "All hell broke loose."

Kermit Duck's family has been fishing in the waters of St. Mary Parish for as long as he can recall. Kermit, age thirty-five, and his wife, Stephanie, brought the oil to the meeting in a ziplock bag to prove that it was out there. Kermit had applied for but had not been hired by VOO, although he had spent two months on a waiting list. He had been unemployed for four months and was desperate to see a real cleanup effort take place so that one day he might be able to fish again. Stephanie's job was not enough to support their entire family, which includes their two young children, ages just one and five. Desperation covered Stephanie's face; frustration dominated Kermit's.

The next morning, Kermit took me to see where he had collected the oil. During the five hours we spent on the water traveling between Oyster Bayou and Taylor's Bayou, we encountered a lot of oil. Other than a small amount of boom outside the mouth of East Bay Juno, we saw no boom, skimmers, absorbents, or cleanup crews. The Juno boom was coated with oil, and so was the area behind it.

We did see plenty of grass and beach freshly soaked with oil, however. The strong harsh smell of crude filled the air as we neared the shore. The oil had washed up in waves, covering a large patch of grass here, leaving a clean patch beside it there. Fields of oil glistened as the sun picked up the oil's sheen.

We walked along a shell beach on the south end of Oyster Bayou. It was speckled throughout with fresh tar balls that reached from the reeds to inside the water's edge. Kermit's friend Buddy used an oar to dig below the beach surface, revealing more oil underneath.

Kermit was desperate for the world to know that there was still oil on these grasslands and beaches—threatening his home, his oysters, and his livelihood. "These grasses are our last line of defense against hurricanes. What are we supposed to do if they die?" he asks. I called a *Washington Post* reporter I know from his work on oil issues. He referred me to the paper's reporter in the Gulf that week, and shortly thereafter, photographer Julie Dermansky arrived to tour the area. Her powerful photographs accompanied the August 14 front-page *Washington Post* story on VOO, and a full slide show appeared on the Web site.

Stephanie's grandmother was sick and Kermit was unable to take Dermansky out on the water; two of his friends, Chuck Frielich and Harry Rebarbie, filled in. Kermit was quoted in the article, however, saying, "It took 100 days to get here, but it's here now. There are no white sandy beaches here, but it's our home. The grass in the marshes has oil mixed in it. I'd be glad to clean it up if I had the right equipment—and if they'd call."[19]

"Rapid Response" Cleanup Crews

Not only were people anxious to have boom installed off their shores to stop the oil from hitting their land and to clean it up out of the water, they were also desperate to clean the oil up off their beaches once it made landfall.

BP provided grants to the states to aid with these efforts. On May 5, 2010, BP gave $25 million each to Florida, Alabama, Mississippi, and Louisiana to "accelerate the implementation of Area Contingency Plans." This money was used to, among other things, acquire boom, build berms to protect shores, and conduct cleanup.

Two weeks later, BP gave Louisiana another $25 million and the other three states $15 million each to promote tourism. It gave smaller sums, around $500,000 to $1 million, directly to the Louisiana parishes.[20]

BP also subcontracted for its own beach cleanup efforts. One of the more jarring photographs on BP's Web site is of a beach cleanup crew in the "Making It Right" section. The picture could be straight out of a science fiction movie about some sort of postapocalyptic vacation. There is a bright blue, crystal-clear sky and an immaculately clean, light brown, sandy beach. The beach is so smooth that it looks more like cement than sand.

Marching across the beach are workers, their heads down, in spotless antiseptically white protective body suits, yellow hats, yellow shoe booties, and gloves. These are BP's "rapid response teams," deployed to beaches throughout the Gulf "whenever oil is spotted." Although the gear they wear is to protect them from the toxic oil and chemicals with which they work, not a drop of either is depicted in the photo or mentioned in the caption.

BP's Shoreline Cleanup Assessment Teams (SCATs) "patrolled the coast daily looking for any oil." Then, when oil was found, the task forces were deployed to clean it up. Under court order, BP required that workers receive safety training before being dispatched "to coastal areas that have needed protecting," according to the site.

There are very few statistics on BP's beach cleanup workers, but for a time they were a constant and reassuring presence on the beaches hardest hit by oil. With a few exceptions, press reports record either the frustration of journalists trying to acquire this information or the use of the cumulative numbers cited above.[21] The lack of good information is most likely a result of the number of subcontractors involved—some reports put the number at fourteen separate companies—and because BP, not the federal government, hired the workers and, as in so many other aspects of its efforts, did not particularly feel the need to be transparent.

Company spokesman Daren Beaudo did say that BP used "hundreds of contract companies" but did not have a "list of who is contracted to do what and by what companies."[22]

In early June, CNN spoke with Deb Witmer, a BP public affairs officer, and reported, "BP is amassing a small army of 4,500 unemployed workers in Alabama, Mississippi, and Florida to be ready on short notice to clean oil off the Gulf coast beaches. . . . Beach cleaners will be paid $18 an hour and supervisors $32 an hour. . . . BP aims to have at least 1,500 workers in Mississippi, 1,000 in Alabama and 1,600 in Florida."[23] The cleanup workers I spoke with reported pay scales more frequently in the $12 to $15 range. We do not know how much BP spent on the cleanup efforts in total.

On July 28, I walked into a HAZWOPER training class with Mayor Ron Davis of Prichard, Alabama, in the city's Public Safety Building. When they saw the mayor, smiles, waves, and nods of appreciation spread throughout the packed classroom As far as the eye could see, all of the students, like the mayor, were black.

Students in Prichard, Alabama, in July 2010 receive HAZWOPER training in hopes of being hired as BP cleanup workers.

It was the tenth and final class the city would provide. The HAZ-WOPER training is required for all who work on the cleanup—something all of these students were eager, even impatient, to do. When asked about the potential health hazards, the long hours, and the heat, one student's response, "It's worth it for the money," is greeted with nods.

Mayor Davis's classes have been packed since they began on May 1. This is not only because Prichard is desperate. The town, with a population of twenty-six thousand, had a 40 percent poverty rate and a 14 percent unemployment rate when the *Deepwater Horizon* exploded. It is also because Mayor Davis's classes are unique. They are free, taught jointly by the city's fire chief and chief fire marshal, and as such, make graduates eligible for a variety of future employment opportunities beyond the oil spill cleanup. The mayor could offer the classes for free thanks to a preexisting grant with the EPA.

Most other classes, offered by less reputable contractors, can cost as much as $500, with some trainers offering to garnish students' wages to cover the cost. The students report that far too often the classes are simply scams that fail to provide the proper training and never lead to the $15-an-hour jobs with BP.

The oil "is a blessing and a curse, for sure," one student says. The disaster has brought great devastation to the Gulf coast, but it has also offered these students a rare opportunity for advancement.

Mayor Davis beams with pride at his students and is eager to have their stories told. More than 1,000 people have applied for his classes, more than 250 have been trained, and more than half of those have been placed with a BP contractor for beach cleanup. The mayor did not even have to advertise; there were already too many people signing up just from word of mouth.

"They're making $15 an hour, working twelve-hour days, seven days a week," the mayor explains. "The first class worked three weeks before they got a day off." He doesn't say this as a complaint; rather, it's a selling point. "Can you imagine people bringing in that kind of money, $1,000 a week after taxes, who hadn't had a job in years? People were spending, buying groceries. It's a blessing on this community."

Prichard is located about twelve miles north of downtown Mobile. It is a world apart, however, from the tourist beaches of Pensacola, Gulf Port, and Dauphin Island, which Mayor Davis's students were hired to clean. The town is a poster child for rural postindustrial blight. The main street is shuttered and run-down; the random hair salon and bar the only open businesses. Mills, paper factories, and production plants have closed down year after year. The Dairy Fresh mill, which employed 150, "just closed down, they're moving to Hattiesburg [Mississippi]," the mayor tells me. The town has lost not only jobs but also tax revenue—$100,000 in 2007–2008 alone—which forced the mayor to file for bankruptcy. "These are folks who want to work, but have nowhere to turn," he says.

Mayor Davis is committed to rehabilitating his town. With a degree in criminal justice and social rehabilitation services, he served as an officer for eighteen years with the Mobile County sheriff's office and was going to retire but decided to come back and help the city where he was born. He was elected to the city council in 2000, elected mayor in 2004, and resoundingly reelected in 2008. He is now the interim president for the World Conference of Mayors.

Ron Davis, the mayor of Prichard, Alabama, in his office in July 2010.

Back in his office, he proudly shows me a photograph of himself taken with President Obama, who, he states emphatically, has "taken an incredible, difficult situation and done the best that he could be expected to do" in the wake of the spill.

Mayor Davis is doing his part, and it has not been easy on him or his students.

To be eligible for his classes, the students had to take a physical and submit to a background check. "As you know," the mayor explains, "people going on Dauphin Island and Gulf Shores—you know, you can't put anybody in these communities. I told each class, 'You got a drug problem, extensive criminal background, don't even bother showing up for class.'"

The mayor's concerns were understandable. The cleanup workers were controversial all across the Gulf. Racism was one of the main reasons. I heard more than one white member of a community complain about the skin color of those coming to clean "their" shores. One white shrimper I met told me, "It's just like when Castro emptied his jails and put them all on boats to America!"

The restrictions placed on the workers did not help. With police guarding the beaches where the cleanup crews worked, and supervisors stopping the public or press from approaching the workers, the impression was created that the public was being protected from the workers, when in reality BP was being protected from the public's hearing what the workers had to say.

One typical account of the restriction on reporters to speak to cleanup workers came from PBS reporter Spencer Michels in June. "On a dock at Venice, where workers who had been hired by BP were coming back from [beach] cleanup duties, we couldn't get a comment from anyone," Michels explained. "We saw them sitting around after their boat trip back, but we were told by coordinators on the dock not to talk with any of them."[24]

The same day as this report, BP chief operating officer Doug Suttles released a letter that tried to clear the air by saying that workers were allowed to talk to the press and that no one should restrict them.[25] As described earlier, after this directive, crew leaders began to speak more freely, but members of the crew largely continued to work in silence.

The largest source of public opposition to BP's cleanup efforts, however, was that BP refused to allow concerned citizens to volunteer and participate. BP attributed this decision to the need to provide jobs. However, there were nearly 700 miles of beach oiled by BP's blowout. There were never enough people employed to adequately monitor and clean these shores. There were also more than enough people willing and eager to pay for and acquire the necessary safety training and equipment to do the work as volunteers. Nonetheless, they were denied access. The reason appeared to be BP's desire to maintain control over access to information.

Casi Callaway was livid about this point. "We had five thousand volunteers trained and ready to deploy," she tells me, "but BP wouldn't let them clean the beaches. These are our beaches. We love them. We want to protect them."

Callaway introduced me to Les Switzer, a volunteer who had shown up at her office after first going to BP to offer his help. "I don't want BP's money to fix what they've broken!" he tells me, his frustration rising as he recounts the story. "I just wanted to help." Instead, he was shown a contract with a confidentiality clause and the strongest thought came to his mind. "I realized that it was just like a crime scene," he says. "They killed our Gulf, and now the murderer is in charge of cleaning up the scene of the crime." Les says, "I realized it was crazy, then I broke down and cried."

There was also controversy over the manner in which the crews were deployed, with numerous reports of workers being specially bused in to provide photo ops for visiting dignitaries and the press. On my first trip to Grand Isle, Louisiana, I was shocked to see a pristine beach. President Obama had been there just the day before and, as it turned out, BP had bused in an army of some four hundred oil cleanup workers in advance of his arrival to create a spotless beach.

Jefferson Parish councilman Chris Roberts, whose district encompasses Grand Isle, was irate. Calling it "a sad stunt" by BP "to mislead the president and the American public," he said that BP was trying "to make it appear as though they've been on top of their game in mobilizing assets when that's not been the case at all." He said "the level of cleanup and cooperation we've gotten from BP in the past is in no way consistent to the effort shown on the island today."[26] Moreover, "as soon as the president was en route back to Washington," he said, "the workers were clearing out of Grand Isle, too."[27]

David Gauthe of BISCO, who was with me that day and helped me get on the beach because he knew one of the security guards, told me that he had never seen the beaches so clean since the spill. Shortly thereafter, of course, the beaches were oiled as the waves brought in fresh oil and then went back out to sea for more.

Like the VOO workers, the cleanup crews also reported health problems. There was also a catch-22 with their gear: it protected them from toxic exposure but greatly exaggerated their exposure to the heat, which readily topped a hundred degrees during the summer. Thus, more often than not, the crews would be seen wearing only boots and gloves rather than their full gear, thereby reducing heat exposure, while increasing their exposure to toxic oil and chemicals.

These crews conducted vital and invaluable work. But, as noted earlier, there were many more who were eager to do this work, both for pay and for free, than BP utilized, even though the need was great.

BP cleanup workers scooping up tar balls in July 2010 on Dauphin Island, Alabama, work without protective gear.

Moreover, once the well was capped, the desire to eliminate the appearance of the need for cleanup caused BP to send the cleanup crews home. I personally continued to walk on beaches covered with oil and to travel in boats to oiled grasslands well after the cap was in place. The only difference was that there were no longer cleanup workers, and as the months rolled on, by and large, no more reporters to observe their absence.

The Claims Process

The claims process became one of the great weights around the necks of hundreds of thousands of Gulf residents and the service providers who hoped to help them. The money it provided has been desperately needed and welcomed, but the program has also been inundated with debilitating problems that forced the federal government to take over BP's operations. Even the new independent process, however, remained plagued by deficiencies.

BP reports having its claims process in place within days of the explosion and making its first claim payment by May 3.[28] According to BP, it paid almost $396 million in claims to individuals and businesses between May 5 and August 22, when the claims process began to be administered by Kenneth Feinberg.[29]

According to a *New York Times* analysis of BP's claims processing on July 3, "Since May, it has paid just under a third of the more than 90,000 claims it has received. . . . The payments averaged about $2,500 a month for a deckhand or $5,000 for a fisherman."[30]

Darryl Willis, BP America's vice president of resources, became the public face of the process. His personal story became as familiar as his face to the people of the Gulf as he appeared regularly in TV commercials and print ads that emphasized his strong local ties and commitment to the area. He explained in ads and interviews that he was born and raised in Louisiana. His mother, age seventy, lost her home in Hurricane Katrina, and he knew how frustrating the recovery efforts had been.[31]

BP's process generally looked good on paper, and Willis's desires certainly seemed quite genuine; moreover, people were thankful that there was a program and were hoping for the best. Within weeks, however, it was clear that the process was buried in bureaucracy, was much too slow and strict, had insufficient funds and staff to be run as was needed, was much too complex for claimants, and was provided only in English for much too long.

The financial need across the Gulf cannot be underestimated. I attended numerous community forums where people pleaded with BP to get their claims checks. People were made homeless, hungry, and desperate by the disaster. Small needs, such as school supplies and being able to purchase a birthday gift for a child, made people nearly as distressed as bigger ones, such as rent and food.

Checks were slow, small, and inconsistent. One source of trouble was identified in BP's choice of subcontractors. BP hired ESIS Inc., a global risk-management services company that provides recovery management services and claims management, to provide the backbone of the claims process.[32] ESIS, however, advertises its recovery management services as having the "goal of reducing our clients' loss dollar payouts" and offering "flexible service delivery models that are aimed to reduce your loss costs."[33] Among the many groups that protested ESIS was the New Orleans–based Advocates for Environmental Human Rights, which wrote to Admiral Thad Allen accusing BP and ESIS of minimizing payments to claimants and potentially providing insufficient compensation.[34]

As the chorus of voices raising concerns with the process became ever more vocal throughout May, including both Gulf residents and government officials, BP appointed an independent mediator to "review and assist in the claims process" on May 26.[35]

One of the most often cited problems was the documentation that BP required the claimants to present. In the largely cash-driven fishing industry, few had the required pay stubs, paychecks, tax returns, and other necessary paperwork. Moreover, BP required three years' worth of paperwork, which, in the storm-centered lives of the Gulf coast, was often impossible to provide. "Most of our community's paperwork washed away in 2008 in Hurricane Gustav," Jamie Billiot told me. "What were they supposed to do?"

Tuan Nguyen, the deputy director of the Mary Queen of Vietnam Community Development Corporation in eastern New Orleans, told the Associated Press that "it's a very cash-involved industry" and that he has encountered "hundreds of workers, mostly deckhands, who lack the documentation BP needs from claimants."[36]

On June 8, responding to the criticism of BP's claims process, Admiral Thad Allen wrote to BP CEO Tony Hayward seeking more information. "We need complete, ongoing transparency into BP's claims process, including detailed information on how claims are being evaluated, how payment amounts are being calculated, and how quickly claims are being processed."[37]

In a meeting on June 10 between Allen and BP officials, BP agreed to speed up the payments. Five days later, the White House announced that BP would no longer be in charge of the claims process. Instead, an independent claims facility would be created, with the mandate to be "fairer, faster, and more transparent in paying damage claims by individuals and businesses."[38]

Governor Bobby Jindal of Louisiana summed up the mood of the Gulf a few days later. "The bottom line is that the BP claims system is broken."[39]

The independent claims facility, however, did not begin until August 23. ProPublica, an independent, nonprofit investigative journalism organization, found that in the interim, BP was intentionally deferring the processing of thousands of claims. BP spokesman Daren Beaudo, told ProPublica, "It may be simpler for Mr. Feinberg to take on those" claims. The deferral was not made public until the story broke, however, and the claimants were not made aware of their "special status." Rather, according to one who posted a comment on the ProPublica Web site, "we were given the run around" while being forced to wait in limbo for Feinberg.[40]

Community groups filled in where BP was failing. Many of the organizations supported by the Gulf Coast Fund provided services to their communities thanks not only to emergency grants provided by the fund but also to its organizational support. Filling BP's void was difficult for many reasons. One was trying to simply understand the claims process, no small feat in itself. The other was staff time and their own financial instability. Finally, the groups' goals were not just service provision but also looking out for the health and long-term economic survival of their communities. This meant staying on top of all of BP's actions, as well as local and federal governments'.

I attended a two-day strategy session in Houma, Louisiana, organized by the fund in late July. Prior to Hurricane Katrina, the few small, understaffed, and underfunded advocacy organizations and volunteer networks of the Gulf coast worked in virtual isolation, cut off by the rural landscape, a lack of technology and funds, and a mentality of go-it-aloneness. "What Katrina did was give us a reason to know each other," LaTosha Brown tells me. "Our local issues were now regional and even national. The disaster was simply too big for one local community to address alone. But, together, we could build a common agenda and capacity." The Gulf Coast Fund, which Brown now directs with Annie Ducmanis, the project manager, was one outcome. All that they learned and built in response to Katrina was now put to work to address the BP Gulf oil disaster.

The fund's strategy session focused on everything from how local community groups could help out-of-work families to preparing for the upcoming annual meeting of the five Gulf states' governors to demanding that Congress establish a citizen advisory council to respond to the disaster. Roberta Avila of the STEPS Coalition of Biloxi, Mississippi, for example, discussed how her organization had hosted a huge claims fair in Biloxi focused on language accessibility, service provision, and personal care for the hardest hit groups. She juxtaposed her fair to one set up by BP and the governor's office that was not publicized in the local newspaper but rather through local trade associations and which focused on large Realtors and big business owners.

Sheila Tyson of the National Coalition on Black Civic Participation and Citizen Participation Program of Birmingham, Alabama, discussed how the spill struck right when school was starting. She later recalled helping a family "where the kids needed school supplies, utilities were due, car insurance was due, telephone, water, light, gas, all of this stuff was due, but no one recognized that they had all these needs." Families could go to one of the many food distribution centers set up at churches in response to the spill, but Tyson recalled this family asking, "How am I going to cook it if I don't have money to pay my utilities?" Tyson's center helped them fill out paperwork for claims, with the provision of clothes and food, and even advocated "with Washington, D.C., for new laws to address oil spills." When asked about the future, Tyson responded that "many in her community may never recover."[41]

Many of those seeking services after losing jobs and income were employed in the tourist industry, the third leg—with oil and fish—of the Gulf coast economy. "The summer that wasn't" is a phrase used by many on the Gulf Coast to describe the summer of 2010.

"The Summer That Wasn't": The Toll on Tourism

In November 2010 *National Geographic Traveler* magazine ranked "99 coastal areas worldwide" and "their geotourism quotient." Coming in dead last was the Louisiana Gulf coast. Second to last was the Mississippi Gulf coast. Both were categorized as "catastrophic" due to the double whammies of Hurricane Katrina and the BP oil disaster.[42]

By some estimates, tourism directly supports some four hundred thousand jobs in Gulf coast counties along the shore, representing 15 percent of total private employment. Annual spending by visitors alone along the shore is estimated at more than $20 billion in Florida, $7.2 billion in Texas, $3.6 billion in Louisiana, just under $2 billion in Mississippi, and $1.4 billion in Alabama.[43]

Aggressive spending by BP and the Gulf states to lure tourists back to the Gulf and the absurdly premature declarations by many that the beaches and waters were safe ensured that the worst economic loss predictions were not met.

The oil spill also brought in its own revenue stream. Some of the hardest hit areas were often also those where cleanup workers, state and federal agency employees, military personnel, journalists, and advocates spilled in. Gulf oil disaster paraphernalia also sold extremely well. In New Orleans, stores along Bourbon Street were filled with T-shirts, baseball caps, and refrigerator magnets reading FUBP (F*ck You BP). Images of BP CEO Tony Hayward with oil dripping from his mouth and blood from his fingertips were also very popular.

Nonetheless, tourists by and large took their business elsewhere, and those hardest hit were the small businesses, already operating on a shoestring from years of hurricanes, that had no cushion on which to withstand the latest disaster.

The effect on the tourist industry ran from hotel owners and all who care for their guests to restaurants, bars, gas stations, grocery stores, convenience marts, and souvenir shops.

On a typical June weekend on Grand Isle, for example, Artie's Sports Bar would be crammed with more than a thousand patrons peering at the ocean, reported Frankie Marullo, whose brother owns Artie's. But he reported only a few dozen customers a day since the explosion. "The tourism business is shot," he said. "This place is wiped out. It's going to kill this little island."

Amy Martin, whose family firm, McGuire's, owns four eateries in the Pensacola area, said the number of customers was down 50 percent in June compared with last year, while sales have dropped 25 percent. "It's worse and worse every day," she said.

Renee Lefeaux and her family only recently had gotten their beach-front condo on Alabama's Orange Beach into rental condition after it was severely damaged by Hurricane Ivan in 2004. In early June, she and her daughter went parasailing. Instead of the seaweed tumbling and rolling, it was clumping together in a way that didn't seem natural, she recalled. "We knew firsthand it was coming," Lefeaux, who lives in Baton Rouge, said of the oil. "It was only a matter of days. . . . It was very sad." The guests who had planned to rent the condo were dropping away. "I had booked through August, and now there's not a soul booked the rest of the year," Lefeaux explained.[44]

Some of the hardest hit were the rental properties, known as "cabins," in the Gulf. These small wooden vacation homes dot islands and bayous and are most often found elevated on stilts because of their close proximity to the water.

July is usually a prime month for rentals, especially the July 4 weekend. Not so for the summer of 2010, during which fireworks were canceled both on Dauphin Island and Grand Isle. Amy Vice, a real estate agent on Dauphin Island, reported, "Last year this time I had 111 [rentals]. "We've got ten coming in tomorrow." Vice said that most of her day is spent sifting through cancellations. "I'm afraid that what's going to happen is once it is capped and it's finished and cleaned up, people won't come back," Vice said. "We have a lot of repeat customers and they're trying to keep with us, but they are afraid to bring their families down here." Overall for the holiday weekend, Dauphin Island rentals were off by a reported 80 percent.

"We're doing a fraction of what we did last year," said Adam Almord, the owner of the Common Loon cafe on Dauphin Island. "The support we're getting is strictly through locals" and cleanup crews.[45]

TripAdvisor, "the world's largest travel website," revealed significant declines in U.S. page views for Gulf coast destinations for the twenty days leading up to May 20 and to July 18 compared to the same twenty-day periods one year earlier. For example, consumers searched 52 percent less for Pensacola, Florida, in July, 65 percent less for Gulf Shores, Alabama, and 48 percent less for Destin, Florida. Searches increased significantly, however, in some cities outside of the "spill zone." For example, searches for Miami and West Palm Beach increased by some 15 percent in May over the previous year.[46]

In June, Jim Desai, the manager of Beachside Resort Hotel in Gulf Shores, Alabama, reported, "Most of our cancellations are related to the oil spill—90 percent, in fact. . . . Guests are saying the beach is getting worse, the oil is on the land, and the water is getting browner and browner."[47]

Smith Travel Research reported that hotel occupancy in July was down or flat from a year earlier in four of five Florida west coast destinations, while eight markets on Florida's east coast were all up, "most by double-digit percentages."[48]

The *Pensacola Business Journal* reported, "Revenues to hotels, restaurants, shops and rental agencies on Pensacola Beach and Perdido Key fell between 25 and 75 percent during June, July and August. In July, for example, of the 130-plus businesses on Pensacola Beach, only six had positive sales numbers."[49]

All of these business owners and those they employ then turned to the claims process in an attempt to stay afloat.

BP's $20 Billion

At the White House on June 16, in the one and only meeting between President Obama and CEO Hayward, BP agreed to establish a $20 billion escrow account that would cover costs, including claims to be paid by the newly established independent claims facility. The $20 billion was not a lump sum but would be contributed throughout a four-year period at a rate of $5 billion per year. BP agreed to "provide assurance for these commitments by setting aside $20 billion in U.S. assets."

In addition to the claims, the escrow funds would cover all "judgments and settlements, natural resource damage costs, and state and local response costs." Thus, although at first glance $20 billion seems like a large figure, the inclusion of all these potential costs makes it more like a drop in the bucket. The agreement also mandated, however, that BP would not "assert any liability cap under OPA to avoid liability" and that the account "is neither a floor nor a ceiling on liability."

The OPA includes a $75 million liability cap unless the responsible party is found to have acted with gross negligence, willful misconduct, or violated applicable federal regulations.

In addition, BP would contribute $100 million to a foundation to support unemployed oil rig workers and another $500 million for a "ten-year Gulf of Mexico Research Initiative to improve understanding of the impacts of and ways to mitigate oil and gas pollution."[50]

Kenneth Feinberg was appointed as the independent claims administrator. Feinberg, a Washington D.C.–based attorney, was best known prior to the BP disaster as the head of the September 11 Victim Compensation Fund, which distributed nearly $7 billion to more than five thousand victims and families of victims of 9/11. More recently, he served as the special master for compensation, or "pay czar," appointed by President Obama in 2009 to establish CEO salaries at companies being bailed out by the federal government under the Toxic Assets Relief Program (TARP).[51]

The Gulf was thrilled with both the establishment of the independent facility and the appointment of Feinberg. The people eagerly awaited his arrival.

The Independent Claims Facility was renamed the Gulf Coast Claims Facility and became formally operational on August 23. All claims previously filed with BP were transferred to the new process, but the claimants had to file new paperwork with the facility in order to receive payments.[52]

The facility established two forms of payments and gave claimants until November 23 to choose between the two. The Emergency Advance Payment provided up to six months of lost income in advance and had to be filed for before November 23. Accepting this payment would still allow the claimant to sue. The other option, the Final Payment, aimed to provide compensation for long-term damages, and the claimant had three years to estimate and claim these damages. In accepting the Final Payment, claimants gave up their right to sue BP or any of the other companies involved in the disaster at any time.[53]

In launching the facility, Feinberg said that "the goal will be to get the emergency six-month payment checks out the door, within 48 hours for individuals, after receipt of the claim form and sufficient supporting documentation."[54] Feinberg also stressed the benefits of the Final Payment.

The right to sue is critical to ensuring justice for those harmed by the disaster. People were very quick, however, to compare their situation to that of Alaskans in the wake of the *Exxon Valdez*. As James Billiot tells me in Dulac, "If you don't take the settlement, you'll spend twenty to thirty years in court like the people in Alaska. And you'll have to pay for a lawyer. You lose either way. BP's lawyers will be big, rich, and powerful. We'll lose."

Legal suits by affected communities began immediately in the wake of the *Valdez* and were brought under one class action suit representing nearly thirty-three thousand individuals in 1994. That year, a court ruled in the class action case that punitive damages "necessary in this case to achieve punishment and deterrence" should be imposed against Exxon in the amount of $5 billion, a year's average profits. Exxon appealed. In 2008, after nearly twenty years and during which time more than three thousand of the claimants died, the U.S. Supreme Court ruled in Exxon's favor and imposed a highly restrictive limit on punitive damages—a one-to-one ratio—yielding damages for Exxon of a measly $507.5 million. I heard the cautionary tale of the *Valdez* told repeatedly in the Gulf as people considered how best to address their financial needs and Feinberg's options.

While there was a great deal of hope placed in Feinberg, unfortunately, the process continued to be encumbered by extreme operational failure under his oversight. Filing remained incredibly complicated, particularly in navigating the switch from BP to the facility. Complaints continued to come from residents and elected officials alike.

In September, associate U.S. attorney general Thomas J. Perrelli sent a stern letter to Feinberg noting that the current pace was "unacceptable" and that he "needs to devote whatever additional resources—or make whatever administrative changes—are necessary in order to speed up this process."[55]

"Under Feinberg the process is still the same," Daniel Nguyen of Mary Queen of Vietnam described in an interview. "Every time we call the 1-800 number we get someone new. So every time we call we have to reexplain everything. It's a really time-consuming process. It's also very inconsistent because each person that answers the phone provides a different answer to the same question. We've actually had an afternoon where we sat down, came out with a question and we just sat there for like an hour calling over and over and recording all these different answers we would get to the same question." Even when they could get through, problems persisted. "We'll have people working the same business earning the same salary and they'll get two completely different claims checks. It's incredibly frustrating," Nguyen said.[56]

Feinberg conducted a series of town hall meetings throughout the Gulf in October that were headline news. The events were packed with participants eager to hear from, or more often, confront the administrator. At one, a woman conveyed the sentiments of many when she held a sign that read FEINBERG, YOU MAKE ME MISS FEMA, a reference to the Federal Emergency Management Administration, which failed so miserably under the Bush administration in the handling of Hurricane Katrina.

The associate attorney general sent another letter to Feinberg in November, saying that he continued to have concerns with the pace of the claims process and asking Feinberg to finish processing all emergency claims by December 15. He also made suggestions about the next phase of the facility so that "it does not limit the rights of claimants and does not slow their receipt of payment."[57]

Feinberg did not make the deadline. By the end of 2010, the facility had received more than 450,000 claim requests but had paid out just over one-third, or only about 150,000, for a total payout to date of approximately $2.3 billion. An additional nearly 80,000 claims were reported to "require more documentation," and more than 170,000 claim requests had been denied.[58] Feinberg has also reported that 10,000 claimants have applied for Final Payments, abdicating their rights to sue BP, Transocean, or any other company that may be found at fault for the disaster.[59]

BP's failures, which led to the explosion of the *Deepwater Horizon*, the release of the BP Macondo oil well monster, and the inability to clean up after the monster and to adequately compensate its victims, all combined to make the company CEO, Tony Hayward, one of the most hated corporate CEOs in history. Unfortunately, neither Hayward nor BP was acting alone.

6

Big Oil Plays Defense

BP and the Oil Industry Respond to Disaster

BP's corporate record in the Gulf oil disaster is flush with nontransparency and disinformation. There was at least one person at BP, however, who was not afraid to tell it exactly how he saw it, at least in terms of how the catastrophe affected him personally: CEO Tony Hayward.

As Hayward asked his fellow BP executives one week after the *Deepwater Horizon* exploded, "What the hell did we do to deserve this?"[1]

Within eight months, Hayward would be dubbed "the most hated man in America" by his hometown newspaper, lose his job, and earn the title of worst CEO of 2010.[2] What happened to the man who at the outset seemed the perfect CEO to handle the disaster?

There was, of course, the endless string of heartless gaffes Hayward seemed incapable of slowing. Being second only to Saddam Hussein in presiding over the worst oil spill in world history certainly did not help. But, likely the most important reasons Hayward fell so far are found in the response to the disaster: the shocking unpreparedness, the deadly doses of dispersant, and the heavy cocktail of cover-ups and disinformation tossed out by Hayward and BP all along the way.

Even though Hayward makes a convenient villain, however, he is far from a lone actor within his company, within this crisis, and within the oil industry. Just as we should not take solace that BP is a reformed company now that Hayward is no longer at its helm, neither should we believe that offshore drilling is now safe because BP has been publicly flogged.

There are many critical ways that BP stands out as a cautionary tale for putting profits over people's lives. However, its operations are so intimately intertwined with those of all the other major oil and oil services companies that it would be a tragic mistake to isolate BP. All of the companies that work with BP must share responsibility for its failures. Moreover, the failures that led to the blowout of the Macondo well and the disastrously long road to capping it are endemic in the industry as a whole and, more important, in the failure of the government to even attempt to regulate it.

Every major oil company was involved in the effort to try to contain and clean up the Macondo well oil gusher, and not a single one knew what to do, nor did the government regulators whose job it was to oversee them and ensure the safety of their operations to their workers, the public, and the environment. No company had invested any significant dollars into cleanup research or preparedness. Ships to contain the oil were not ready nor were adequate boom or skimmers. And while all of their applications to drill deepwater wells state their preparedness for even much larger blowouts than that at the Macondo, the companies were not prepared. Not a single one had actually developed the technology, much less the equipment, with which to address a deepwater blowout. Instead, both the companies and those who regulate them learned on the fly for eighty-seven long days.

In fact, none seem to have learned from the disaster, as pumping oil anew from the reluctant, and very dangerous, Macondo oil well still remains an open possibility.

BP: "Beyond Propaganda"

BP was the largest oil and gas producer not only in the Gulf of Mexico but also in the entire United States before the *Deepwater Horizon* exploded, and it remains so today. In 2009, it was the fourth-largest corporation in the world (by revenue), the fourth most profitable company with nearly $16.6 billion in profits, and the fourth-largest company operating in the United States (after Wal-Mart, Shell, and ExxonMobil). Its 2009 revenues of $239 billion were larger than the gross national products of all but thirty-two of the world's countries.[3]

Formed in 1908 from the British acquisition of Iranian (then Persian) oil, it was first called the Anglo-Persian Oil Company and then, in 1935, the Anglo-Iranian Oil Company. In 1914, the British government became its majority owner. During World War I, the British government then seized the assets of British Petroleum, a German-owned gasoline company operating in Britain, sold them to Anglo-Persian, and in 1954 took the name British Petroleum Company, or BP. BP remained a government-owned company until 1987, when the British government sold the last of its shares.

BP bought its way into the United States through a series of mergers and acquisitions. In 1987, the same year it fully privatized, BP launched its U.S. operations, BP America, with the purchase of Standard Oil of Ohio. In 1998, BP spent $53 billion to purchase the American Oil Company (Amoco), one of the largest corporate mergers in history. The purchase made BP both the largest producer of both oil and natural gas in the United States and Britain's largest corporation. Two years later, BP bought the Atlantic Richfield Company (ARCO). Although they are marketed separately, the Amoco, ARCO, and BP gasoline brands are all owned by BP.

Following the ARCO merger, in 2001, the company changed its name to BP p.l.c. and spent $200 million on a public relations ad campaign to launch its new tagline, "Beyond Petroleum," and a new logo: the bright yellow and green flowerlike sunburst. Although the campaign made it appear to some that Beyond Petroleum was its new name, BP p.l.c. officially stands for nothing at all.

Meanwhile, a group of BP shareholders found that the company had spent more on the launch of its new name and logo than on renewable energy in all of 1998, leading CorpWatch, a leading consumer watchdog group, to argue that a more appropriate phrase for the company's rebranding effort would be "BP: Beyond Propaganda . . . Beyond Belief."[4]

Hayward Takes the Helm

In many ways, Tony Hayward should have been the perfect person to lead BP through the Gulf oil disaster. Young, good-looking, and comfortable in front of a camera, he is deeply knowledgeable about—and has virtually staked his career on—deep offshore operations. He is well ingrained in BP as an organization, and he is extremely ambitious. However, Hayward is also brash, reckless, and absolutely willing to go anywhere, do anything, work with anyone, and suffer any consequence to get more oil and increase profits. Upon taking the helm of BP, Hayward pushed further cost cutting, less oversight, and greater risk taking. In the Gulf of Mexico, the consequences finally caught up with him.

In 1982, after Hayward earned his Ph.D. in geology, his first job was working for BP. He was a rig geologist in Aberdeen, the heart of Britain's North Slope offshore production—considered, with the Gulf of Mexico, the most difficult offshore production area in the world. Eighteen years later, after working his way up through the company, Hayward was appointed company treasurer in 2000 and then, in 2003, head of BP's exploration and production arm, which earned him a seat on the board of directors. In 2007, at age fifty, he was named CEO.

Hayward's tenure, first as head of BP's exploration and production arm and then as CEO, has taken the company into increasingly risky environments and operations, pushing the envelope of both technology and corporate responsibility. Offshore expansion into ever deeper waters around the globe has been a hallmark of that expansion. His stint as treasurer, moreover, earned him the simultaneous, although seemingly contradictory, reputation as a "bean counter" who puts money above all else.

As CEO, Hayward took BP into the Canadian tar sands, where it partners with the Husky and Devon Energy companies. Tar sand oil extraction is one of the most environmentally destructive methods of extracting oil from the earth and one that harms the land of First Nation, or indigenous, populations through the destruction of Boreal forests and the contamination of drinking and fishing waters.

Hayward shut down BP's alternative energy headquarters in London in 2009, accepted the resignation of its clean-energy boss, and imposed steep budget cuts on the remaining program.[5]

Hayward led BP's efforts to push the government of Iraq to open its markets to foreign oil companies after the 2003 U.S. and British invasion (a topic discussed in great detail in both *The Bush Agenda* and *The Tyranny of Oil*). BP became the first foreign oil company to sign a production contract with the new government of Iraq in 2009 and is partnering with China National Petroleum Company on the project.[6]

Hayward significantly expanded BP's offshore operations all around the world, including in Egypt, Trinidad, and Canada, as well as in Angola and Azerbaijan—among the world's most repressive and corrupt governments.

BP's Azerbaijan operations made front-page news when they were cited in classified U.S. embassy cables released in December 2010 by Wikileaks, a nonprofit organization that leaks documents that "reveal unethical behavior in governments and institutions," according to its Web site. The cables revealed that in September 2008, BP suffered a massive blowout, similar to that in the Gulf of Mexico, at its Azeri-Chirag-Guneshi (ACG) offshore field in the Caspian Sea but largely failed to disclose it.

BP is the operator and largest shareholder in the consortium with partners Chevron, ExxonMobil, Hess, Statoil, and the Azerbaijani state-owned oil company Socar. The U.S. embassy wrote shortly after the incident, "ACG operator BP has been exceptionally circumspect in disseminating information about the ACG gas leak, both to the public and to its ACG partners." It said, "BP would analyze [mud] to help find the cause of the blowout and gas leak." Another cable in January 2009 indicates that BP blamed a "bad cement job" for the blowout. Fortunately, the blowout did not lead to an explosion, and the company was able to evacuate its 212 workers safely. It did, however, lead to two oil fields being shut down and the expelling of "water, mud and gas" into the Caspian.[7]

Without the Wikileak, few in or out of Azerbaijan would have known of the incident. BP's annual report, for example, does not mention the blowout but rather only a "subsurface gas release."[8] BP declined to answer questions about the incident when asked by reporters for the *Guardian* newspaper. When British reporter Greg Palast went to Azerbaijan in December to investigate, he was arrested by a team of BP, Azeri, and military police for filming a BP oil rig. He was told, "Here in Azerbaijan we believe in human rights. Please give us your film!" Though the film was confiscated, Palast was released. The full story of BP's operations, however, remains shrouded in secrecy.[9]

In Alaska, Hayward launched a new potentially record-setting project, drilling two miles down through ocean and earth and then six to eight miles out horizontally to hit oil. The drilling takes place three miles off the coast in the fragile Beaufort Sea, home to many threatened species, including the polar bear. The project was allowed to proceed in 2010 even though a moratorium on new deepwater offshore drilling was in place because BP argued, and government regulators agreed, that it should be considered an "onshore" project. It takes place on an artificial island built by BP, a thirty-one-acre pile of gravel in about twenty-two feet of water.[10]

Hayward also pushed BP into the Gulf of Mexico, where it partners with every major oil company. With minority partners Petrobras of Brazil and ConocoPhillips, BP broke all previous offshore records by drilling the Tiber well off the coast of Texas, the deepest offshore well in the world.

With minority partner ExxonMobil, BP's Thunder Horse platform off the coast of Louisiana is the size of three football fields and is the largest oil and gas production facility in the world.

About midway between the two is BP's Mad Dog field, one of the largest discoveries in the Gulf, with estimated reserves of between 200 and 450 million barrels of oil. BP's Mad Dog minority partners are Chevron and BHP Billiton, one of the world's largest mining and energy companies.

Next door to Mad Dog is the Atlantis field, BP's second-largest Gulf production platform and the world's deepest moored (as opposed to dynamically positioned) production facility. BP's minority partner in Atlantis is BHP Billiton.

Breaking so many records would generally be considered a good thing if it were not for the concern that BP has flouted safety concerns as it has pushed ever deeper into the ocean. Either its partners did not press BP for the necessary information to judge the safety of these operations for themselves, or they willfully ignored BP's failures.

After the explosion of the *Deepwater Horizon*, Kenneth Abbott, a contractor for BP, went public under whistleblower protection laws with charges that BP broke federal laws and violated its own internal procedures by failing to maintain crucial safety and engineering documents related to the Atlantis. He had raised concerns about the facility in November 2008, when he first discovered that the platform had been operating with only about 10 percent of the total engineer-approved documents it needed to run safely, leaving it vulnerable to a catastrophic disaster. He went public following the explosion, which he described as making him "personally sick at heart," because "several different causes for the [Macondo well] blowout have been reported on the news. Many of them would be caused by the same problems I have seen on Atlantis."

A congressional investigation is under way. In testimony before Congress in June 2010, Abbott emphasized the importance of the missing documents and the challenges of offshore drilling broadly, explaining, "A project such as Atlantis is incredibly complex in two ways: First, there are many components produced by many vendors which must all work together. Second, there are many challenges created by the extreme water depth which must be overcome by cutting edge engineering techniques. One of the functions of the owner/operator, BP in this case, is to assure that engineering knowledge and expertise look at the system overall to be sure that all of the parts function together."

Abbott disclosed internal e-mails revealing that BP was well aware of the problems. The e-mails included those from BP manager Barry Duff, then an Atlantis project manager in BP's Houston operations headquarters, warning that incomplete and unapproved design specifications had been given to platform operators. "This could lead to catastrophic operator errors," he argued. He continued, "There are hundreds if not thousands of subsea documents that have never been finalized, yet the facilities have been turned over" to operators who rely on the design specifications to safely operate Atlantis.[11]

Abbott called his findings "astounding" when he learned that out of over seven thousand drawings and documents, almost 90 percent never received any approval of any kind, including for design. At one point BP vetoed a corrective plan because it would have cost $2 million, Abbott alleged.[12]

Mike Sawyer, a Texas-based engineer who works for Apex Safety Consultants and who evaluated BP's Atlantis documents, supports Abbott's allegations. Sawyer found BP operations "reckless" and said that BP's "widespread pattern of unapproved design, testing and inspection documentation on the Atlantis subsea project creates a risk of a catastrophic incident." He called the documents in question—welding records, inspections, and safety shutdown logistics materials—"extremely critical to the safe operation of the platform and its subsea components."[13]

In a statement in May, BP said, "The investigation found that the operators on the platform had full access to the accurate up-to-date drawings necessary to operate the platform safely."

Nonetheless, along with the nonprofit advocacy organization Food and Water Watch, in September 2010, Abbott launched a lawsuit against BP demanding that Atlantis be shut down until BP can prove it meets U.S. safety and engineering standards.[14]

The reason Hayward pushed BP to the depths of the ocean and the ends of the earth to find new oil is simple: BP had been the object of takeover rumors for years because its reserve-replacement ratio—how much oil it has in relation to how much it uses—had been falling steadily since 2001. Hayward was supposed to turn this crippling trend around.

Hayward's expansion was successful in the ways that counted: he reversed the trend and by 2010 raised the reserve-replacement ratio to the highest it had been in six years. BP is also the second-largest owner of booked reserves—how much oil still in the ground a company can call its own and a primary measure of an oil company's financial worth. With 10.5 billion barrels in 2009, BP had the second-largest oil reserves in the world after ExxonMobil, which had 11.7 billion. More than 50 percent of BP's oil and gas reserves are in the United States.

Hayward was also promoted to clean up BP's image and, supposedly, its operations as well, after a series of catastrophic and deadly incidents in the United States.

"I personally believe that BP, with its corporate culture of greed over profits, murdered my parents," Eva Rowe testified before Congress in 2007.[15] Congress was investigating the worst workplace accident in the

United States in more than fifteen years: a massive explosion at BP's Texas City Refinery in March 2005 that killed 15 workers, including Rowe's parents, and injured 180 more.

As a result, BP was forced to pay the two largest fines in the history of the Occupational Safety and Health Administration (OSHA). The first was a direct result of the explosion, and the second, in October 2009, under Hayward's tenure, arose from BP's noncompliance with the settlement agreement, with OSHA finding 270 "notifications of failure to abate" and 439 new willful violations.[16]

The U.S. Chemical Safety Board, an independent federal agency, investigated the blast and released a devastating indictment of BP. "The Texas City disaster was caused by organizational and safety deficiencies at all levels of the BP corporation," the 2007 report found. "The combination of cost-cutting, production pressures and failure to invest caused a progressive deterioration of safety at the refinery."[17]

Perhaps most important, as we consider the incidents' relevance to the Gulf oil disaster, was the board's warning that such dangerous operations were not limited to the Texas City Refinery or to BP but could be industry-wide. The board found a pervasive "complacency towards serious safety risks" across the leading oil companies. It called on OSHA to "require these corporations to evaluate the safety impact of mergers, reorganizations, downsizing, and budget cuts."[18]

Just one year after the Texas City explosion, in March 2006, BP released 200,000 gallons of crude oil onto the Alaskan tundra—the largest oil spill in the history of the North Slope. The spill happened at the company's giant Prudhoe Bay field, which it manages for partners ExxonMobil and ConocoPhillips. Investigators determined that a buildup of sediment in a pipe caused a massive leak and that BP failed to properly inspect or clean the pipeline, procedures required by law to prevent pipeline corrosion. The investigation revealed that in 2004, the company had become aware of increased corrosion in the pipeline. Court documents cited BP's "failure to allocate sufficient resources to ensure safe and environmentally protective operation of the pipelines that leaked." Prosecutors estimated that BP saved $9.6 million by choosing not to regularly clean and inspect the pipelines. "I think we need to put a particular emphasis on the need to give high priority to maintenance and maybe a little less priority to profits," said the judge in the case, Ralph Beistline. "In this particular case, the need to protect the environment should be our ultimate priority."[19]

Even after the ruling, BP once again refused to mend its ways. In March 2009, under Hayward's tenure, the Department of Justice filed a civil lawsuit against BP for failing "to comply in a timely manner with a Corrective Action Order" involving the Alaska oil spill.[20]

Rather than clean up BP's operations, Hayward pushed the envelope even further, cutting staff, grading employees every year based on how much money they saved the company, and admonishing caution.

In his first meeting as new CEO in Houston in September 2007, Hayward used "straight talk" to tell his new staff that it had been over-cautious and that "assurance is killing us." Hayward said that too many people were engaged in decision making that was leading to excessive caution. He would be instigating a thorough shake-up by slashing man-agement and eliminating staff. He admonished the Houston staff for delays in bringing on high-profile field start-ups in the Gulf of Mexico, "such as *Thunder Horse* and *Atlantis*," and noted that "refinery difficulties" at Texas City "meant third-quarter revenues would be heavily reduced."[21]

In November 2010, Professor Robert Bea's *Deepwater Horizon* Study Group concluded that "at the time of the Macondo blowout, BP's cor-porate culture remained one that was embedded in risk-taking and cost-cutting—it was like that in 2005 (Texas City), in 2006 (Alaska North Slope Spill), and in 2010 (the *Deepwater Horizon*)."[22]

At the beginning of the Gulf oil disaster, it seemed that Hayward and BP would do the right thing and handle the situation with speed and responsibility. Rather than hide behind Transocean, BP released a statement within twenty-four hours of the explosion of the *Deepwater Horizon*. Hayward shared his concern and pledged assistance, saying, "Our concern and thoughts are with the rig personnel and their families. We are also very focused on providing every possible assistance in the effort to deal with the consequences of the incident."[23]

The same day, BP vice president David Rainey participated in a press conference in New Orleans with Transocean and the coast guard.

After the sinking of the rig, Hayward released an even stronger state-ment, which acknowledged not only that an oil spill was happening but that BP would do all it could to clean it up and protect the environ-ment. "We are determined to do everything in our power to contain this oil spill and resolve the situation as rapidly, safely and effectively as possible. . . . We have assembled and are now deploying world-class facilities, resources and expertise, and can call on more if needed. There should be no doubt of our resolve to limit the escape of oil and protect the marine and coastal environments from its effects," said the April 22 statement.[24]

As coast guard rear admiral Mary Landry went on the TV morning talk shows on April 23 stressing that all was well and no oil was leaking from the disaster site, Hayward flew to BP's U.S. corporate headquarters. The massive three-building campus—lake included—sits in a nondescript corporate suburb of Houston. BP's global exploration and production business is based there, and its seven thousand employees are tucked away in giant glistening buildings of steel, concrete, and glass. The United States is the heart of BP's oil production, almost 60 percent of which takes place on the Gulf coast.

Inside the building, BP, with Transocean and other oil companies, made decisions that were okayed by U.S. government staff. Even though BP's operational and managerial failures caused the disaster, the federal government did not take charge; it had neither the staff nor the expertise to tell BP what to do. Unfortunately, no one—not BP, the federal government, or any other company—knew what to do in response to a deepwater blowout. Moreover, just as the federal government was unable to regulate BP prior to the incident, so, too, was it unable to regulate its activities in response to the disaster.

The National Oil Spill Commission concluded in November that "on April 20, the oil and gas industry was unprepared to respond to a deepwater blowout, and the federal government was similarly unprepared to provide meaningful supervision. . . . BP had to construct novel devices, and the government had to mobilize personnel on the fly, because neither was ready for a disaster of this nature in deepwater."[25]

The Regulators

From the earliest days of the spill response, officials from the Interior Department's Minerals Management Service (MMS) described themselves as observers of the operations in Houston; the personnel from BP,

Transocean, and Cameron (the company that manufactured the BOP) were making the decisions, informed by experts from every major oil company. A *Washington Post* reporter was granted rare access to the Houston "war room" in early July. He reported that of 569 people working that day, only 18 were federal employees. "Decisions are reviewed and approved by Government officials," but they were made by BP.[26]

One MMS staffer told the National Oil Spill Commission that despite working more than eighty hours a week, he had to miss more than half of the meetings he was supposed to attend each day. Another asserted that BP, and the oil industry more broadly, "possessed ten times the expertise that MMS could bring to bear." Two MMS employees recalled high-level officials at the Department of the Interior asking what they would do if the U.S. government took over the containment effort. Both said they would hire one of the major oil companies.[27]

The lack of regulatory capacity of the MMS, the agency tasked with overseeing most aspects of offshore drilling, may be best understood by looking at its "parents." President Ronald Reagan and his interior secretary, James Watt, created the agency in 1982. Neither believed in regulation, but both believed in the oil industry. Watt stated that he intended to "mine more" and "drill more" as head of Interior.[28] The agency principally focused, therefore, not on providing oversight but rather on granting access: leases for private companies to work the natural resources found on public lands. At the same time, there was not all that much offshore production to be regulated.

Deepwater technology, particularly ultra-deepwater, is still a relatively new phenomenon, and actual *production* of oil from such wells is even newer. Although oil derricks were built on lakes as early as the 1930s, the first production from water "well beyond the sight of land" happened in 1947, when Kerr-McGee Oil Company's Kermac 16 derrick began producing oil in eighteen feet of water, ten and half miles from the shore of Morgan City, Louisiana. Though a historic development, offshore drilling would take another thirty years to contribute more than 10 percent of the global oil supply and another twenty years before it topped 30 percent—the same percent it accounts for today.[29] The rigs inched their way out to sea in fits and starts, and regulation followed accordingly.

In 1969, at a time when a "deepwater" well was one drilled at 400 feet of water, the very first annual offshore technology conference was held in Houston. That year, the American Petroleum Institute (API), the leading lobby for the oil and gas industry, published its first "Recommended Practice" document for the design, fabrication, and installation of offshore platforms.[30] The government then adopted the regulations as its own. Although the industry has undergone astonishing technological changes in the intervening four decades, most of the regulations used to govern offshore operations come from this period. What updates have been made continue to be done by the industry itself through the API.

The first discovery of oil at a depth of more than 1,000 feet, today's definition of a "deepwater well," was in 1975, but this well did not start producing until the 1980s. Similarly, the first ultra-deepwater hole was drilled in 1986 at 5,000 feet, but it did not start producing until the 1990s. In 1997, there were just seventeen rigs drilling in deepwater in the Gulf. Today there are forty-two. The number of ultra-deepwater projects, meanwhile, is still just 17 in the Gulf and 148 in the world.[31]

The greatest advances in offshore drilling have all occurred in the last decade. Anyone in the business will tell you that drilling at such depths remains incredibly risky, even with the most conscientious oversight. The Chevron Corporation has written, "Navigating uncertain weather conditions, freezing water and crushing pressure, deepwater drilling is one of the most technologically challenging ways of finding and extracting oil."[32] Micky Driver, a Chevron spokesman, has admitted, "It's lots of money, it's lots of equipment, and it's a total crapshoot."[33]

Unfortunately, the approach of the regulators, not to mention the regulations themselves, changed little following the model set by Watt, particularly during the eight years of the George W. Bush administration, when two Watt devotees headed the Department of the Interior.

Even during the Clinton administration, however, the push was to reduce oversight and increase trust in the industry to police itself. Bruce Babbitt, Clinton's interior secretary, recently said in an interview that the Clinton administration's efforts to reduce the overall size of the federal government left him with nearly 10 percent less staff and forced him to increase dependence on the industry. Interior shifted to a model of oversight euphemistically called "performance-based regulation," which means that regulators step in only after the industry has already messed up.

"That is a mistake for which I shoulder part of the blame," Babbitt says. "It was not a good decision. My belief, with considerable hindsight, is there is no place for performance-based regulation [in offshore drilling] because of the high risk."[34]

The Clinton administration was a friend of industry, but the George W. Bush administration was part and parcel of the oil industry. In 2000, the oil industry spent more money than it had ever spent on any election in order to put George Bush and Dick Cheney into office. When Bush and Cheney won, the industry could stop lobbying and start legislating.

The George W. Bush Administration

Gale Norton served as President Bush's interior secretary from 2000 until 2006, when she resigned, tarnished by the Jack Abramoff influence-peddling scandal. Under Norton, the Interior Department's top priorities were to increase domestic oil and gas production, offer more incentives to drillers on the Gulf coast, and open the Arctic National Wildlife Refuge and other wilderness areas to drilling. The department reduced spending on enforcement, cut back on auditors, and sped up approvals for drilling applications. In 2001 alone, the Interior Department opened four million new acres of public land to oil, natural gas, and coal mining.[35]

Much of this plan was written by the American Petroleum Institute and then implemented, often word-for-word, by the secretary.[36]

On February 26, 2005, Norton and MMS director Johnnie Burton joined BP, at its expense, at the dedication of its Thunder Horse oil platform in the Gulf, then the largest oil rig in the world. Standing on the rig, Norton declared that it was "created to protect the blue waters that it stands in, no matter how great the storm. . . . *Thunder Horse* is a dramatic embodiment of how far technology has progressed." Less than five months later, the thirteen-story structure nearly capsized, the result of equipment failures during Hurricane Dennis.[37]

In November 2006, the Interior Department awarded Shell new leases for oil shale development in Colorado. Two months later, Norton was hired by Shell as general counsel to the unit working on oil shale development.

Norton's immediate subordinate at Interior, J. Stephen Griles, had in the course of his career lobbied on behalf of Chevron, Sunoco, Unocal, Occidental Petroleum, and the American Petroleum Institute, among others. Like Norton, Griles was an early Watt devotee who began his government career as deputy director of Surface Mining in Watt's Interior Department. Griles left the Reagan administration to cash in on his successful deregulatory efforts. He launched his own D.C. lobbying firm, J. Stephen Griles and Associates, and represented the nation's largest oil and mining companies before the federal government.[38]

Bush tapped Griles for the number two spot at Interior in 2001 and gave him the keys to the nation's oil and mineral holdings. We have a unique view of Griles's activities at Interior because of the many investigations launched against him, including the federal inspector general's report from 2004 describing his tenure at Interior as "an ethical quagmire." Griles's logs show that in just two years, from 2001 to 2003, he met on at least thirty-two occasions with other administration officials to discuss pending regulatory matters that were a concern to his former clients.[39]

These meetings flout federal ethics rules that prohibit executive branch officials from participating in any matter that could advance their own financial interests or that involves former employers or clients. Griles pushed for rollbacks in environmental standards for air and water and tried to exempt the oil industry from royalty payments.

After passing billions of dollars over to the industry, Griles left the Bush administration in January 2005 to form Lundquist, Nethercutt, and Griles. The lobbying firm's clients included the American Petroleum Institute and BP. In March 2007, Griles pleaded guilty to felony charges of obstruction of justice by lying on four separate occasions about his connections to Jack Abramoff.[40]

In September 2008, the Interior Department's inspector general found that staff of the MMS had accepted thousands of dollars in industry gifts, used cocaine, and engaged in sex with oil industry representatives, noting that "a culture of ethical failure" had pervaded the agency. Government officials handling billions of dollars in oil royalties partied and had sex with and accepted golf and ski outings from employees of the energy companies they were dealing with.

Thirteen former and current Interior employees in Denver and

Washington were accused of rigging contracts, working part-time as private oil consultants, and having sexual relationships with—and accepting improper gifts from—oil company employees. The investigations revealed a "culture of substance abuse and promiscuity . . . wholly lacking in acceptance of or adherence to government ethical standards." Four companies were accused of, among other things, providing gifts to Interior employees, in what the report described as "a textbook example of improperly receiving gifts from prohibited sources."[41]

During Norton and Griles's tenure, offshore drilling expanded, as did reports commissioned or performed by the Interior Department to identify problems with key elements of offshore operations, including the blowout preventer (BOP). None of the potential problems identified was addressed, and each played a role in the explosion of, and the inability to repair, the *Deepwater Horizon*. Moreover, not only did the MMS refuse to act on these warnings, it increasingly reduced oversight.

In February 2000, a rig spilled oil into the Gulf of Mexico after a crew member accidentally pushed the wrong button and severed the connection between the rig and the BOP. A study for the MMS identified the need for a second blind shear ram for backup, but the agency refused to require it. Instead, with the industry's support, the MMS reduced the frequency of inspections from once a week to every two weeks, citing the disruptions that these tests caused to oil drilling and extraction efforts.

All of the deepwater oil drilling rigs currently working in the Gulf of Mexico rely on remotely controlled submersible vehicles to turn on the BOPs if the primary controls fail. However, a consultant hired by the MMS in 2003 warned that these machines were frequently unreliable during blowouts, moving too slowly and often lacking power to do the job. The same consultant concluded in a federally financed study that even if the rig crew managed to turn the BOP on, the most critical safety component inside these machines—the shear ram, which is meant to cut quickly through the well pipe to stop the flow of oil and gas—was often not strong enough to cut through the modern pipes that drilling rigs use.

"This grim snapshot illustrates the lack of preparedness in the industry to shear and seal a well with the last line of defense against a blowout," said the September 2004 report, written by West Engineering.[42]

The MMS did little to address the concern. Instead, in 2005, it finalized a policy from the end of the Clinton era, telling companies that they did not have to provide detailed blowout and response scenarios for each exploration plan.

David Abraham served as the Office of Management and Budget's examiner for offshore programs from 2003 to 2005. When he asked MMS officials about their plans for keeping up with expanded drilling, he said they routinely gave the same answer. "They said, 'Our processes work,'" Abraham told the *Washington Post* in August. "I said, 'It's like an airplane—everyone always says the wheels will come down, but what happens when they don't?' They should have had people who could say, 'This is what we do when the wheels don't come down.' What they said is, 'Don't worry. Our regulatory regime works.'"[43]

Enter the Obama Administration

When President Obama came into office, he brought several plans to increase regulation and taxation on the oil industry. But he did not come with an agenda focused on cleaning house at the MMS. Although he changed the leadership, the vast majority of the staff stayed the same, as did the rules governing its operations and the financial resources available to it. The dominant relationship might best be described as a grandparent (the oil industry) to a grandchild (the regulators)—and a rather sickly grandchild, at that.

As President Obama took office and prepared to dramatically expand offshore drilling, just sixty-two inspectors were responsible for overseeing four thousand production wells in the Gulf of Mexico. The MMS's track record, moreover, was far from perfect. From 2001 to 2007, the MMS reported nearly 1,500 serious drilling accidents in offshore operations in U.S. waters, leading to 41 deaths, 302 injuries, and 356 oil spills.[44]

Perhaps most important in terms of the Gulf oil disaster, President Obama left in place the Bush administration's associate director of offshore programs, Chris Oynes, appointed in 2007. Oynes, age sixty-three, had spent more than thirty years working for the federal government, including thirteen as MMS regional director for the Gulf of Mexico from 1995 to 2007. As the *New York Times* wrote, in the world of oil and gas in the Gulf, "few people have mattered more" than Oynes.[45]

Oynes was a central figure in what Earl E. Devaney, the Interior Department's inspector general, told senators in 2007 was a "jaw-dropping example of bureaucratic bungling."[46] In 1998 and 1999, the Clinton Interior Department waived royalty payments on eleven hundred offshore leases. At the time the leases were signed, the price of oil was between $9 and $15 per barrel. The companies were supposed to start paying royalties when oil topped $34 per barrel. But the $34 clause was omitted. Oynes signed off on seven hundred of the leases (all of those in the Gulf) with missing price thresholds.[47] Midlevel officials at Interior spotted the omission in 2000 and quietly made sure to include the clause in all subsequent leases. But no one tried to fix the leases that had already been signed, and almost no one talked about them until oil prices started to climb above $34 per barrel in 2004. This is when the oil companies reportedly began pressing the Bush administration to clarify how it would treat the leases, trying to ensure that they would not have to pay any royalties. The cost to the American taxpayer? An estimated $53 billion.

Instead of being fired by Bush, Oynes was promoted. Upon his promotion, Congresswoman Carolyn Maloney of New York said, "It is completely ridiculous that MMS would take the person most likely responsible for the royalty rip-off and put him in charge of the whole show. . . . This is just one more in a long line of actions showing that MMS sides with Big Oil over the American taxpayers. If it isn't the revolving door that brings oil and gas industry folks positions of power at MMS, then it's the promotion of an insider responsible for one of the government's costliest and most ridiculous missteps in recent history."[48]

More of Oynes's failures as the regulator in the Gulf came to light in May 2010, when the Department of Interior inspector general released a scathing report of the MMS Lake Charles, Louisiana, district office, which is one of five with authority over the entire Gulf coast. The inspection covered 2005 to 2007, when the office was under Oynes's direction. It revealed a very similar relationship to the oil industry reported in the Denver and Washington offices above. The report found a widespread culture of abuse and dangerously cozy relationships between the regulators and those they were supposed to be regulating. Said acting inspector general Mary Kendall in her cover letter to Secretary Ken Salazar, "Of greatest concern to me is the environment in which these inspectors operate—particularly the ease with which they move between industry and government."[49]

The report cited the common practice of accepting gifts from oil companies, including hunting, fishing, and skeet shooting trips; Christmas parties; and even free tickets to see Louisiana State University beat the University of Miami in the 2005 Peach Bowl in Atlanta.

Two employees admitted illegal drug use. One was reported to have possibly used crystal methamphetamine while working on offshore platforms. Another admitted to using the drug and that "he might have been under the influence of the drug at work."

The investigation found that in mid-2008, an MMS employee conducted four inspections on drilling platforms when he was also negotiating, and later accepted, employment with the drilling company owner.

"Obviously, we're all oil industry," MMS Lake Charles district manager Larry Williamson told the investigators. "We're all from the same part of the country. Almost all of our inspectors have worked for oil companies out on these same platforms. . . . They've been with these people since they were kids. They've hunted together. They fish together. They skeet shoot together. . . . They do this all the time."

One confidential source told investigators that some inspectors allowed oil and gas company personnel to fill out their own inspection forms, which would then be completed or signed by the inspector and turned in for review. The inspectors could not verify the allegation.

"The Inspector General report describes reprehensible activities," said Secretary Salazar. "This deeply disturbing report is further evidence of the cozy relationship between some elements of MMS and the oil and gas industry."[50]

On May 17, just before the report was released, Oynes notified his superiors at the Interior Department of his decision to accelerate his retirement and leave on May 31. Oynes departed with full government benefits, including pension, in hand.

The Interior Department shares some regulatory oversight of deepwater drilling with the coast guard. Whereas the Interior Department regulates everything to do with the actual oil and gas operations (for example, drills, pipes, and BOPs), the coast guard regulates the aspects of offshore drilling that move around in the ocean (for example, rigs).

Within both the Interior Department and the coast guard are

employees who have tried for decades to implement and enforce meaningful regulations over this industry. They have, however, been successfully beaten back by the industry at virtually every turn, and the industry has been able to largely regulate itself.

In June 2009, for example, the MMS tried to implement stricter safety and environmental standards and more frequent inspections at offshore rigs. Every major oil company, including BP, lobbied against the proposed rules. In a letter to the agency, BP said that "extensive, prescriptive" regulations were not needed for offshore drilling and urged the MMS to allow the operators to define the steps, largely on their own, that they would take to ensure safety.

Chevron's harshly worded letter, written by Sandi Fury, called the new rules an "abrupt change from past direction of the MMS and directives to Industry." She emphasized Chevron's own role in crafting the existing voluntary rules, which had been written by the American Petroleum Institute, and their "flexibility and versatility," which would be undermined by the MMS's "prescriptive requirement."[51]

Exxon also supported the American Petroleum Institute rule, API RP 75, and argued that new rules would be onerous, writing that "the proposed rule prescribes stricter requirements than the approach on which it is based—API Recommended Practice, and may generate significant difficulties for operators and contractors to abide by the rule."[52]

As usual, the industry won, and no new rules were implemented.

One of the most shocking practices that has continued through the Obama administration is the granting of "categorical exclusions" exempting oil companies from having to adhere to environmental laws in their offshore operations. The exclusions are supposed to be granted only if the proposed activity does not have "a significant effect on the human environment." They were intended to allow companies to not have to file new environmental assessments every time they added a door to a rig, for example. Instead, they became a means to evade the law altogether, ballooning to encompass waivers for entire operations, including the drilling of the Macondo well.

Under the Clinton administration, categorical exclusions granted in the central and western Gulf rose from 3 in 1997 to 795 in 2000. During the George W. Bush administration, they exploded: the MMS granted an average of 650 categorical exclusions per year in the region. The number then fell to just 220 during the Obama administration's first year, but included all new Gulf deepwater wells, including BP's Macondo well. Granting the exclusion meant that the Obama administration waived BP from having to perform the legally prescribed assessment of how its operations would and could potentially impact the environment.[53]

On May 12, 2010, coast guard captain Hung Nguyen, cochairman of the Marine Board Investigation into the *Deepwater Horizon* disaster, asked coast guard commander Michael Odom, an offshore drilling inspector, if the regulations "are adequate to support your activity as an inspector."

"No," responded Odom, "the pace of the technology [has] definitely outrun the current regulations, and the variety of systems that we're seeing, a lot of them are not even addressed in the regulations." The regulations were written twenty to thirty years ago, before the ultra-deepwater drilling done by the *Deepwater Horizon* and seventeen other rigs currently in the Gulf was even taking place.

The captain also grilled Interior Department regulator Michael Saucier on oversight of the BOP. The captain concluded, "So, my understanding is that it is designed to industry standard, manufactured by industry, installed by industry with no government witnessing oversight of the construction or the installation. Is that correct?"

"That would be correct," Saucier replied.

The same industry self-certification, the captain later concluded, applied to virtually all operations on the rig.[54]

No Plan

The regulators did not require the industry to implement best practices to ensure the safety of their operations, and thus the operations were not safe. The regulators also did not require the companies to prepare for a blowout in deepwater, and thus they were not prepared.

When Hayward arrived in Houston, he worked with all of the major oil and oil services companies to try to stop the Macondo blowout. In one meeting in early May, thirty-five representatives of these companies, including Exxon, Shell, and ConocoPhillips, sat around a table to try to find a solution.[55] They may have held in their hands their "disaster preparedness plans" required of all companies in order to receive permits for offshore operations.

Unfortunately, all five of the major oil producers in the Gulf of Mexico—BP, Chevron, Exxon, ConocoPhillips, and Shell—used virtually identical and tragically inadequate plans. They also all used the same shoddy subcontractor, the Response Group, to write the plans. Three of the companies' 2009 plans, including BP's, listed as a consultant biologist Peter Lutz, who died in February 2005. Four companies ensured that their plans addressed the need to protect walruses, sea lions, and seals, although none of these live in the Gulf, revealing that the reports were not only cut and pasted among the companies but also originally written for Arctic operations. The BP plan even lists a Japanese shopping and search Web site as a link to one of its "primary equipment providers" for rapid deployment in the event of a spill in the Gulf.[56]

Most important, the plans absolutely do not work, as BP's response to the *Deepwater Horizon* explosion has made horrifically clear. Nonetheless, each plan received the government's approval. All of the plans used the same reassuring language that a spill was not likely, but even if one happened, the impact would be small because of the distance from shore. The companies assured the government that they could handle oil spills much larger than the Macondo blowout.

Given the incredible complexity of drilling in the Gulf, one of the more shocking aspects of the plans is that rather than conduct an analysis of each individual well and the potential impact of a blowout occurring there, the companies were permitted to use common assumptions for the entire region. Thus, BP did not do a specific analysis for the Macondo well, aka, "the Well from Hell." Rather, BP used the same model for all its wells in the Gulf, based on a location about thirty-three miles away from where the blowout actually occurred.[57]

At a June 15 hearing, House Energy and Commerce chairman Henry Waxman (D-CA) said that the "cookie-cutter" plans show that "none of the five oil companies has an adequate response plan." Congressman Bart Stupak (D-MI) said, "It could be said that BP is the one bad apple in the bunch. But unfortunately they appear to have plenty of company. Exxon and the other oil companies are just as unprepared to respond to a major oil spill in the Gulf as BP."

ExxonMobil chief executive Rex Tillerson said it was "an embarrassment" that his plan included provisions for walrus protection in the Gulf. Grilled by Congressman Stupak about how the companies could say they could handle spills many times the size of the current one, Tillerson conceded, "We are not well-equipped to handle them."[58]

Containment

The lack of preparation became apparent within days of the Macondo well blowout. Other than the lengthy process of drilling a relief well, accurately estimated by BP to take several months, there was no available, tested technique to stop a deepwater blowout. BP mobilized two rigs to drill separate relief wells, a primary well and a backup insisted upon by Secretary of the Interior Ken Salazar. The first rig began drilling on May 2, and the second began on May 17.

Each relief well involved drilling a standard hole deep into the ocean floor near the original *Deepwater Horizon* pipe. Then it gets tricky, as the new pipes need to intersect the original one. The target is slender— about eight to ten inches in diameter (the size of a Frisbee)—and the distance is great. One expert described it as "like threading a needle from across the room. With your eyes closed."[59] Electromagnetic fields guide the new pipes, homing in ever closer to the original. Once the intersection is reached, cement is pumped through the relief pipe into the blown-out well to close it in.

With the relief drilling under way, the oil companies tried to shift devices used in shallow waters—100 to 400 feet—to the Macondo well 5,000 feet below the ocean's surface. Predictably, all of the initial efforts failed. While Hayward was bouncing around on beaches assuring Americans that all was under control, in private the companies knew the situation was far graver.[60]

BP enlisted Transocean, Shell, Exxon, Chevron, Anadarko, and other companies to aid in its ultimately fruitless efforts to close the BOP stack.[61] These efforts are described in chapter 2.

On April 25, BP began to consider placing a large containment dome, known as a cofferdam, over the larger of the two leaks from the broken riser. At the top of the cofferdam, a pipe would channel hydrocarbons to the *Discoverer Enterprise*, a containment ship on the surface. BP chief operating officer Doug Suttles told the National Oil Spill Commission staff that according to BP engineers, the chance of success was at best 50 percent.

By this time, crews were well aware of all three leaks and the likelihood that some 15,000 barrels of oil a day were spewing from the well. Nonetheless, BP released a statement on April 26, which was not well reported at the time but circulated widely in the months to follow, in which Hayward assured that "the improved weather has created better conditions for our response. This, combined with the light, thin oil we are dealing with has further increased our confidence that we can tackle this spill offshore."[62]

As crews started to maneuver the cofferdam over the leak on May 7, Hayward conducted one of his first post-blowout television interviews. After seven days of thick, oil-drenched waves slopping up onto Gulf beaches, there was hope that BP had a solution. It sounded straightforward—BP would simply lower a cap on top of the leak and stop it. BP officials told the *New York Times* that the cofferdam would remove about 85 percent of the oil spilling into the sea.[63]

This was the fourth-largest corporation on the planet, supported by the second-, the third-, and the fifth-largest corporations in the world, in turn supported by the wealthiest government the world has ever known. Surely they all knew what they were doing, didn't they? Surely they'd prepared for such an event, hadn't they?

Hayward crushed everyone's hope, however, and everyone's venom found a fresh target. For it wasn't just *what* he said, it was *how* he said it: very calmly and without apology or apparent concern, Hayward let us know that BP had no clue, no plan, and no hope of a prompt solution.

"It's a technology first," Hayward said, standing on a dock in Venice, Louisiana, in an interview with CNN's David Mattingly. "It works in three to four hundred feet of water, but the pressures and temperatures are very different here," Hayward said. "So that is why we cannot be confident that it will work and that is why we continue with other significant interventions." Mattingly asked incredulously, "Do you have something ready to go just in case this fails?" Hayward responded that BP was trying to contain the spill "to the maximum extent possible" and that there are further operations on the BOP, but these "will probably take two to three weeks to put into place."

You can just hear the words reverberate around the nation: "A few weeks? Wasn't this horrific tragedy supposed to be fixed a week ago?"

Mattingly pressed Hayward: "But you are drilling for oil a mile under the ocean. Was something . . . why wasn't something already ready to go in the event of this kind of a disaster?" Hayward answered, incorrectly, that "this has never happened in 25 years in the industry" and then corrected himself, saying, "It was considered to be an extremely low, low probability, and what we have implemented is the spill response plan."[64]

This may well have been the interview that sealed Hayward's fate with the public. Little did we know that the lack of preparedness and the lack of consideration for risk were endemic to the entire industry. Moreover, Hayward was actually overplaying the other options BP had. While Hayward told Mattingly that BP had "further operations on the BOP," in fact, BP's internal documents from May 7 state that the company had concluded, "[t]he possibility of closing the BOP has now been essentially exhausted."[65]

From here, Hayward only went downhill. As the weeks rolled into

BP CEO Tony Hayward flanked by security guard and press on an oil-soaked beach in Louisiana in May 2010.

Paul Horsman, a marine biologist and oil spill specialist for Greenpeace, and an activist stand in the oil-soaked waters of South Pass, Louisiana, in May 2010, protesting BP's spill.

months and each effort not only failed but seemed increasingly ridiculous, Hayward took the heat and withered under the sun.

The dome failed. Hydrates formed before the cofferdam could be put in place, clogging the opening through which the oil was to be funneled and causing it to float up toward the ocean surface. One high-level government official recalled Andy Inglis, BP's chief executive of exploration and production, saying, "If we had tried to make a hydrate collection contraption, we couldn't have done a better job."[66]

On May 13, another week's worth of oil had attacked the waters, wildlife, and beaches of the Gulf, and another 194,000 gallons of dispersant were sprayed from airplanes and helicopters and dumped at the bottom of the ocean. The disaster was already identified as the largest oil spill in U.S. history, with independent estimates reaching to 50,000 barrels of oil a day. Nonetheless, Hayward assured us once again not to worry, telling the *Guardian* of London, "The Gulf of Mexico is a very big ocean. The amount of volume of oil and dispersant we are putting into it is tiny in relation to the total water volume."[67]

Dr. Joye and the rest of the *Pelican* team's discovery of the existence of huge underwater plumes of Macondo well oil were front-page news the next day. A few days later, Hayward offered this gem to *Sky News*: "I think the environmental impact of this disaster is likely to have been very, very modest."[68] When asked if he was sleeping at night in light of the oil spill's disastrous effects, Hayward said, "Yeah . . . of course I am."[69]

The public was outraged. Mocking Hayward's sense of "modesty," on May 24, protesters swarmed outside his Houston headquarters. Exposing the "naked truth" of BP's deadly operations, lies, and harm, women from across the United States including the Gulf coast, wearing little more than poster board, marched from Transocean's offices just down the road to BP's Houston headquarters. The strategically placed signs on the nearly naked women read EXPOSE BP'S NAKED GREED, EXPOSE BP'S NAKED LIES, EXPOSE THE NAKED TRUTH OF "DRILL BABY DRILL"—SPILL BABY SPILL!, EXPOSE THE TRUTH ABOUT BIG OIL.

Rae Abileah and Dana Balicki of Code Pink expose
the "naked truth" of BP's deadly operations at a protest
at BP's Houston headquarters in May 2010.

Rae Abileah and Dana Balicki said that the protesters were there to mourn the deaths of the eleven workers and the livelihoods of thousands of people throughout the Gulf. They led the chants, asking, "Hey, BP, what do you say? How many fish did you kill today? Hey, BP, what do you say? How many millions did you make today?"

As news crews filmed, snapped photographs, and asked for interviews, thick black oil (fake) and dead fish (real and smelly) were doused over women who were dressed as silvery mermaids and bright colorful birds as they writhed on the cement. Greedy "BP executives" swooped in, snatching $100 bills out of the muck. The protesters ended by cordoning off the headquarters with yellow police tape, protecting the public from further harm by BP.

Television hosts had a field day with Hayward's comments. Jimmy Fallon said, "In a new interview, BP's CEO said that the Gulf coast oil spill is relatively tiny compared to the 'very big ocean.' That's like telling someone who's just been shot not to worry about the bullet because they're really, really fat." Jay Leno quipped, "I love this. On the news today, the CEO of British Petroleum says he believes the overall environmental impact of this oil spill will be very, very modest. Yeah. If you live in England!"

Nearly one month into the crisis, things were getting out of hand. Public outrage was growing, the Russians were suggesting we should nuke the ocean in order to seal off the flow of oil, and plans for the "junk shot" and "top kill" got underway.

Top Kills, Junk Shots, and Top Hats

Well, folks, here's the latest update. I guess this is good news. BP officials say the "top kill" plan is working. The bad news— BP officials are a bunch of lying weasels.

—Jay Leno

The top kill and junk shot may be standard industry procedures for stopping blown-out wells in shallow waters, but they sounded utterly ridiculous to the public—and the public was right. The methods involve throwing mud, golf balls, and tire rubber onto a blowout to stop it up.

More precisely, a *top kill* involves pumping heavy drilling mud into the top of a well through the BOP's choke and kill lines, at rates and pressures significant enough to force escaping hydrocarbons back down the well. A *junk shot* complements a top kill. It involves pumping bridging materials—including pieces of tire rubber and golf balls—into the bottom of a BOP. Those bridging materials ideally get caught on obstructions in the flow path for hydrocarbons—such as pieces of drill pipe and partially deployed BOP rams—and further impede the flow. By slowing or stopping the flow of hydrocarbons, a successful junk shot makes it easier to execute a top kill.

BP and federal government officials knew the procedures would work only on a blowout spewing, at most, 13,000 to 15,000 barrels of oil a day. At this time, however, government officials estimated the rate of flow to be 12,000 to 25,000 barrels a day.[70] Weeks earlier, Dr. Ian MacDonald had publicly released his estimate of 26,500 barrels of oil a day, followed by Dr. Crone's estimate of as much as 50,000 barrels a day. All the while, BP had refused to publicly release the spill cam footage, refused access to experts for independent measurements, and continually declared publicly that the size of the flow was irrelevant because it would not affect the efforts to contain the spill.

It turned out this latter argument was, in a sense, truthful. BP was going to try the same tactics regardless of the flow rate because it did not have anything else to do. It simply was not ready to address the actual magnitude of the spill. For example, until June 16, BP had just one containment ship capable of capturing just 15,000 barrels of oil a day. Even when the second ship finally came on line, combined, they were still able to capture just 25,000 barrels of oil a day—only about a third of the oil spewing from the well at its height. Thus, it wouldn't have mattered if BP had succeeded in inserting pipes to capture more oil, as it had nowhere for the oil to go. BP did, however, want the public to think that it was trying to do something.

BP was most likely borrowing a script from Exxon, whose executives were, after all, advising the company in its efforts. In court proceedings from the *Exxon Valdez* disaster, an audiotape captured an Exxon official demanding cleanup equipment as follows: "I don't care so much whether it's working or not but . . . it needs to be something out there that looks like an effort is being made. . . . I don't care if it picks up 2 gallons a week. Get that shit out there . . . and . . . standing around where people can see it."[71]

After three days, the junk shot and top kill failed, predictably.

Hayward apparently oversaw the operations himself, sitting like Jigsaw from the horror movie *Saw* watching videos of his torture victims. The London *Telegraph* reported, "Hayward has remained largely inscrutable. Holed up inside the energy giant's dimly lit Houston command centre, watching grainy footage of attempts by robot submarines to plug the gushing well, overseeing the 'Top Kill' operation that may or may not have been successful, Hayward has retained his composure even as the pressure has mounted from President Obama."[72]

While things were going downhill publicly, behind the scenes the U.S. government started to push its way into the decision-making process. Immediately after the failure of the top kill, BP teams in Houston met throughout the night of May 28 to assess the operation. Some meetings occurred behind closed doors, without government participation.

At one point, Lars Herbst of the MMS and Admiral Kevin Cook of the coast guard, who had been dispatched by National Incident Commander admiral Thad Allen to be his representative in Houston, entered a meeting and stated that they had a right to be present. Apparently, according to the National Oil Spill Commission, government officials had not previously insisted on joining these types of meetings, and BP personnel were surprised by the interruption. Asserting the right to be present for BP's top kill analysis was a turning point for the government team. After the failure of the top kill, the government significantly increased its oversight of the containment effort.[73]

Hayward was feeling the pressure from all sides. Code Pink protesters doused his logo in oil, Dr. MacDonald started calling Hayward "Tiny," Jay Leno was calling him Tony "Haywire," and the government was barging in on his "private" meetings.[74] Hayward was having a tough time of it, and he wasn't afraid to let the public know it.

On May 30, standing in front of piles of bright orange oil booms, a weary Hayward told the nation in a TV interview, "We're sorry for the massive disruption it's caused their lives. There's no one who wants this over more than I do. I would like my life back."[75]

It became a battle cry. Fishers all around the Gulf started wearing

T-shirts reading HEY TONY, I WANT MY LIFE BACK TOO! Photographs of oil-soaked brown pelicans were printed with the banner I WANT MY LIFE BACK. Protesters all around the world scrawled it on posters. There was even a song, "Hey Tony," by New Orleans–based musician Louie Ludwig, with the plaintive chorus, "Hey Tony, I'd like my life back too."[76]

Within days, the *Sunday Times* of London declared Hayward "the most hated man in America."[77]

Next came a citizens' arrest in Washington, D.C. On June 4, hundreds of people gathered in front of BP's D.C. headquarters, demanding the CEO's arrest. Louisiana native Reverend Lennox Yearwood, one of the protest leaders, is president of the Hip Hop Caucus, a nonprofit organization focused on engaging young people and people of color in the policy-making process by linking with famous hip-hop artists such as Jay Z and Sean "P. Diddy" Combs. With leaders of public interest and environmental organizations, including Public Citizen, Greenpeace, and 350.org, present, the reverend pronounced Hayward guilty of disregarding worker and environmental safety, price-gouging consumers, and destroying the Gulf, culminating in a finding of criminal negligence. The protesters shouted, "No respect for workers. No respect for earth. BP can't see what life's really worth!"[78]

BP employees on the building's upper floors watched as the protest moved from their lawn to their front doors. "I see you in the windows. Your greed is killing my people!" Reverend Yearwood called up. He then led protesters in prayer for those who have lost their lives to the fossil fuel industry and for elected leaders to find the wisdom to recognize "creation's bountiful and powerful gift of the sun and wind." They prayed, "Let BP know that not only are we fighting for our children, but we are fighting for their children as well."

No BP executives appeared outside, so Reverend Yearwood tied the black and white 1920s-style jail suit the protesters had brought for Hayward to the doors of the company's building.

Lightly Skewered

Perhaps the pressure worked. Just a few days later, on June 16, President Obama and CEO Hayward had their first and only meeting. They met in the White House and hammered out BP's $20 billion escrow deal. In addition to responding to the public outrage, the deal was likely intended to soften things up before Hayward faced off against some of his harshest critics: the United States Congress.

On June 17, Hayward testified for the first and only time on Capitol Hill. He sat alone at the long brown wooden table. The rest of the oil industry had had its opportunity a few days before, when they were grilled about their identical spill response plans. The House Committee on Energy and Commerce called both hearings, forcing the executives to face Congressmen Edward Markey and Henry Waxman, the two men who had led the most exhaustive congressional investigations into the disaster, its causes, and its consequences.

Congressman Edward Markey has represented Massachusetts's seventh congressional district since 1976. At the time of this hearing, he was chairman of the Select Committee on Energy Independence and Global Warming and the Energy and Environment Subcommittee of the Energy and Commerce Committee. On April 29, Congressman Markey began demanding that the executives appear before the committee. Nearly seven weeks later, they conceded.

Congressman Henry Waxman has represented California's thirtieth district since 2003. At the time of this hearing, he was chairman of the Energy and Commerce Committee. Formerly the chairman of the Committee on Oversight and Government Reform, the principal investigative committee in the House of Representatives, Congressman Waxman led in-depth investigations into the operations of corporations such as Halliburton and Bechtel.

Hayward sat before the committee in a dark suit, red tie, and white shirt, showing little emotion. In his opening statement, Congressman Waxman admonished Hayward, "When you became CEO of BP, you promised to focus like a laser on safe and reliable operations. We could find no evidence that you paid any attention to the tremendous risks BP was taking. We have reviewed 30,000 pages of documents from BP, including your e-mails. There is not a single e-mail or document that shows you paid even the slightest attention to the dangers at this well."

He called "BP's corporate complacency astonishing" and said that the warnings from BP employees on the Macondo well "fell on deaf ears."

"There is a complete contradiction between BP's words and deeds,"

Waxman said. "You were brought in to make safety the top priority of BP. But under your leadership, BP has taken the most extreme risks. BP cut corner after corner to save a million dollars here and a few hours there. And now the whole Gulf Coast is paying the price."[79]

Before Hayward could get started, Congressman Joe Barton (R-TX), had an apology to offer the CEO. Barton apologized for the "$20 billion shakedown" Hayward was forced to suffer. Long one of the most well-subsidized members of Congress by the oil industry, Barton said, "I'm ashamed of what happened in the White House yesterday. I think it is a tragedy of the first proportion that a private corporation can be subjected to what I would characterize as a shakedown. I apologize. I do not want to live in a country where any time a citizen or a corporation does something that is legitimately wrong, it is subject to some sort of political pressure that is, again, in my words—amounts to a shakedown, so I apologize."[80]

Hayward sat calmly as Congressman Waxman and other members recited a litany of his abuses, he was unfazed when Barton apologized, and he remained cool even when fourth-generation shrimper Diane Wilson from Texas jumped up from the back of the hearing, her hands and face smeared in fake oil, and yelled, "Arrest this man! He is a criminal! His company is stealing the livelihoods of thousands of us in the Gulf and killing our coasts! Tony Hayward should be arrested!" Hayward kept speaking over her, so the cochairman of the hearing, Congressman Stupak, had to ask him to stop while the police swarmed. Wilson was arrested, held briefly, and released.

Diane Wilson grew up on the water. She loves shrimp almost as much as she loves shrimpers. A few weeks after the arrest she told me that this crisis had gripped people all across the country, even those who had never been on a boat, because "the waters of the Gulf are as much a part of national heritage as is the Grand Canyon." The images on people's televisions touched them as though "blood was poured down Niagara Falls." But most people do not know how to channel those feelings into action, she maintains. "I am cursed. I'm a supreme optimist," she says. "I believe in impossible things happening if the people seize the moment."

After her action appeared on evening news programs and was covered in newspapers and on the radio, Wilson received phone calls and e-mails from people asking what they could do. "It was certainly worth it," she says, "even if Hayward never saw me."

After Wilson's "commotion" was over, Hayward's testimony got under way. He offered a short opening statement, apologizing for the "tragedy" and accounting for BP's efforts to contain and stop the oil spill, its compensation efforts, and the $20 billion escrow fund. He then waited for questions.

His responses, regardless of the questions, consisted primarily of one of three answers: "I was not part of the decision-making process on this well," "I had no prior knowledge," and "It's too early to reach conclusions." The members of Congress were angry and accused Hayward of stonewalling. "This is an investigation," Congressman Waxman reminded him. "That is what this committee is doing. It is an investigatory committee. And we expect you to cooperate with us. Are you failing to cooperate with other investigators as well? Because they are going to have a hard time reaching conclusions if you stonewall them, which is what we seem to be getting today." Waxman said that rather than taking responsibility, Hayward was "kicking the can down the road and acting as if you had nothing to do with this company."[81]

The criticism was harsh, not just in Congress but also back home in London, where the *Daily Telegraph* said, "Hayward has the communication skills of a tax inspector, dry and arrogant." The *Times* said that Hayward resembled "a weary registrar in a south London crematorium. Americans say he looked like Mr. Bean. Make that Mr. Has Bean." The conservative *Daily Mail*, however, struck out at Congress: "BP's chief executive was subjected to a grilling so savage, it was more like ancient Rome than Capitol Hill."[82]

Clearly in need of a vacation after his "savage" grilling, Hayward took off two days later to the southern coast of England, where he spent the day watching his yacht compete in the J. P. Morgan Asset Management Round the Island race.

Jamie Billiot was furious. "Tony Hayward says, 'I want my life back,' then he goes on vacation to a yacht race! People were mad. You want your life back? What about people who actually live here? Grandchildren will not grow up the same way as their grandparents! It blows my mind!"

On June 21, David Letterman offered the "Top Ten Ways Tony Hayward Can Improve his Image" on his late-night show. Number one was, "Dial it back from arrogant bastard to smug son x*@." The audience erupted in laughter and applause.

Closure

By mid-June, the federal government had had about enough of Tony Hayward and BP. Principally under Secretary of Energy Steven Chu's guidance, BP was now forced to submit its plans to the federal government for authorization before moving forward. BP would submit its plans to Chu's on-site scientists, who would then prepare their own analyses of BP's plans. Based upon those analyses, the science advisers would force BP to evaluate worst-case scenarios and plan for contingencies.

In essence, the National Oil Spill Commission concluded, "They played 'devil's advocate,' questioning BP's proposals to ensure that BP had fully considered and mitigated even low-probability risks." A senior government official characterized BP's attitude prior to the increased supervision as "hope for the best, plan for the best, expect the best."

While BP expressed "frustration to Commission staff about the nature of the science team's pushback," arguing that theoretical scientists consider risk differently from engineers, the government team was "skeptical of BP's risk management practices, given that BP's well had just blown out."

On June 19, coast guard rear admiral James Watson, who had replaced Rear Admiral Mary Landry as the federal on-scene coordinator, issued a letter to BP formalizing the more extensive government review process.

The new efforts would build on the few successes BP had experienced in capturing some of the oil flowing from the Macondo well. On May 16, BP succeeded in inserting a tube into the riser that was able to carry oil and gas up to the *Discoverer Enterprise* on the surface. But nine days later, when it was replaced by the top hat, the tool had collected just 22,000 barrels of oil, the equivalent of one-third of one day's flow.

On June 1, BP implemented a plan to cut off the portion of the riser still attached to the BOP stack and install a collection device in its place. The cap, or top hat, would connect through a riser to the *Discoverer*

Enterprise. BP had, by this time, constructed seven new different top hats—none of which had been ready prior to the disaster. A remotely operated vehicle (ROV) used large hydraulic shears to cut the riser at a small distance from the top of the BOP stack, the cap was successfully installed, and by June 8, the *Discoverer Enterprise* was collecting nearly 15,000 barrels of oil a day through the top hat.

On June 16, another vessel, the *Q4000*, was specially outfitted to process and burn, rather than collect, up to 10,000 barrels of oil a day. The *Helix Producer*, a production ship that connected to the BOP through a freestanding riser, was the final collection system that BP was able to deploy. It was a key addition to BP's collection capacity, which BP envisioned would eventually reach 90,000 barrels a day. The *Helix Producer* became operational only on July 12, however, and thus only collected oil for the three days remaining before BP shut the well on July 15.

BP did not have vessels ready to go for this purpose. It had to gather them and then specially outfit them with products from around the world. "Had BP obtained another production vessel sooner, it might have been able to collect oil through the BOP's kill line at a rate comparable to the collection rate of the *Q4000*," the National Oil Spill Commission concluded.

In early July, Dr. Tom Hunter from Dr. Chu's science team and James Dupree of BP traveled to Washington, D.C., to brief a group of high-ranking government officials about BP's plans to use a capping stack in an attempt to close in the well. The next day, Secretary Chu and Dr. Hunter briefed the president, who gave his approval as well.[83]

The capping stack was essentially a smaller version of a BOP. Designed to connect to the top of the BOP stack, it contained three rams capable of shutting off the flow of hydrocarbons as well as its own choke and kill lines. The government also pushed BP to install two pressure sensors in the capping stack, which later proved critical in generating accurate flow-rate estimates.

On July 10, after removing the top hat from the top of the riser, ROVs unbolted the stub of the riser connected to the top of the BOP stack, removed the stub, slid the capping stack into place, and bolted the capping stack to the top of the BOP stack. On July 12, the cap was successfully installed.

Three days later, on July 15, with the guidance of Exxon, Shell, and government scientists, BP tested the well. For the first time in eighty-seven days, no oil flowed into the Gulf of Mexico. The BP Macondo oil and gas monster was cut off. The cap worked.

Just when things were going so well, BP sought to nullify public concern about the capping process in the best way it knew how: it misled the public. On July 16, John Aravosis revealed on his Americablog.com site that BP had doctored photos on its Web site of the Houston command center that was overseeing the operation, filling in three video screens that had been blank in the original photo with false images. Scott Dean, a spokesman for BP, confirmed the "alteration" and said, "We've instructed our post-production team to refrain from doing this in the future."[84]

The BP oil monster was not so easily tamed, however. The work was not yet complete. The cap was just a temporary measure. The hole still needed to be completely shut in. There were two methods to do this: the relief wells and a static kill. Like the top kill, the static kill involved pumping heavy drilling mud into the well.

As the static kill was getting under way, Tropical Storm Bonnie pushed through the Gulf. At the same time, another storm was blowing through BP's corporate offices.

Bye-bye, Tony

Following a daylong meeting of its board of directors in London, BP announced in a written statement released on July 27 that Tony Hayward would be stepping down as CEO.[85] Hayward was not fired. The decision was by "mutual agreement." Hayward would receive about $17 million in severance and pension and was appointed to the board of BP's Russian partnership, TNK-BP. Congressman Markey called Hayward's $17 million a "golden parachute" and said that BP should instead "be dedicating its resources to compensating the residents of the Gulf Coast who are the victims of this tragedy."[86]

Hayward's replacement, BP managing director Robert Dudley, came as little surprise. Dudley had already taken over as BP's front man in the Gulf following Hayward's crushing "I want my life back" gaffe and his disastrous congressional testimony. A Chicago native, Dudley had worked his way up, first through Amoco and then continued to rise following Amoco's merger with BP. Although he would be BP's first non-British CEO, he had spent most of his career outside of the United States, working in countries such as Angola, Algeria, Egypt, China, and Russia, and in the Caspian region and at BP's London headquarters.[87]

BP chairman Carl-Henric Svanberg said in the statement, "The BP board is deeply saddened to lose a CEO whose success over some three years in driving the performance of the company was so widely and deservedly admired."[88] At the same time, however, the company announced its worst quarter in history and one of the largest losses in British corporate history, a jaw-dropping $17 billion loss.[89] More losses were inevitable as, on the same day, Greenpeace activists shut down fifty BP gas stations in London.

Hayward is also quoted in the statement, saying, "The Gulf of Mexico explosion was a terrible tragedy for which—as the man in charge of BP when it happened—I will always feel a deep responsibility, regardless of where blame is ultimately found to lie."

As usual, however, Hayward could not hold his tongue. In interviews following his departure he said, "I became a villain for doing the right thing."[90] And that if he had a degree in acting from "RADA [the Royal Academy of Dramatic Art] rather than a degree in geology I may have done better" in handling the fallout from the disaster.[91]

On August 3, with Hayward out the door, BP began the static kill of the Macondo well.

At 11 p.m. it achieved control of the well. BP followed the mud with cement.

No one was quite breathing a sigh of relief yet, however. "You want to make sure it's really dead dead dead. Don't want anything to rise out of the grave," Secretary Chu said. "Death" meant completion of at least one relief well.

On September 19, 152 days after the April 20 blowout, the needle was threaded.

"BP has successfully completed the relief well," Admiral Thad Allen announced. "With this development, we can finally announce that the Macondo 252 well is effectively dead."[92]

Mission accomplished. Or was it?

Not Dead, Just Sleeping

The hole that BP, Transocean, Halliburton, and the other companies drilled into Macondo 252 has been sealed. But that is just one route into an enormous reservoir. BP still owns the Macondo well, and no one has told BP, nor any other company for that matter, that it is prohibited from drilling there. In fact, BP has officially left open the possibility of going back into the reservoir. Asked whether it should, Admiral Allen said that would be a policy decision between BP and the Department of the Interior. "Frankly, it's above my pay grade," he said.[93]

Whether the Obama administration has the ability to withstand the pressure of the oil industry to grant this and its many other wishes remains an open question.

7

Obama Steps Up

But Is the Disaster Over and
Could It Happen Again?

On May 2, 2010, President Barack Obama stood on a dock in Venice, Louisiana, fifty miles from the site of the explosion of the *Deepwater Horizon* and twenty miles from where BP's oil monster had first made landfall just three days before. It was the president's first visit to the Gulf and his second address to the nation since disaster struck.

As an increasingly driving rain fell on his face and pushed at the pages of his speech, the president, dressed in shirtsleeves and a light windbreaker, joked that he was told it was supposed to just be drizzling outside. "Is this Louisiana drizzle?" he quipped. Quickly turning to business, the president stood firm and resolute in the face of the storm.

One word, never said aloud, hung over the event like a swirling whisper: *Katrina*. Everything in President Obama's tone, words, and deeds was designed to demonstrate that this would not be a repeat of the catastrophic failures of the Bush administration in response to the deadly 2005 hurricane.

This was no flyover with the president peering out an airplane window from high above, as President Bush had done two days after Hurricane Katrina hit New Orleans. President Obama's speech hit every key failure of his predecessor and made clear that he would do the opposite. The president assured us that he took this event with the utmost seriousness, his administration was immediately and thoroughly committed to resolving it all the way through to the end and would spare no resource in doing so, he was already working closely with local government officials, and even though BP was responsible "for the leak" and would be "paying the bill," as "president of the United States" (a phrase he repeated multiple times), he was committed to the full restoration of the Gulf and its people.

The president concluded that "this is one of the richest and most beautiful ecosystems on the planet, and for centuries its residents have enjoyed and made a living off the fish that swim in these waters and the wildlife that inhabit these shores. This is also the heartbeat of the region's economic life. And we're going to do everything in our power to protect our natural resources, compensate those who have been harmed, rebuild what has been damaged, and help this region persevere like it has done so many times before."[1]

As the crisis has unfolded, it turned out not to be a repeat of Katrina, and President Obama has kept many of his pledges. But it is also far from a government success story. Part of the failure is evident in the president's speech, which was pitch-perfect but for a few glaring problems. The most significant came when the president stated, "Because this leak is unique and unprecedented, *it could take many days to stop*" (emphasis added). Hindsight is certainly 20-20, and this was when BP was still working with the BOP in the hope that it could be jump-started to close the well. But it was also the same day that BP America chairman and president, Lamar McKay, very pessimistically told ABC News that efforts to trigger the BOP were like performing "open heart surgery at 5,000 feet in the dark."[2]

It was also a time when the entire oil industry was gathered in Houston simultaneously shrugging its collective shoulders in acknowledgment that none had planned for such an event, knew what to do, nor what their next move would be. Finally, the statement came just days after the administration acknowledged that the well was releasing five times more oil than previously stated—5,000 barrels a day—while experts reported the amount to be five times larger still.

Whether the idea that the leak could be stopped in days was purposely an understatement or what he actually believed, it certainly provides an excellent example of the administration's ultimate failure in its response to this disaster: an attempt to make the whole thing disappear from public sight and mind—whether through propaganda or dispersants. As that failed to happen and the extent of the spill got progressively worse with every passing day, week, and month, the administration consistently misrepresented the actual size, devastating immediate impact, and the potential long-term harm of the disaster, while making catastrophic choices that made it all far worse. The misrepresentations began with Admiral Landry's announcement that no oil was leaking from the Macondo well, followed by months of inaccurate statements about the size of the flow, and crescendoed with White House "climate czar" Carol Browner's declaration that "virtually all the oil is gone" from the Gulf long before this was so.

"I was terribly disappointed by the Obama administration's lack of transparency," Stan Jones, whose award-winning coverage of the *Exxon Valdez* disaster for the *Anchorage Daily News* was so critical to understand the crisis, told me. "Why they continued to deny facts that were manifest even when looked at by a layman. I guess he didn't want another crisis, and he thought he could spin it into insignificance."

Obama's speech was also a clear representation of what one can only dub a "pathological relationship" between the administration and BP.

In the speech, the president walked a thin line evident throughout the disaster. He argued that "BP is responsible" and "the federal government is throwing all it's got at the disaster," but he never answered the key question, "Who's in charge?"

Publicly, the answer fluctuated largely with the public mood. In reality, the answer was BP. There were areas where this made grudgingly good sense due to a lack of government know-how. But in the end, a corporation was allowed to make life-and-death decisions about public health, worker safety, economic livelihoods, the environment, and entire ecosystems that it should never have been allowed to make.

When BP had already demonstrated its long history and current practice of putting money over safety, the environment, and so on; its lack of preparation and capacity to address either the blowout or the subsequent oil and gas release; and motivations directly at odds with the public's on so many fronts, putting BP in charge was a tragic mistake.

The pathology is certainly not limited to Barack Obama and BP. It is endemic to the relationship of the oil industry to the United States in general and to the Gulf coast in particular. Obama is not an oil president as George W. Bush was, and his implementation of a moratorium on offshore drilling in the Gulf was decidedly not what the oil industry wanted in response to the disaster. However, this is an "oil nation": the power of the wealthiest industry on earth is virtually impossible to withstand, both for the public and for our elected officials.

"Here is an industry increasingly obliged to engage in ultrahazardous activity," Philip Zelikow, who served as executive director of the commission that investigated the September 11, 2001, terrorist attacks, testified. "I wouldn't look just at BP. They are functioning in a much larger institutional culture. Everyone who worked on this rig has worked on other projects, maybe at other companies. What you want to discover is if there is something distinct and pathological about all of these institutions, or something distinctly pathological about BP."[3]

By minimizing the size and significance of the disaster, the Obama administration played right into the hands of an oil industry working to ensure that no meaningful long-term policy changes emerge. The lack of transparency reduced both public concern and the public platform required for change to occur. The administration's reasons were obvious: polls showed that the more public attention focused on the disaster, the lower the president's approval ratings fell. With midterm elections looming and quite a bit more on the president's plate with two wars, an economic recession, health-care legislation, and Republicans in Congress determined to kill every legislative proposal he put forward, the White House needed all the good public will it could muster. Getting the oil spill out of the public's mind was one way to try to accomplish this.

In the long term, the ultimate failure is that the nation may emerge from the tragedy without meaningful policy changes to ensure that such an event never happens again.

Obama is certainly not alone in blame. The oil industry released one of the most powerful and successful lobbying and public relations campaigns in its long history. Thus, the Obama administration was responding to public polls that were in themselves being driven by the oil industry to isolate BP, end the moratorium, and ensure that Congress would pass no new legislation related to the oil spill.

Unfortunately, the counterweight was not strong enough. The environmental, human rights, worker, and public safety groups were simply not organized enough, funded enough, or in enough agreement to ensure that the response to the BP oil spill would be equal to the response to either the 1969 Unocal oil blowout in California or the 1989 *Exxon Valdez* disaster in Alaska.

The oil industry appears to have won—at least for now—but the struggle is far from over.

The Tyranny of Oil

In 2008, I published a book about the oil companies that compose the world's most powerful industry. It opens with then senator Barack Obama's victory speech upon winning the Iowa caucus in January 2008 in which he forcefully declares that he will free the United States once and for all from "the tyranny of oil." I found the words so poignant that I used them as the title of my book.

Throughout the 2008 presidential campaign, candidate Obama increasingly went on the attack against Big Oil, denouncing the industry's profits and pledging to impose a windfall profit tax, cut industry subsidies and tax breaks, and investigate oil and gas price manipulation. Obama pledged also to address global warming, protect the environment, and expand investments in alternative energy and green jobs. In ranking his lifetime record on environmental issues, the League of Conservation Voters found that Senator Obama voted "with the environment" 86 percent of the time. In the general election, Senator Obama ran against Republican senator John McCain, who was found to vote with the environment just 27 percent of the time and whose running mate, Alaska governor Sarah Palin, denied that climate change is manmade while chanting the oil industry's favorite election mantra, "Drill, baby, drill!" every chance she got.

The 2008 presidential election was the most expensive in history, with the oil industry spending more money than it had spent on any previous election. In the end, Obama and his running mate, Senator Joe Biden—neither of whom had records of receiving significant oil industry backing—received less than a third of the political contributions from the oil industry of their challengers. Nonetheless, they did receive $884,000 from the oil industry. One of the ways they likely achieved this amount of money was in a dramatic flip mid-campaign in support of lifting the moratorium on offshore oil drilling.

The Moratorium

The moratorium on offshore drilling was imposed following the January 28, 1969, blowout on the offshore oil rig *Platform Alpha* owned by Unocal (now Chevron) five miles off the coast of Summerland, California. Three million gallons of oil spilled directly into the Santa Barbara Channel, coating thirty-five miles of shoreline with oil up to six inches thick.

The community was already organized against offshore drilling and they were ready to take action. Millions of supporters and advocates were enlisted, spawning a massive environmental movement that within just a few years achieved the establishment of the U.S. Environmental Protection Agency and the passage of the U.S. Clean Air and Clean Water acts. After thirteen years of activist organizing, in 1981, Congress finally implemented the Outer Continental Shelf (OCS) moratorium, which prevented new leases for oil and gas development off the Pacific and Atlantic coasts as well as in Bristol Bay, Alaska.

The Gulf of Mexico was omitted from the moratorium under the extreme influence of the states' powerful oil-funded legislators. In an added victory for the oil industry, the legislation included a provision that the moratorium would automatically expire unless renewed annually. In 1990, George H. W. Bush added an additional level of presidential protection that deferred new leasing until 2002. Bill Clinton extended the presidential deferral to 2012.

The moratorium affected only the granting of new leases. Therefore, twenty-three oil and gas production facilities already in place off the coast of California remain active today, and hundreds of rigs operate off the coast of Alaska. There was no drilling off the Atlantic coast prior to 1981, and there is none today. In the Gulf of Mexico, off the coasts of

Texas, Louisiana, Mississippi, Alabama, and western Florida, where there has never been a moratorium, drilling exploded, with production rising roughly 70 percent in just the last twelve years.

Speaking recently about offshore drilling continuing in the Gulf while being banned in most other areas, Kieran Suckling of the Center for Biological Diversity said, "The environmental movement was either so far removed from it that it was unaware, or it was aware and afraid to challenge it because of local politics. . . . Or it was unwilling to challenge because it has written off the Gulf as America's dumping ground."[4]

"We've always been a plantation state," Oliver Houck, an environmental law professor at Tulane University, said of Louisiana. "What oil and gas did is replace the agricultural plantation culture with an oil and gas plantation culture."[5]

The reason for wanting to maintain and even expand the moratorium is not just fears of another massive blowout. It is that, among other dangers, the environmental impacts of offshore drilling are significant even under normal operations. From 1998 through 2007, offshore producers released an average of more than 6,500 barrels of oil a year into U.S. waters—64 percent more than the annual average during the previous ten years.[6]

The first half of 2008 alone brought more than 1,100 barrels of oil spilled in five incidents.[7] In just the Gulf of Mexico, the Mineral Management Service (MMS) reports that oil spills dumped nearly 520,000 barrels of oil from 1964 to 2009.[8]

However, even these numbers are suspect. As Professor Robert Bea of the *Deepwater Horizon* Study Group told the *Washington Post*, "I see the numbers, and I shrug my shoulders." MMS statistics say, for example, that a 1970 blowout on a Shell Oil well that killed four people triggered a spill of 53,000 barrels. But Professor Bea, who at that time worked for Shell tracking the oil spill, says that the spill was ten times that size and contaminated shorelines on the Yucatan Peninsula as well as the U.S. Gulf coast. The 1970 Shell blowout happened on a production platform, he noted. "We knew what the production rates were," he said.[9]

Even under normal operations, it is estimated that every offshore oil platform generates approximately 214,000 pounds of air pollutants each year, including some 50 tons of nitrogen oxides, 13 tons of carbon monoxide, 6 tons of sulfur dioxide, and 5 tons of volatile organic hydrocarbons.

According to the National Academy of Sciences, a single offshore well produces between 1,500 and 2,000 tons of waste material. Debris includes drill cuttings, which is rock ground into pieces by the bit, and drilling mud brought up during the drilling process, which contains toxic metals such as lead, cadmium, and mercury. Other pollutants, such as benzene, arsenic, zinc, and other known carcinogens and radioactive materials, are routinely released in "produced water," which emerges when water is brought up from a well along with the oil or gas.[10]

Drilling is also harmful to marine life, including species of commercial fish. The first step to drilling any offshore well involves taking an inventory of estimated resources. Every technology employed for this purpose harms marine ecosystems and species.

The seismic survey, for example, involves ships towing multiple air-gun arrays with tens of thousands of high-decibel explosive impulses. These air-gun arrays fire regular bursts of sound at frequencies in the range of 20 to 150 hertz, which is within the auditory range of many marine species, including whales. The auditory organs of fish are particularly vulnerable to loud sounds such as those produced by survey air guns, as fish rely on their ability to hear to find mates, locate prey, avoid predators, and communicate. The sounds have even been known to kill some species outright, including salmon, whose swim bladders have ruptured from exposure to intense sounds.

Both dart core and grab sampling, additional survey techniques, are extremely destructive to seafloor organisms and fish habitat, discharging silt plumes that are transported on ocean currents and smother nearby life on the seabed.[11]

It is widely estimated that oil and gas operations are responsible for some 60 percent of wetland loss and coastal erosion in the Gulf.[12] Oil operations are made possible by digging canals and channels throughout the wetlands, which allows saltwater to intrude inland. The saline in the water causes the dieback—the gradual dying of plant shoots, starting at the tips—of freshwater vegetation, which ultimately leads to wetland erosion. At the same time, the spoil banks, or piles of

waste, created during construction impede natural freshwater flow, leading to increased periods of flooding and drying.[13] Finally, U.S. Geological
Survey scientists have blamed the extraction of oil and gas for subsidence,
the sinking of the surface level; when fluids are pumped out of the
ground, air pressure under the surface diminishes and the surface grad-
ually sinks.[14]

In spite of all these dangers, the oil industry has never forgiven the
American public for the existing moratoriums and has lobbied relent-
lessly to have them lifted. As the then CEO of Chevron, David O'Reilly,
said in 2007, "Eighty-five percent of our coastlines are off-limits to exploration . . . what's wrong with our country?"[15]

For twenty-seven years, however, the industry failed, and the moratorium held. In June 2008, candidate Obama made a speech giving
some excellent reasons that it should. Standing on the scenic Riverwalk
along the St. Johns River in Jacksonville, Florida, presidential candidate
Obama wore a sharp red tie and crisp white shirt with his sleeves rolled
up for action. Birds chirped and flew in the sky, grass blew in the
wind, water flowed down the river, and cars honked and drove on the
overpass above his head. The senator looked young, energetic, and
determined while expressing his emphatic opposition to expanding offshore drilling.

"When I'm president, I intend to keep in place the moratorium here
in Florida and around the country," he said. "That's how we can protect
our coastline and still make the investments that will reduce our dependence on foreign oil and bring down gas prices for good." Such drilling,
the candidate emphasized, "would have long-term consequences for our
coastlines, but no short-term benefits." At a time of skyrocketing gas
prices, the senator explained, "Now believe me, if I thought there was
any evidence at all that drilling could save people money who are struggling to fill up their gas tanks by this summer, or this year, or even the
next few years, I would consider it. But it won't."[16]

Obama was moved to make this speech because just days earlier,
Republican presidential candidate John McCain had announced the
reversal of his long-held opposition to new offshore drilling in a major
speech delivered right before heading to Texas for a series of fund-raisers
with energy industry executives. The next day, Senator McCain raised
$1.3 million at just one luncheon.[17]

In response, in his Jacksonville speech, Senator Obama said, "So let me just repeat, John McCain's proposal, George Bush's proposal to drill offshore here in Florida and other places around the country would not provide families with any relief. . . . That may not poll well. . . . My job is not to go with the polls; my job is to tell the American people the truth about what's going to work when it comes to our long-term energy future. . . . That's what I intend to do as President as well."[18]

One month later, George W. Bush lifted the presidential moratorium. A month after that, Senator Obama flipped as well.

In August Obama said he would now accept additional offshore drilling and the lifting of the moratorium. The reason he cited, however, was not a new belief that offshore drilling was suddenly a good idea, but rather it was a trade-off he was willing to give to the oil industry in order to overcome gridlock in Washington on energy legislation. "My interest is in making sure we've got the kind of comprehensive energy policy that can bring down gas prices," Obama said in an interview. "If, in order to get that passed, we have to compromise in terms of a careful, well-thought-out drilling strategy that was carefully circumscribed to avoid significant environmental damage—I don't want to be so rigid that we can't get something done."[19]

In September, Congress, with the help of senators McCain and Obama, allowed the congressional moratorium to expire. House majority leader Steny Hoyer (D-MD) assured the public that month that restoring the ban "will be a top priority for discussion next year."[20] The goal, they all said, was passage of a comprehensive energy policy. Yet no such policy passed. Just two weeks after the election, moreover, Hoyer reversed himself completely, saying that House Democrats would not try to reimpose the moratorium, and they did not.[21]

The reason the legislation failed and everybody flipped on the moratorium had far more to do with the incredible escalating wealth of the oil industry that had taken place over the eight years of the Bush administration. As recently as 2004, only four of the top ten global corporations by revenue were oil companies. In 2009, for the first time in history, seven of the ten largest corporations in the world were oil companies. Only twenty-three countries had GDPs larger than the revenues of either of the world's largest oil companies: Royal Dutch Shell or Exxon-Mobil. Meanwhile, ExxonMobil's profits, at $45 billion for 2008, were more than three times those of Wal-Mart, the second-largest U.S. corporation.

As *Fortune* magazine reported in its ranking of the world's largest corporations for 2009, "Oil companies are the biggest money makers on this year's Global 500." The five most profitable oil companies operating in the United States raked in a combined $136 billion in pure profits for the year; there simply was no comparable industry.[22]

The industry turned this money into a spending tsunami in the 2008 election. To be sure, the vast majority of this oil money went, as it always does, to Republican candidates and their supporting issue groups. Moreover, the industry spent more money than it has spent on any other presidential election to try to get McCain and Palin into office. However, it also gave enough to the Democrats and the Obama-Biden ticket to ensure that its influence remained intact. When its chosen candidates lost the presidency and its chosen party lost the House and Senate, the oil industry lost the ability to legislate that it held under the Bush administration.

In response, it broke all its previous lobbying records and showered the U.S. Capitol and state capitols with unprecedented amounts of money, lobbyists, and influence. It did the same in the public sphere, launching media blitzes that filled the airwaves with pro-industry propaganda. Neither the Obama administration nor the Democrat-controlled Congress nor the American public has thus far proven to be immune from the onslaught.

The industry had many goals for 2009; two of the most important were killing meaningful climate-change and energy legislation and seeing the lifting of the moratorium on offshore drilling implemented by the signing of new leases and the authorization of new offshore activities.

The environmental movement was torn in half by this process. The desire to see legislation passed addressing the world's deteriorating climate had been a central (and seemingly perpetually out of reach) goal for everyone from the smallest local grassroots environmental organizations to the largest national groups. With President Obama in office and the Democrats in control of both houses, they thought there was finally a fighting chance. Little did they estimate the crushing power that the U.S. oil industry and its staunchest allies in Congress could bring to bear, even without the "climate-change deniers" in power.

As 2009 progressed, the vision of what "meaningful" legislation entailed was slowly whittled down, with supporters falling off with each new cut. The environmental community was splintering between those who accepted and those who rejected the deal making that succeeded in bringing some energy industry representatives to the table but was yielding bills that many believed were at best watered-down reforms and at worst would exacerbate climate change and its deadly effects. One section of the Kerry/Lieberman bill in the Senate, for example, is titled "Insuring Coal's Future," even though coal is the most climate destructive energy source, emitting 29 and 80 percent more carbon dioxide per unit of energy than oil and gasoline, respectively.

In October 2009, Senator Lindsey Graham (R-SC) demonstrated his support for climate-change legislation with a *New York Times* op-ed piece coauthored with Senator John Kerry (D-MA). The deal was historic, as Graham was the only Republican senator who supported such legislation. But in order to achieve their goals, the Democrats agreed to many more energy industry favors, including the inclusion of an expansion of offshore drilling and the implementation of the lifting of the moratorium.

More cleavage occurred within the environmental movement as some groups agreed to the deal, believing that just about any legislation was better than none, while most others joined the opposition. Passage remained elusive nonetheless as industry representatives, led by the American Petroleum Institute and the U.S. Chamber of Commerce, along with their representatives in Congress, continued to block the legislation.

In an effort to show the oil industry and pro-industry senators that even if Congress could not pass legislation, he was serious about supporting their interests, President Obama very publicly proclaimed his determination to expand offshore drilling.

At Andrews Air Force Base on March 31, 2010, the president stood before the *Green Hornet*, an F-18 fighter jet that runs on biofuel. It would be flown "in just a few days on Earth day," the president effused. "If tests go as planned it will be the first plane ever to fly faster than the speed of sound on a fuel mix that is half biomass."

He then announced that he would implement the end of the moratorium and start new leasing off America's Atlantic coast and parts of Alaska. "So today we are announcing the expansion of offshore oil and gas exploration," the president declared. "Under the leadership of Secretary Salazar we'll employ new technologies that reduce the impact of oil exploration. We'll protect areas that are vital to tourism, the environment and our national security. And we will be guided not by political ideology but by scientific evidence."[23]

Two days later, the president added, "It turns out, by the way, that oil rigs today generally don't cause spills. They are technologically very advanced. Even during Katrina, the spills didn't come from the oil rigs, they came from the refineries onshore."[24] These statements, including those regarding Katrina, were tragically inaccurate. Before both hurricanes Katrina and Rita touched ground, they destroyed or damaged 167 offshore platforms and more than 450 pipelines connecting the platforms to land, causing nine major oil spills that released at least 7 million gallons of oil and other pollutants into the water.[25]

Sitting front and center at the Andrews Air Force Base ceremony were all of President Obama's key cabinet secretaries involved in both energy and the environment, including Interior secretary Ken Salazar. A former Colorado attorney general and then senator, Salazar's appointment as secretary was met with mixed feelings from the U.S. environmental movement. Among those voicing opposition were organizations most focused on offshore drilling, including the Center for Biological Diversity. Kieran Suckling, the center's cofounder and executive director, told me in an interview of Salazar, "He was one of the top offshore oil drilling boosters in Congress, and so as soon as he was tapped for the [Interior] job we really knew that, wow, we're likely to see a big expansion of offshore oil drilling, and in fact that's exactly what we saw."

The president closed his speech by urging Democrats and Republicans to come together to pass comprehensive energy and climate legislation.

The die was cast. Those who supported the legislation also had to support offshore drilling, or at least remain silent on its harm.

Twenty days later, the *Deepwater Horizon* exploded.

Two days after that, the *Green Hornet* made its successful Earth Day maiden voyage.

The Administration Responds

On April 20, when the *Deepwater Horizon* exploded, President Obama was "alerted to the event" and began "actively monitoring the situation," according to the official White House blog.[26]

On April 22, the president convened a "principal level meeting" in the Oval Office to discuss the situation and ongoing response efforts with senior officials, including Department of Homeland secretary Janet Napolitano, Admiral Thad Allen, Secretary Salazar, FEMA administrator Craig Fugate, Chief of Staff Rahm Emanuel, assistant to the president for Energy and Climate Change Carol Browner, and National Economic Council director Lawrence Summers. The White House released a press statement following the meeting, saying, "The President and First Lady's thoughts and prayers are with the family members and loved ones facing the tragic situation in the Gulf of Mexico." The president "made sure that the entire federal government was offering all assistance needed in the rescue effort as well as in mitigating and responding to the environmental impact and that this response was being treated as the number one priority. The president asked the responding departments to devote every resource needed to respond to this incident and investigate its cause." Finally, the president thanked "the bravery of the professionals across the government who have worked non-stop over the last two days to save lives and protect the environment."[27]

The next morning, Rear Admiral Landry hit the airwaves, making back-to-back appearances on every news program declaring that no oil was leaking from the Macondo well. On April 24, using BP's estimates, the rear admiral then said that just 1,000 barrels of oil a day were leaking from the well.

On April 27, top administration officials, including secretaries Napolitano and Salazar and Director Browner, met with BP executives in Washington, D.C. That same day, and only after John Amos went public with his 5,000–20,000 barrels of oil a day estimate, did the administration release its 5,000 barrels a day estimate. Following this meeting, on April 28, the administration formally designated BP and Transocean as the "responsible parties" for the spill. The designation has great legal consequences for the companies. However, the determination as to who the responsible parties are is very straightforward. In the case of offshore facilities, the "lessee or permittee of the area in which the facility is located" is the responsible party.[28] There was thus absolutely no question that BP was a responsible party. The delay may have been due to determining Transocean's role, the role of BP's minority partners in the Macondo field, and negotiations related to the determination of the spill as one of "national significance."

As described in chapter 2, the federal government's oil spill response is laid out in the National Oil and Hazardous Substances Pollution Contingency Plan. Much of the plan is dependent upon the designation of the spill as one of national significance. It took the administration nine long days to make this designation, during which time 607,500 barrels of oil (a conservative estimate) roared into the Gulf.[29]

Underscoring well the intimate relationship of the U.S. oil industry with even a decidedly non-oil White House, on April 29, the White House released a photograph of the president meeting in the Oval Office with key senior administration officials regarding the spill. In the photo, Secretary Napolitano and national security advisor general James Jones are briefing the president.

After retiring from a forty-year career in the Marine Corps, General Jones served as president and CEO of the U.S. Chamber of Commerce's Institute for 21st Century Energy, a position he held from the institute's founding in 2007 until he was appointed Obama's national security advisor. He also served on Chevron's board of directors during 2008. The U.S. Chamber of Commerce is the preeminent lobbying body for U.S. business and the most powerful lobbying force before the federal government. To focus its energy-related lobbying, the chamber founded the institute with "key oil company backing." In speeches on behalf of the institute in 2008, Jones expressed his strong support for the need to repeal "remaining moratoria on domestic energy production and exploration," expand leasing for oil and gas, and reduce "burdensome regulations" that stymie energy production and industry innovation.[30]

With the designation of the spill as one of national significance, the role of the national incident commander was implemented, putting coast guard admiral Thad Allen in charge of the government's response effort. This was the first time the designation had been made and there was little meaningful guidance in the plan for the actual role of the National Incident Command.

This lack of guidance meant that the White House and Admiral Allen could define the role themselves. Allen knew the Gulf—and disaster in the Gulf—well, having served as the federal on-scene coordinator in New Orleans during Hurricane Katrina. Although he was very proud of the coast guard's ability to save "between 33,000 and 34,000 people," he was not afraid to criticize the response. What we "were really dealing with was the equivalent of a weapon of mass destruction being used on the city without criminality." Nonetheless, "we treated it as a regular run-of-the-mill hurricane, which was not the right response," Allen said.[31]

Allen also believed that there was a key difference between hurricanes and oil storms: while the coast guard has the expertise to deal with the former, the "means to control the source" in an oil spill "lies in the private sector," Allen argued.[32] Allen and the administration did not only put BP in charge "of the source" of the spill, they put it in charge of just about every aspect of the disaster except PR.

The role of the National Incident Command, "as envisioned by Admiral Allen, primarily functioned as a national coordination and communications center to deal with high-level political and media inquiries," the National Oil Spill Commission concluded. Allen did not believe his job was to direct tactics or response operations, "but to deal with political and high-level strategy issues associated with the response." Similarly, the goal of the federal on-scene coordinator (Allen's immediate subordinate) and the Unified Area Command was "to coordinate with the responsible party."[33]

The level of entwinement was made clear in late August, when Allen was asked whether a recent pressure test was conceived by the government scientists or BP's engineers. He replied, "It's hard to say anymore, we've been together so long."[34]

There was also no differentiation made between BP's response efforts and those of the federal government. Thus, one of BP's responsibilities was to provide the public with data on the response efforts. The federal government then used BP's numbers as its own, with the only difference being that the federal government also provided numbers for National Guard deployments. Nowhere are these numbers broken down into which are BP and which are federal responders or equipment, much less by other service branches, separate federal agencies, or by what the people and equipment were doing.

Thus, for instance, on July 15, the day the oil stopped flowing and the height of the response effort, the federal government's Web site for issues related to the *Deepwater Horizon* disaster, www.restorethegulf.gov, stated that the response involved approximately 44,000 responders, more than 6,870 vessels (including skimmers, tugs, barges, and recovery vessels), approximately 4.12 million feet of boom, and 17,500 National Guard troops from Gulf coast states. These are the same numbers provided by BP on its Web site (excluding the National Guard).

Although we do not have exact numbers, we know that in reaction to the largest oil spill in the nation's history, the Obama administration released an equally unprecedented response. It included senior-level personnel, such as cabinet secretaries and agency directors, who made frequent visits to the Gulf. It also included personnel from agencies such as Wildlife and Fisheries, who searched for and rescued harmed animals; the EPA, who monitored the air, soil, and water; NOAA, who monitored the waters and fisheries; the Department of Energy, who ultimately participated in the closing of the well; OSHA, who monitored workers involved in the cleanup; the Small Business Administration, who provided loans; the Justice Department, who conducted investigations; and all branches of the military, with portions of Grand Isle often looking like full military encampments.

There were numerous problems with BP and the federal government working so closely together. First, the very company whose failure, likely to be found as criminal negligence, had led to the crisis was now being asked to direct the response. Second, the federal government, which had failed so miserably to regulate BP's activities, was now being expected to supervise the company's response. Finally, BP's interests, moreover, like those of any similarly placed company, are fundamentally at odds with the public's interests. BP has no interest in sharing information or being transparent in its operations; of having its potential negligence exposed; of ensuring a full accounting of the costs, impact, and extent of a spill; of mitigating the harm on people and the environment; or of ensuring that all who have been harmed are cared for. All of these are the concerns of government, but the government had largely abdicated its job to BP.

The public definitely discerned a pathology taking hold in the Gulf. As public attention became riveted and as the crisis continued well beyond the mere "days" assured by the president, criticism arose about

BP's handling of the mess and the government's handling—or lack thereof—of BP.

Polls revealed a disturbing trend for the White House: as public attention to the disaster increased, the president's approval ratings for his handling of the crisis declined. Media coverage of the disaster built rapidly, and after accounting for about 6 percent of the total news coverage at the time of the explosion, it grew to account for more than 30 percent in early June.[35]

Over the same period, the president's approval rating for his handling of the spill fell from about 50 percent at the start of the crisis to approximately 40 percent in the first week of June—the same week that the administration disavowed the *Pelican* cruise's findings and the existence of oil plumes while simultaneously announcing that the oil was gushing from the Macondo well at a rate of 20,000 to 40,000 barrels a day, four to eight times larger than previously reported.[36]

As a comparison frequently noted in the press, at its height, approximately 64 percent of people had disapproved of President Bush's handling of the aftermath of Hurricane Katrina, a virtually identical low point for President Obama.[37]

The tone of the public debate also experienced a marked change at this time. Rather than focus on what the administration was *doing*, a national dialogue began as to the president's emotional state while he was doing it; namely, whether the president was expressing adequate rage against BP. *Newsweek* described "a demand for a show of presidential fury," *New York Times* columnist Frank Rich discussed "the frantic and fruitless nationwide search for the president's temper,"[38] and newspapers, pundits, and entire cable shows focused on why the president wasn't displaying rage. CBS news correspondent Chip Reid even asked press secretary Robert Gibbs whether he had "really seen rage from the president" and, if so, to "describe it."[39]

Some of the president's most publicly active weeks in response to the disaster followed. From May 28 to June 8, the president took two additional trips to the Gulf of Mexico, spoke on the spill from the Rose Garden, gave a weekly address on the disaster, and appeared on *Larry King Live* and then on *Today*. By the end of the week, his poll numbers would reach their lowest point since the start of the disaster.

On May 28, President Obama was on his second trip to the Gulf coast, touring affected areas and meeting with local politicians and affected community members. He declared for the first time, "I ultimately take responsibility for solving this crisis," and "I'm the president, and the buck stops with me."[40]

When the president appeared on *Larry King Live*, he told the nation that he was "furious at this entire situation" in the Gulf of Mexico because "somebody didn't think through the consequences of their actions." He continued, "I would love to just spend a lot of my time venting and yelling at people. But that's not the job I was hired to do. My job is to solve this problem and ultimately this isn't about me and how angry I am. Ultimately this is about the people down in the Gulf who are being impacted and what am I doing to make sure that they're able to salvage their way of life. And that's going to be the main focus that I've got in the weeks and months ahead."[41]

Many commentators noted the racial undercurrent of the tenor of the pressure on the president. "What are we really after in this collective longing for the President to 'go off'?" asked Ruth King, a writer for *O, the Oprah Magazine*. "What would President Obama's rage prove to us? We have a history of tying black and rage together, and when this doesn't happen, we don't know what to do with ourselves."[42]

Perhaps another part of the problem was that the deadly creature in this very public story was an oil blob that made for exciting footage but not much else. When the two leading human protagonists, Tony Hayward and Barack Obama, refused to get riled, the media increased the pressure. Hayward's response was to become a bit of a blundering idiot. Obama finally took the bait and tried to "man up."

On June 8, the president sat down with Matt Lauer on NBC's *Today* and said, "I don't sit around just talking to experts because this is a college seminar; we talk to these folks because they potentially have the best answers—so I know whose ass to kick."

The administration did make substantive changes as well in order to change public perception about who was in charge. In late May, Admiral Allen became the sole public face of the administration in response to the spill. Admiral Allen is tall, heavyset, and gruff—in other words, stereotypically "manly." He not only ended the joint press conferences with BP, he also replaced the "various Cabinet secretaries who were serving as the spokesperson du jour."[43]

Allen left his post as coast guard commandant so that he could instead focus "full time on the Gulf." He began holding daily press briefings either on location at the beach or at a hotel in downtown New Orleans, far away from the incident command centers where the press briefings had been taking place and far away from BP. At the same time, Rear Admiral Mary Landry was replaced as federal on-scene coordinator and rotated back to her role as Eighth Coast Guard district commander.[44]

Under pressure from Congressman Markey and Congresswoman Lois Capps (D-CA), who had introduced legislation for an independent commission with subpoena power to investigate the disaster, President Obama signed an executive order on May 21 establishing the National Commission on the BP *Deepwater Horizon* Oil Spill and Offshore Drilling. The commission, whose interviews and efforts have been critical to understanding this disaster and whose preliminary reports have served as a key source of information, was restricted in its efforts, however, due to its lack of subpoena power.

On June 1, U.S. attorney general Eric Holder announced that the administration had begun civil and criminal investigations into the disaster and that he planned to "prosecute to the fullest extent of the law" any person or entity that the Justice Department determined had broken the law in connection with the oil spill.[45]

Impatient with investigations, the Center for Biological Diversity took action against the companies into its own hands. The Clean Water Act allows citizens to sue on behalf of the government. Thus, in the largest citizen enforcement action ever taken under the Clean Water Act, on June 18, the center sued BP and Transocean for illegally spewing oil and other toxic pollutants into the Gulf of Mexico. The $19 billion suit seeks the maximum possible penalty. If BP's violations are found to have been the result of gross negligence or willful misconduct, the maximum fine is $4,300 per barrel of oil spilled. The penalties would be paid to the U.S. Treasury and will be available for Gulf coast restoration efforts.

The most significant changes made by the Obama administration were the restructuring of the MMS and the implementation of an interim moratorium on certain deepwater operations in the Gulf.

Restructuring the MMS and the "Now You See It, Now You Don't" Offshore Drilling Moratorium

Having so recently and so publicly wrapped its arms around the safety and desirability of offshore production put the Obama administration in a precarious position regarding offshore oil operations in the wake of the disaster. The administration moved cautiously, therefore, in its response. On April 26, Secretary Salazar directed the MMS to conduct physical inspections of all offshore rigs and platforms, followed a few days later by the president giving the secretary thirty days to complete a report with "recommendations on what, if any, additional safety measures should be required for offshore operations."[46] At this point, there was no mention of changes at the MMS, a halt to existing offshore operations, or a repeal of the lifting of the moratoriums.

The push for a more aggressive response came from many corners, but one of the loudest and best organized was from the Center for Biological Diversity and its executive director, Kieran Suckling.

The center has been called "pound for pound, dollar for dollar, the most effective conservation organization in the country" by *LA Weekly*. Founded in 1989, the center employs over seventy people in field offices from Alaska to Washington, D.C., and has over 220,000 members. It uses a variety of tactics to protect all species (including humans) from extinction, but the dominant one is the law. The center records an over 90 percent success rate for lawsuits to protect species, habitats, and wild places, including a landmark case in 2007 challenging the federal suppression of climate science in the Bush administration. The center is one of a mere handful of U.S. environmental organizations—including Environment America, Oceana, the Sierra Club, the Surfrider Foundation, and Greenpeace—that had been working on offshore drilling prior to the Gulf oil disaster.

Suckling holds a master's degree in philosophy and once taught Taoism, Buddhism, and Hinduism to freshmen at the State University of New York at Stony Brook. Suckling grew up Catholic and now, at age fifty-six, calls himself "something of a philosophical animist with a lingering affinity for the saints." He describes his childhood as "guerrilla middle class." After his parents divorced, his mother, a nurse, tried to keep his family in the upper-middle-class town his engineer father's career had previously afforded them.

"We lived like refugees among the wealthy," Suckling tells me. "No heat in the winter, hitchhiking to school. . . . [It] gave me a very skeptical view of what passes for middle-class norms."

On the day the *Deepwater Horizon* exploded, Suckling had already spent much of the day on work related to the Gulf coast states and offshore oil drilling. On that day, the center had filed a scientific petition to protect 404 southeastern freshwater species under the Endangered Species Act, including species in all the Gulf states affected by the BP spill.

"Midday I worked on an initial draft of a press statement attacking the Kerry, Graham, Lieberman climate bill, which we expected to be publicly announced around Earth Day. One [of] our core complaints was that the bill would encourage more offshore oil drilling," he tells me.

Later in the day, Suckling worked on a "plan to distribute 350,000 Endangered Species condoms at Earth Day events across the country."

In response to the disaster, the center unleashed a tide of critical research, lawsuits, and member activism while Suckling personally released a storm of commentary to the media taking the administration to task.

The first step was a focus on the MMS, the primary regulatory agency that had so critically failed in its functions. The center made sure the public was aware that BP's drilling operation at the Macondo well had been approved by the MMS with no environmental review, but rather, in March 2009, was granted a categorical exclusion from National Environmental Policy Act provisions.[47] As Suckling told the *Washington Post*, the federal waiver "put BP entirely in control" of the way it conducted its drilling. "The agency's oversight role has devolved to little more than rubber-stamping British Petroleum's self-serving drilling plans," Suckling said.[48]

On May 7, while the categorical exclusion made headlines, the center next uncovered that since the explosion of the *Deepwater Horizon*, the MMS had approved nearly twenty-five new offshore drilling projects, including a new plan by BP approved on May 5, all using the same exemption from environmental review as BP's Macondo well.

"This oil spill has had absolutely no effect on MMS behavior at all," Suckling told the *Guardian* of London.[49] Elsewhere he stated, "It is inconceivable that MMS could look out its window at what is likely the worst oil spill in American history, then rubber-stamp new BP drilling permits based on BP's patently false statements that an oil spill cannot occur and would not be dangerous if it did. Heads need to start rolling at MMS."[50]

In response, the administration announced that no new applications for drilling permits would go forward for any new offshore drilling activity until "the Department of the Interior completes the safety review process that President Obama requested."[51] To this Suckling responded, "The only thing Salazar has stopped is the final, technical check-off which comes long after the environmental review. . . . It is inconceivable that his attention is apparently on providing BP with new environmentally exempted offshore oil wells instead of shutting down the process which put billions of dollars into BP's pocket and millions of gallons of oil into the Gulf of Mexico."

Suckling argued that it was not enough to stop issuing new permits; the existing exemptions should be revoked. "MMS needs to formally revoke all 400 environmental waivers given out in the past 18 months and redo each and every decision."[52]

"MMS used to stand for Minerals Management Service. It now stands for misconduct, management and spills," said Congressman Edward Markey, during a House hearing in May.[53]

As public pressure grew to take action at the MMS, on May 13, Secretary Salazar announced that the Interior Department would begin overseeing a restructuring of the MMS to separate leasing and royalty collection from its regulatory functions. Within days, MMS associate director of offshore programs, Chris Oynes, announced his early resignation in advance of the Interior Department inspector general's scathing report of Gulf of Mexico inspections under his watch, followed shortly thereafter by the ouster of MMS chief Elizabeth Birnbaum.

Appointed in 2009, Birnbaum quickly came under criticism at the MMS. As the *New York Times* reported, "Agency scientists and other employees complained that since taking the post in July, Ms. Birnbaum has done almost nothing to fix problems that have plagued the minerals agency for over a decade. She rarely visited the agency's far-flung offices, so few staff members have ever seen her. The same agency managers who during the Bush administration ignored or suppressed scientists' concerns about the safety and environmental risks of some off-shore drilling plans are still there doing the same things, they said."[54]

The MMS is no longer; it has been replaced by a new Bureau of Ocean Energy Management, Regulation and Enforcement, or BOEMRE, headed by former Justice Department inspector general Michael R. Bromwich. BOEMRE is responsible for conventional and renewable offshore energy development. About seven hundred of the current seventeen hundred employees will be shifted to this agency. The Bureau of Safety and Environmental Enforcement, with roughly three hundred employees, will carry out oversight, inspections, safety, and environmental protection in all offshore energy activities. The Office of Natural Resources Revenue, with about seven hundred employees, will handle both onshore and offshore royalty and revenue functions, including the collection and distribution of revenue, auditing and compliance, and asset management.

Response to the change has been muted. While separating out these highly conflicted functions is unquestionably a vital and necessary improvement, the move does not go far enough as the staff members and staffing levels remain the same, as are the regulations they are given to enforce. Suckling, who jokingly refers to the new agency as "bummer," said that Salazar was making a baby step forward at a time when dramatic action was needed, particularly substantive regulatory reforms.

On May 30, the administration went from baby step to adult stride when it made its most significant policy response to the spill by implementing a six-month moratorium on some offshore drilling in the Gulf of Mexico.

Secretary Salazar made the announcement, explaining that the six-month moratorium would "provide time to implement new safety requirements and to allow the Presidential Commission to complete its work. . . . Deepwater production from the Gulf of Mexico will continue subject to close oversight and safety requirements, but deepwater drilling operations must safely come to a halt. With the BP oil spill still growing in the Gulf, and investigations and reviews still underway, a six-month pause in drilling is needed, appropriate, and prudent."[55]

The moratorium was implemented at a time when the polls reflected increased opposition to offshore drilling. In April 2009, 68 percent of respondents said they "favor allowing offshore drilling in U.S. waters." The number fell to 54 percent in May 2010, the lowest rate in years.[56]

At the same time, the public was now enthralled by events in the Gulf as the "spill cam" had gone public on May 20 and the administration once again revised its estimate of the amount of oil being released, from 5,000 to as much as 25,000 barrels a day. In a poll taken in mid-June, respondents were asked to rank the importance of a series of issues to them. In a list that included the wars in Iraq and Afghanistan, unemployment, terrorism, health care, and the federal deficit, only the economy ranked higher than the oil spill in the Gulf.[57]

The moratorium also came at a time of increased public action. In addition to protests by groups such as Code Pink in Houston and the Hip Hop Caucus in Washington, D.C., a direct action by seven Greenpeace activists in Port Fourchon, Louisiana, grabbed headlines.

On May 24, the group went out to a Shell Oil vessel that was preparing to depart for the Arctic to begin new offshore drilling. After sneaking aboard the rig, they used climbing gear to hang off the side of the vessel. Using oil spilled from the Macondo oil well, they painted the words ARCTIC NEXT? on the side of the rig. All seven activists were arrested.

"I've been documenting the BP Deepwater Disaster in the Gulf with Greenpeace for the past month," stated Greenpeace senior campaigner Lindsey Allen. "Having witnessed the destruction that has been unleashed on this fragile, irreplaceable ecosystem, it is unfathomable to allow Shell to drill in the pristine waters of the Arctic. Secretary Salazar must immediately stop plans for new drilling in the Arctic or any U.S. waters. We cannot afford another BP Disaster."

John Hocevar, Greenpeace's oceans campaign director, added, "As long as we continue to rely on dirty and dangerous fossil fuels and offshore drilling, we can't prevent future disasters from destroying our oceans and the industries and wildlife that depend upon them."[58]

Isolating BP

The oil industry was not amused by either Greenpeace or the Obama administration. The industry was in a bind. It needed to act but was

Port Fourchon Harbor Police talk to Greenpeace activist Scott Cardiff as he
dangles from the bridge of the *Harvey Explorer* in May 2010.

unsure how. Eleven men were dead, millions of gallons of oil, gas, and
dispersants were spewing across the Gulf, and the public was not only
furious, it was also paying very close attention to the industry's movements and operations and developing the most negative attitude toward
offshore drilling in years. Dozens of bills were moving through Congress
in response to the disaster; they would do everything from lifting the
$75 million cap on oil company liability for a spill imposed by the Oil
Pollution Act to banning all offshore drilling outright.

Keith Jones, the father of Gordon Jones, who died on the *Deepwater Horizon*, was leading a fight with other families of the deceased to make sure that further limits on oil company financial obligations were lifted. The administration was also still seeking to implement a campaign pledge made by Obama to lift the estimated $36 billion in tax breaks and subsidies granted to the oil industry in the Bush years. All the while, the effort to implement climate and energy legislation kept moving, and midterm congressional elections were coming up in November. What was a poor oil industry to do?

The solution? Isolate BP—circle in, squeeze, and then push it out like pus from an ugly pimple scarring the industry's otherwise pristine face.

On occasion, oil industry executives were forced to implement the strategy publicly, such as when they were called before Congress on June 15.

One by one, they hung BP out to dry.

"We would not have drilled the well the way they did," said Rex W. Tillerson, chief executive of ExxonMobil.

"It certainly appears that not all the standards that we would recommend or that we would employ were in place," said John S. Watson, chairman of Chevron.

"It's not a well that we would have drilled in that mechanical setup," said Marvin E. Odum, president of Shell.[59]

"I sort of shrug my shoulders when I hear these executives speak this way," Professor Robert Bea, who worked for Shell for sixteen years as an engineer and a manager, tells me. "Yes, there are always different methods, different backups, but all of these backups can fail. You tell me how you're going to handle the consequences of failure. It is an industry problem. It's got two parts: keeping the failure from happening; and two, if it does happen, can you fix it? If you can't fix it, then 'Houston, we have a problem.' If you can't tell me how you're going to handle failure, I don't think I want you to do it."

More often, however, the strategy to isolate BP was done at arm's length by the oil companies through proxies such as the American Petroleum Institute and the American Energy Alliance. The goal was to make the *Deepwater Horizon* appear to be a freak, isolated incident caused by one rotten apple. The result was that it worked.

Of course, this simply was not the case. Professor Bea tells me of the *Deepwater Horizon* Study Group, which is composed of, among others, longtime oil industry engineers: "We have come to a unwavering conclusion. This is an industry problem. It is not just BP. BP just got to the finish line first. They know this is an endemic systemic problem."

Nonetheless, the strategy of isolating BP was not at odds with the interests of the administration. Although polls showed increasing upset at the administration's handling of the crisis, they showed even greater anger at BP. The administration surely reasoned that the more the public outrage could be focused on BP, the less outrage would be directed at the president.

Of course, we cannot know all that happened between the oil industry and the Obama administration in response to the spill. We could know more, but the administration has thus far refused to respond to a Freedom of Information Act request that the Center for Biological Diversity submitted in May for "e-mails, phone logs, and meeting notes documenting his [Secretary Salazar's] interaction with oil-industry lobbyists since becoming secretary of the interior."[60] After two months of waiting, the center sued Secretary Salazar for not turning over the requested documents. It is still awaiting a response.

We do know that lobbying and campaign dollars were breaking records throughout the unfolding disaster. The oil industry spent a whopping $112 million between January and July 2010 (the latest date for which data are available) lobbying the federal government—more money than it had spent in any comparable period with the exception of 2008.[61] The industry also teamed up with the U.S. Chamber of Commerce to help the latter meet its goal of spending another $75 million on the November midterm elections—more than double what it spent in 2008—with a goal of heading off new regulation.[62]

The doubling was made possible by the Supreme Court's January 21, 2010, ruling in *Citizens United v. Federal Election Commission* that ended limitations on the amount of money that corporations can spend on elections. The unprecedented war chest was not only a means to elect candidates but also a threat to any who might choose to act against it: "Vote against us now and we'll throw our money behind the other guy in the fall."

Anna Aurilio was in the middle of this storm as the director of the Washington, D.C., office of Environment America, a federation of twenty-nine state-level environmental advocacy organizations originally founded as the Public Interest Research Group (PIRG) in 1970. While pursuing a Ph.D., Aurilio had a realization in her first semester that "I couldn't save the world through physics." In 1987, she became a staff scientist at PIRG, and in 1993, after earning a master's degree in environmental engineering at the Massachusetts Institute of Technology, served first as PIRG's research director and then as its legislative director.

After nearly twenty years on the front lines of the fight over offshore drilling, Aurilio took the explosion of the *Deepwater Horizon* very personally. "It was horrifying. It was a failure for the environmental movement that something like this could happen in 2010. We had not done our job as advocates to make sure that if any of this [offshore drilling] was happening, it was being done safely. I took it very personally as a failure."

Aurilio teamed up with colleagues from groups such as Friends of the Earth, the Sierra Club, Defenders of Wildlife, Greenpeace, Public Citizen, Oceana, MoveOn, and others to launch as strong a response as possible. They ran headlong into the fury of the industry. "You cannot overestimate the stranglehold that big oil has on Congress," she tells me. "I've seen this over seventeen years, but this is the worst. It's incredible how much power they have: the money, the contributions, the lobbyists. It's crushing."

Add to this the ongoing dispute within the environmental movement over how to proceed in the fight for a climate bill in Congress, and a movement already weak in relation to its oil industry challengers was now seriously divided and, for some large national environmental groups, all but silenced.

On June 2, the administration did a quick about-face on the moratorium, announcing that it would be significantly scaled back; it would now apply not to all offshore drilling operations in the Gulf but just to certain deepwater ones. The moratorium halted the work of thirty-three of the forty-six deepwater rigs operating in the Gulf but did not touch the fifty-seven that operated in shallow waters.

The administration "took advantage of the fact that the media was all focused on deep water," Suckling tells me. "That's a real problem, because the safety record of shallow-water drilling is actually worse than for deepwater drilling, and some of the biggest oil spills in the world have been in shallow water, including last year's oil explosion off of Australia."

In what became one of the worst oil-related disasters in Australian history, in August 2009, the Montara wellhead drilling unit operating in just 250 feet of water and about 140 miles off Australia's northwestern coast experienced a massive blowout. Oil spilled from the well for over seventy days. Estimates of the amount of oil released range wildly, from 400 barrels per day (according to the company, PTTEP Australasia) to 3,000 barrels a day (independent experts). The disaster is still under investigation.[63]

In fact, until the *Deepwater Horizon*, the largest offshore blowout occurred in less than 165 feet of water when the *Ixtoc* well, off the Mexican coastline, suffered a blowout in June 1979. It poured nearly 150 million gallons of oil into the ocean and lasted for 290 days, until it was finally capped in March 1980.[64]

The United States got its own shallow-water wake-up call when in September 2010 a Mariner Energy platform standing in 320 feet of water about 80 miles south of Vermilion Bay, Louisiana, had an explosion and fire, throwing thirteen men into the water, all of whom were rescued by the coast guard.

When Keith Jones heard the initial news of the explosion, he went home immediately. "It affected me so much. It was awful," he tells me. "I watched CNN all day. And then I just cried and cried. My brother called it 'picking a scab' and maybe that's a pretty good description. It was just a reminder of everything. I don't know what I'd felt if someone had died."

With the moratorium whittled down to just thirty-three rigs in the Gulf, the industry's next step was a legal challenge to the remaining moratorium. This was harder than you might think in Louisiana. After seven district court judges recused themselves from the case for potential bias because of oil industry ties, one judge was finally found who, although he held stock in Transocean, did not believe that this was a reason for recusal.[65]

The legal challenge worked to a degree. The judge, Martin Feldman, ruled that the moratorium should be lifted on June 22 and then denied the administration's request to stay the ruling pending an appeal two days later. The Fifth Circuit Court of appeals similarly ruled against the administration on July 8.[66] So the White House appealed again.

With the new appeal pending and the moratorium still in place, conditions on Capitol Hill and across the country were looking progressively worse for the industry. On June 26, hundreds of thousands of people all across the world gathered on thousands of separate beaches holding hands and creating a giant "line in the sand" in opposition to offshore drilling in a protest called Hands Across the Sand. At the same time, Environment America, the Sierra Club, MoveOn, and others, delivered four hundred thousand separate statements to the Interior Department from people calling for an end to new offshore drilling.

In July, Keith Jones's SPILL Act, Securing Protections for the Injured from Limitations on Liability, sailed through the House faster, he assures me, "than any bill in U.S. history except the declaration of war against Japan and the authorization for military force after 9/11." The House then passed the CLEAR Act, Consolidated Land, Energy, and Aquatic Resources, an omnibus package pooling together key pieces of legislation in response to the disaster, which eliminated the $75 million cap on oil company liabilities related to spills, tightened worker and environmental safety regulations, and increased funding to pay for more and better regulators.

In early July, a national poll found that after only jobs and the deficit, people still ranked the oil spill in the Gulf as the most important issue facing the country.[67] In response, the companies unleashed their proxies to go on the offensive. The winning strategy in addition to isolating BP was to take concern number one on the public's mind (jobs), and turn it into the reason not to act on concern number three (the oil spill).

Leading the charge was Jack Gerard of the American Petroleum Institute and Thomas Pyle and Robert Bradley of the American Energy Alliance.

Attack by Proxy

The American Petroleum Institute (API) is the leading lobby for the oil and gas industry. It has four hundred corporate members and seven hundred committees and task forces. Its board is composed of the CEOs of every major oil company. Former ExxonMobil CEO Lee Raymond served nearly twenty years as an API board member, including two terms as chairman. When Dick Cheney was Halliburton's CEO, he served on API's board and on its public policy committee.

The API spent $7.3 million on federal lobbying in 2009, an all-time institutional record. By July 2010, it had already spent as much on federal lobbying (over $4.8 million) as it had spent in all of 2008.[68] Its chief economist, lobbyist, and spokesperson, John Felmy, told me in 2008 that the API's primary job on behalf of its oil company members is to eliminate barriers to oil production, including all moratoriums on oil production "off the shores of the east and west coasts and parts of the Gulf of Mexico, Alaska, of course," and Iran.

Jack Gerard became president and CEO of API in November 2008, when the lobby needed a new face to go along with the new Congress and White House. Gerard was seen as less combative and partisan than his predecessor, Red Cavaney. In an interview with the *Mormon Times* a few months after taking the helm of API, the father of eight who is also president of the McLean, Virginia, Stake of the Church of Jesus Christ of Latter-day Saints and chairman of the National Capital Area Council of the Boy Scouts of America, said that at API, "we have our work cut out for us."[69]

Gerard's baby face belies his experience. As past president and CEO of the National Mining Association and of the American Chemistry Council—the chemical industry's leading lobby—he knows Washington, and he knows how to lobby on behalf of unpopular yet well-funded industries.

"I give him a lot of credit," Tyson Slocum, the director of the energy program at Public Citizen, says of Gerard. "He can snap his fingers and cause senators to drop what they're doing and take notice. That says a lot, given that this was an industry that was on the ropes in May."

Slocum has been working on behalf of consumer interests since cutting his teeth exposing Enron's manipulation of the west coast electricity market in 2000 as a policy analyst at the Institute on Taxation and Economic Policy. Slocum has uncovered the same kind of manipulation by oil companies in the oil, gasoline, and even propane markets and exposed the costs to consumers of overconcentration in the oil industry and the high costs (and low consumer benefits) of offshore drilling. He has been going head-to-head against the oil industry in Washington, D.C., in all of these fights for nearly a decade.

Gerard led the oil industry's response to the Gulf oil disaster "such that the industry might even come out stronger as a result," Slocum tells me. "Can you believe it? It's remarkable. That's the story of corporate lobbying in America today—that you can be an industry responsible for mass destruction and still come out on top."

API issued its first public statement on the subject of the moratorium on June 22, welcoming Judge Feldman's decision and heralding the administration for acting "appropriately in its immediate steps to inspect every rig in the Gulf following the *Deepwater Horizon* explosion. Those inspections were necessary to assure Americans that offshore operations were safe and subject to appropriate oversight."[70]

API then lashed out against the moratorium when the administration failed to budge. In July, Gerard began a new line of attack, going after the alleged effect of the moratorium on jobs. He claimed that the moratorium "places the jobs of tens of thousands of workers in serious and immediate jeopardy and promises a substantial reduction in domestic energy production."[71]

In July, Gerard said that the CLEAR Act in the House "seeks to penalize the entire oil industry for the Deepwater Horizon spill," and "if passed, would threaten American jobs, the nation's economy and its energy security."[72]

At the same time, the American Energy Alliance got to work. The alliance is the advocacy arm of the Institute for Energy Research (IER). Thomas J. Pyle is the president of both organizations, which share the same D.C. office. All of the alliance's staff people are former aides to Republican members of Congress. Among his past jobs, Pyle was once a lobbyist for the National Petrochemical and Refiners Association and the director of federal affairs for Koch Industries, the largest privately owned oil company in the United States. Robert Bradley is the founder of IER. He was also the director of public policy analysis at Enron, where he worked from 1985 to 2001, although he rarely includes that on his résumé these days.

As nonprofit organizations, API, IER, and the American Energy Alliance are not required to make the sources of their funding available to the public. However, National Public Radio reported that the alliance has "strong ties to the oil industry," including having received $160,000 from ExxonMobil.[73] While the alliance is registered as a nonprofit advocacy organization, virtually its entire 2008 expenditures—$1.61 million out of $1.67 million—were paid to Crossroads Media, "the premier Republican media services firm," according to the company.[74]

In July, the alliance launched U.S. Energy Jobs and the Web site www.saveusenergyjobs.com to "help educate voters about the unfortunate divergence in safety and health approaches between BP and the remainder of the industry." In its inaugural press release, the new organization announced, "To tarnish an entire industry because of the continuing incompetence of one company is not only wholly unfair, it is a misrepresentation of the facts." It added, "This is not an industry problem. This is a BP problem."

Then, stating that "the numbers speak for themselves," it released the following misleading and erroneous statistics: "760 'egregious, willful' safety violations administered to BP by OSHA compared to Sunoco's eight, two for Conoco-Phillips and CITGO, and one for ExxonMobil, the industry's safety leader. Other companies maintained these impeccable records while drilling over 50,000 wells safely in federal waters."[75]

The Web site provides no supporting data, date range, or documentation, and requests for this information have gone unanswered. The data are both deceptive and incorrect. OSHA has no jurisdiction over offshore drilling vessels. The 760 violations are from BP's refineries and virtually all from BP's Texas City refinery, the site of the most deadly U.S. oil industry disaster until the *Deepwater Horizon*. While BP certainly acted far more egregiously than other companies in these instances, the others are not without violation. Recall the U.S. Chemical Safety Board's warning in its ruling on the Texas City explosion that it found a pervasive "complacency towards serious safety risks" across all the leading oil companies at their refineries.

Moreover, ConocoPhillips faces over a hundred OSHA violations at its U.S. refineries for incidents occurring since 2006. OSHA cites more than twenty violations at ExxonMobil refineries, including a serious violation at its Baton Rouge Refinery involving "the control of hazardous energy" just six days before the *Deepwater Horizon* exploded. Three years earlier, in October 2007, OSHA cited the company for eleven serious violations at its Baytown, Texas, refinery for "management of highly dangerous chemicals."[76]

OSHA violations are not the only concern. In just one example close to my home: Chevron operates a refinery in Richmond, California, about fifteen miles from where I live in San Francisco. The refinery has been in "high priority violation"—the most serious level of violation noted by the EPA—of Clean Air Act compliance standards every year since at least 2006.

In January 2007, a giant explosion rocked the refinery. A leaking corroded pipe "that should have been detached two decades ago," according to investigators, was to blame. The five-alarm fire and hundred-foot flames burned for nine hours. Almost three thousand people in nearby neighborhoods received telephone calls instructing them to stay inside with their doors and windows shut to avoid breathing the toxic fumes.[77]

Perhaps most important, however, is the dangerous claim that the fifty thousand other wells maintained "impeccable records."

In addition to the failures cited above, the *Houston Chronicle* found that in just the five years before the *Deepwater Horizon* exploded, federal investigators documented nearly two hundred safety and environmental violations in accidents on platforms and rigs in the Gulf. The paper described "a stunning array of hazards that resulted in few penalties," including "workers plunged dozens of feet through open unmarked holes. Welding sparked flash fires. Overloaded cranes dropped heavy loads that smashed equipment and pinned workers. Oil and drilling mud fouled Gulf waters. Compressors exploded. Wells blew out." All the major companies were cited. While BP led the others with at least forty-seven accidents or blowouts, Chevron was a very close second at forty-six, and Shell had twenty-two.[78]

The alliance statistics were nonetheless widely repeated in publications across the country and on cable TV programs—rarely, if ever, with any indication of their source and no recognition of their inaccuracy.

The alliance then focused on jobs. "But while oil continues to spill into ocean waters, there is another crisis developing in the Gulf—the death of the 'energy coast' economy and the jobs the oil and gas industry supports."[79]

Back over at API, Jack Gerard was reigniting his successful "Astroturf" rally strategy begun in 2009. That summer, Gerard spearheaded a

series of rallies across the country that sought to create the impression
that there was a grassroots public movement against climate legislation.
Dubbed "Astroturf" as opposed to "grassroots" rallies by their critics, a
leaked internal API memo revealed that Jack Gerard had contacted his
oil company patrons, including Chevron, ConocoPhillips, and Halliburton, to organize the events all around the country, bus in their employees,
and furnish T-shirts and signs.[80]

This time around, Gerard started an online movement through
another proxy, Energy Citizens. The group introduced itself in an e-mail
blast in July 2010 as a "new movement of citizens focused on countering
reactionary policies and restoring a common-sense perspective" that is
"supported by" API. The e-mail was from the Partnership for America's
Energy Security and signed by Deryck Spooner, the executive director
of Energy Citizens and API's "external mobilization director."[81] Its primary "take action" item listed on its Web site states, "Tell Congress to
Get Domestic Energy Production Back on Track."

API was also providing statistics, stating, "America's oil and natural
gas industry supports over 9 million jobs," and "adds more than $1 tril-
lion to the national economy. That's 7.5% of our nation's wealth." At least
API provided a definition for its data, but it is so broad that it could
include any industry that uses oil, which is virtually every business in
the country. Thus, the key word is *supports*.

API's definition reads, "The industry supports jobs not just in explor-
ing, producing, refining, transporting and marketing oil and natural gas,
but also through the purchases it makes of other goods and services that
support the industry's operations. Equipment suppliers, construction
companies, management specialists, and food service businesses are all
a strong link in the industry. These businesses, in turn, purchase other
goods and services that support other jobs throughout the nation."

API then breaks down the numbers by state, repeating the same line
each time: "In Louisiana alone, 330,053 jobs are supported by the industry"; for Texas, 1,772,335 jobs; for Mississippi, 83,820 jobs; for Alabama,
94,732 jobs; and for Florida, 267,277 jobs."[82]

These figures were widely reported in the press and used by political
leaders without a caveat about how broad a net the numbers were cast-
ing. For example, at a congressional hearing in June, Senator Mary
Landrieu (D-LA)—one of the staunchest and most well-funded allies

of the oil industry in Congress and the leading congressional opponent of the moratorium, told Interior Secretary Ken Salazar that the moratorium "could affect 330,000 people in Louisiana alone."[83]

The reason industry advocates exaggerate the numbers is that while oil and gas industry jobs are certainly vital for those whose livelihoods depend on them, and particularly for communities with a high concentration of oil industry workers, there simply are not that many people employed by the industry in the United States.

There are several reasons for this, including the increasing reliance on technology instead of people in oil production. In addition, it is extremely dangerous work, and only a portion of it is unionized. While only about 64 percent of refinery workers are unionized, according to the Seafarers International Union, none of the offshore exploration, production, and service industry in the Gulf of Mexico is unionized.[84] Statistically, unionized workers have higher pay and are more likely to have health insurance, pensions, paid sick leave, and other benefits and safer workplaces than nonunionized workers in similar jobs.[85] Finally, the industry has an aging workforce that is not being replaced with new, younger workers. For example, the University of Houston found that in 2007, "enrollment in U.S. universities for petroleum engineering programs is down by 85 percent from its peak in 1982."[86]

The U.S. Bureau of the Census lists 133,286 people "working in the crude petroleum and natural gas extraction industry in 2007"; of these,12,056 live in Louisiana, approximately 57,600 live in Texas, fewer than 1,000 live in Mississippi, about 230 live in Florida, and 500 to 1,000 live in Alabama.[87] This definition includes only those in exploration, development, and/or the production of oil or gas. It does not include oil refinery workers—approximately 30,000 people nationally, according to the United Steelworkers union, which represents them. Nor does it include support industries.

In a review of the effects of the moratorium released in September, the Department of Commerce analyzed the employment figures in five parishes in Louisiana that are heavily dependent on the deepwater drilling industry: Lafourche, Lafayette, St. Mary, Terrebonne, and Iberia. It found little change in employment between May 2009 and July 2010, whereas employment actually increased slightly between April and July 2010, most likely due to people employed in efforts related to the oil spill.

The department also reported that fewer than 2,000 of 9,700 rig workers had been laid off or had left the Gulf to work elsewhere. The report estimated that the "six-month moratorium may temporarily result in up to 8,000 to 12,000 fewer jobs in the Gulf Coast."

Finally, the report found that most rig workers held on to their jobs during the moratorium, some by doing regular maintenance work and some because it is difficult and expensive to recruit such highly skilled workers, so the companies were hesitant to lay them off.[88]

Many analysts have found the administration's numbers to most likely be overestimates as well. BP established and then heavily publicized a $100 million fund to help workers laid off by the ban. The fund's administrator said, "We were expecting hundreds of applications per day when we opened up on Sept. 1," but as of mid-September, only 220 people had applied for the grants.[89]

API then resurrected the Astroturf rallies on the ground in September, organizing a series of "Rallies for Jobs" in cities all across the country, urging Congress to block legislation proposed in the wake of the BP oil disaster. Rallies held across Texas drew an estimated fifty-five hundred people. But, as in the past, the attendees were largely employees from Texas-based oil industry firms that were hosting the event. About two hundred employees of Houston-based ConocoPhillips attended one rally, for example, with transportation provided courtesy of API.[90]

Before completing the story of the fight over the moratorium and legislation, we must return to the most important fight, that over the fate of the oil in the Gulf.

The Disappearing Macondo Oil

On July 15, BP successfully capped the well. On August 3, it completed the static kill and successfully closed in the well. These were two historic and monumental events worthy of great celebration and tribute. The White House was not content, however, to herald these achievements. It wanted more. It wanted the whole event to be over, or at least to be over in the public's mind. To do away with the entire affair, Carol Browner, director of the White House Office of Energy and Climate

Change Policy, was dispatched onto the airwaves not to advocate for new regulation, the continuation of the moratorium, or legislation in Congress to ensure that such a disaster never happened again, but to complete what Admiral Landry had begun three months earlier.

On August 4, the banner headlines on televisions and newspapers all across America declared: WHITE HOUSE: NO MORE OIL and WHITE HOUSE: 75% OF OIL GONE. Director Browner swept the morning news programs, appearing on ABC, NBC, CBS, MSNBC, and even Fox, to proclaim the good news. "The good news is that the vast majority of the oil appears to be gone," she announced. "As our scientists are saying, the vast majority of the oil has either been cleaned, skimmed, contained. . . . Our scientists have done an initial assessment, and more than three-quarters of the oil is gone. The vast majority of the oil is gone."[91]

"I don't want to be pessimistic here," *Today* host Matt Lauer said politely, "but NOAA was also the group that grossly underestimated the size of the leak in the first place. Can we trust their numbers?"

Rather than defend her sister agency, Director Browner instead emphasized that a "panel of scientists" from in and outside of the government looked at all of the information and reached this initial assessment, but "we think it's very encouraging, you know, that the vast majority of the oil was contained, was cleaned up, and as I said, Mother Nature did its part, too." The director said that we may "see some tar balls," maybe even "some sheen," but that was it. Host Robin Roberts on ABC's *Good Morning America* responded calmly given the magnitude of the announcement. "Very encouraging," she said.

The news would have been encouraging, if it were accurate. Unfortunately, the scientists had come to no such conclusion, and the information the director was providing was false.

It got worse, as that afternoon the director was joined by NOAA administrator Jane Lubchenco at a White House press briefing, who went on to say that the report was "peer reviewed" when it was not.

The report the White House released, "BP Deepwater Horizon Oil Budget: What Happened to the Oil?," did not provide methodology, define terms, reflect the findings of the scientists who worked on it, or support the assertions made by the director or administrator.[92]

The budget put the oil into several different categories: residual, 26 percent; evaporated or dissolved, 25 percent; naturally dispersed, 16 percent; chemically dispersed, 8 percent; and directly recovered from the wellhead, burned, or skimmed, 25 percent. The director counted everything other than that directly recovered from the wellhead, burned, or skimmed as gone. However, oil mixed with dispersant is far from gone, and oil that is dissolved can still remain present in water. Even evaporated oil remains in the atmosphere where it can cause serious harm. From these round numbers, it was only accurate to say that 25 percent of the oil was "gone."

"There's a lot of . . . smoke and mirrors in this report," said Dr. Ian MacDonald of Florida State University. "It seems very reassuring, but the data aren't there to actually bear out the assurances that were made."[93]

The White House particularly unnerved scientists who worked on the report. The National Oil Spill Commission interviewed many of these scientists and found that they were critical of the report and the way it was presented. Many emphasized the large degree of uncertainty in their work and their impression that they were assisting in the development of an operational tool rather than a public government report.

For example, the estimate of how much oil evaporated was calculated using a formula designed for spills near the surface, not 5,000 feet underwater. As for an assertion by Dr. Lubchenco that the oil that has been dispersed is "rapidly degrading," Bill Lehr, a NOAA scientist and an author of the report, said the analysis did not include an actual calculation or measurement of what's happening in the Gulf. "We haven't attempted yet to calculate that rate," he said, and instead relied on assumptions based on past spills in the Gulf.

Given such uncertainties, one researcher who collaborated on the report said he would not have given out exact figures. "We don't have the foggiest idea [about how to measure the oil] with that precision," said Ed Overton, a professor at Louisiana State University.[94]

"Indeed," the commission wrote, "it is unclear whether any of the independent scientists actually reviewed the final report prior to its release. In the words of consulting expert Ed Overton, '[t]o a scientist, peer review means something . . . . Clearly it wasn't a peer review from a scientific perspective.'"

"If an academic scientist put something like this out there, it would get torpedoed into a billion pieces," Dr. Samantha Joye said of the report.[95]

Dr. Joye and her colleagues did their own analysis of the government's data. Georgia Sea Grant, of the University of Georgia, released the analysis on August 16. They classified the oil into categories relevant to discussions of recovery and environmental impact: burned, skimmed, evaporated, degraded, and remaining. Thus, starting with the report's figure for how much oil entered the water, they estimated how much oil could have conceivably degraded and evaporated.

They concluded that of the oil that was released into the Gulf (i.e., not captured directly from the wellhead in containment ships), 10 percent was burned or skimmed, 4 percent degraded, and 7 percent evaporated. Based on their analysis, which has become the standard since its release, as much as 79 percent of the oil remained as a threat to the ecosystem as of early August. They also stressed, moreover, that oil evaporated into the atmosphere can also have environmental and health-related effects for communities living downwind of the Gulf of Mexico.[96]

While oil continued to wash up in waves on beaches and marshes across the coast, float beneath the ocean in plumes, and rest in blobs on the ocean floor, the White House in essence told the public, "Nothing to see here, folks, move along!" And that's exactly what happened, that is, everywhere but in the Gulf.

"[It] just suddenly came up Friday and it's along the beach for miles and miles, and drifting inside in some spots," Ryan Lambert, a resident of Buras, Louisiana, said of a new wave of oil hitting the beaches in September. "Everyone thinks this is over, but it's not—not if we can still get soakings like this."[97]

"All the brown spots and patches you'll see on this beach for the next nine miles is oil," said Forrest Travirca of Fourchon Beach in September. "And if you dig down a few inches or a few feet, you'll see oil, too. And if you walk into that marsh back there, you'll find oil. . . . So don't tell me we dodged any bullets. Or that it wasn't so bad. 'Cause I've been out there every day since May dealing with all that oil we dodged. It just makes my blood boil. There's so much oil in some of these sands that when they heat up, the oil starts bubbling to the surface. That's one reason the cleanup crews have to wear those [protective] shoes. They're literally walking on oil sometimes. So when I hear people say 'It's over,' I just feel like screaming. I want them to come spend a day on this beach, and then tell me it wasn't so bad, or it's over."[98]

While the oil was declared "gone," the fight over the moratorium continued. And although the economic impact of the moratorium on the Gulf may not have been extensive, the political fallout was. Nowhere was this felt more powerfully than in Louisiana. Oil is built into not just the economic but also the historic, social, and political fabric of the state.

When the Unocal oil blowout happened off the coast of California in 1969, California was one of the largest producers of oil in the United States. Nonetheless, Californians responded by sacrificing new offshore drilling. As a result, other industries, such as tourism, grew. Even without this drilling, moreover, the state remains today the third-largest oil producer in the United States, and its largest oil companies continue to dominate the industry, with California-based Chevron the fifth-largest corporation in the world.

Louisianans, in contrast, vehemently rejected the moratorium and those who created and supported it. I got a good taste for why on Labor Day weekend.

The Seventy-fifth Annual Louisiana Shrimp and Petroleum Festival

On Sunday, September 5, a special mass was held in the central park of Morgan City, Louisiana. The park was standing room only as perhaps a thousand people gathered to receive communion before the traditional "blessing of the fleet." The white gazebo served as the church, and a priest in thick lime green vestments conducted the service, young altar boys flanking him on either side. On the table before him, along with the traditional missal, anointing oil, and a wooden cross, sat a small crystal bowl filled with bright pink shrimp, a miniature light brown wooden replica of an oil derrick, and two faux jewel–encrusted crowns. The crowns were made of glistening white crystals with green stones in the form of an oil derrick sprouting up the center, wrapped from base to tip by a giant pink shrimp sporting a yellow hard hat. The crowns would later adorn the 2010 king and queen of the Seventy-fifth Annual Louisiana Shrimp and Petroleum Festival.

The same image that is on the crowns is found on T-shirts, baseball caps, flags, and banners hanging all around the park and town. People in Morgan City find neither the image nor the festival ironic. "Nobody thought it was strange till you all showed up," Nathalie Weber, the president of the festival (and Miss Louisiana Shrimp and Petroleum Festival 1968) tells me, shrugging derogatorily in the direction of the gathered out-of-town media. "We've always celebrated both industries. They go hand in hand. They support each other."

The festival was originally just the Louisiana Shrimp Festival, founded to celebrate the shrimp harvest, and in "a community strongly rooted in Catholicism and tradition," a blessing of the fleet was held to

The 2010 king and queen of the Seventy-fifth Annual Louisiana Shrimp and Petroleum Festival.

ensure "a safe return and a bountiful harvest." The first year, the celebration traveled to "Egle's Place for a fais-do-do, a Cajun dance," and
the festival was born.

Twelve years later, in 1947, the festival, Louisiana, and the world
were changed forever when Morgan City became the home of the first
producing oil well built on water "well beyond the sight of land," earning
the word *petroleum* a place in the festival name from then on.

Offshore drilling would prove to be a central component of Louisiana's economic, political, environmental, and social fabric. Although just
over 12,000 people in the state are employed directly in production
today, many thousands more work in the industries that support those
operations, as well as "downstream" oil workers, such as those at refineries. Moreover, while nationally, oil and gas production employs just
133,000 people, more than half of them reside in the Gulf states.[99]

The state taxes derived from the industry are not as large as many
would hope. None of the major oil companies is based in Louisiana,
although all of them have operations there. Moreover, thanks to bad
deals struck with federal regulators, prior to 2006, the state received no
royalties from oil and gas projects in the federal waters beyond six miles
from the state's shore. A law passed that year gave the state royalty payments on all *new* projects, and then, beginning in 2017, Louisianans
will get payments on all offshore operations. Louisianans are estimated
to have subsidized the industry by hundreds of billions of dollars in the
interim.[100] Finally, in 2008, Moody's Analytics estimates that oil and gas
extraction-related activities contributed to 8 percent of Louisiana's gross
domestic product.[101]

Whereas Louisiana's shrimpers account for 80 percent of the
nation's shrimp production, offshore oil drilling from the entire Gulf of
Mexico (the numbers are not broken down by state) accounts for just
23 percent of domestic U.S. production. If you include the oil produced
on land in Louisiana and that produced in its state waters, the total from
the Gulf increases to 28 percent of domestic production.[102]

It is very important to note, however, that all of the oil produced in
the Gulf does not necessarily stay there, nor even in the United States.
Rather, oil companies are increasingly exporting oil and other petroleum
products out of the country. After remaining constant at around 1,000
barrels per day throughout the 1990s, exports of oil and petroleum

products, including from the Gulf coast, have been on the rise since 2006, totaling some 2,200 barrels per day.[103] The reason is that Americans have been steadily reducing their gasoline consumption by utilizing more fuel-efficient vehicles and driving less, such that the high point of U.S. of gasoline consumption was in 2007.[104]

After the priest delivered his sermon at the festival, communion was served and the crowd proceeded to the dock for the blessing of the fleet. There were several boats, but by far the most popular was the one carrying the young queens crowned at parish harvest festivals across the state. Ranging in age from sixteen to eighteen, they wore sashes bearing their titles across the front and their names across the back. Krista Lynn, Amy Lynn, and Meagan, for example, bore the titles Strawberry Queen, Oyster Queen, and even Frog Queen. The young women smiled, waved, and granted interviews while their boat was docked. The boats then circled the harbor, with the queen's boat meeting the king's boat bow-to-bow in a uniquely wet "kiss."

Local dignitaries then offered speeches, including interim lieutenant governor Scott Angelle. I spoke with Angelle for some time following his speech. He was eager to talk, affable, and good-humored. Angelle, then a Democrat, had been secretary of the state's Department of Natural Resources when Mitch Landrieu, then lieutenant governor, was elected mayor of New Orleans. Republican governor Bobby Jindal appointed Angelle to the post just six days after the explosion of the *Deepwater Horizon* to fill out Landrieu's term on the condition that Angelle not run for the office in the upcoming November election, at which time he returned to his original post.

While he and I talked, at least two candidates for his office walked by. They were easy to notice; their supporters, dressed in campaign T-shirts, holding signs, and handing out brochures, surrounded them as they passed. A month after our meeting, Angelle switched parties, citing his dissatisfaction with the national Democratic Party's energy policies, particularly the moratorium on drilling in the Gulf of Mexico.

Angelle spoke with me of the incredible resilience of the people of Louisiana to withstand this crisis. He said that they would fish and drill again. "This was an event. A horrible event. We'll get through it. We're in good shape," he reassures me. "It was a group of companies that failed. Not the system."

Scott Angelle (right), lieutenant governor of Louisiana, at the
Seventy-fifth Annual Shrimp and Petroleum Festival in Morgan
City, Louisiana, in September 2010.

Angelle explains that the September 11 terrorist attacks reflected a
"systemic failure. Multiple cities were involved, multiple airlines. But
the airline industry was shut down for just four days, that's it." In the
case of the *Deepwater Horizon*, he notes, "There is no evidence of systemic failure here, yet the Obama administration overreacted. Shutting
down an entire industry? That's not what we do in America."

Angelle surmises that the administration had another agenda: to
move the nation to alternative energy before the market was ready.

"Cheap and available gasoline built the middle class of the United States," he argues. "Look at Florida. You wouldn't have people there without air-conditioning. Wouldn't have air-conditioning without energy."

"I live in the world of probability," Angelle explains. "Academics, like this administration, live in the world of possibility. It's *possible* there will be another incident, but it's not *probable*." In a refrain I was hearing a lot that day, Angelle says, "We're sad about the spill, we're mad about the moratorium."

When I point out that the MMS had not been appropriately regulating the industry and had failed in its duties, Angelle quips, "Well, obviously there's something wrong at DOI [Department of Interior]! These guys are having sex with and doing cocaine with those they're supposed to be regulating!" He then looks down at me with a knowing glint and asks jokingly, "Two great things in life, right?" and laughs at his own joke. Getting back to business, he says, "We had an accident, it happens all the time. I'm confident in these companies."

I then point out that there had been four hundred incidents on offshore drilling rigs in the Gulf in the last five years alone, to which he replies, "Compared to what? I don't know if that's a lot. How many birds are killed by windmills each year?"

Most important, he insists, we need the oil production from the Gulf, or we will become more dependent on foreign oil. I ask, "What about the fact that the oil produced here doesn't stay here and that oil companies are shipping more oil out of the Gulf?" He denies it. "That's not happening. Not happening here. Oil produced here stays here."

Before moving on to his next event, he gives me his cell phone number and tells me to text him anytime for a follow-up interview.

The rest of the day is filled with carnival rides, Cajun food, live music, dancing, and a parade complete with high school marching bands, national elected officials, including Senator David Vitter, and—the highlight—the car pulling the king and queen.

I walked around all day interviewing people. Everyone was having a great time. Once I started asking questions, however, I learned that a lot of anger was boiling just below the surface. The anger was focused on just two people: not BP CEO Tony Hayward and Transocean CEO Steven Newman, but rather President Barack Obama and Speaker of the House Nancy Pelosi. The dislike was not strictly a "party-issue."

Although Louisiana has a Republican governor, its senators are split between one Republican and one Democrat, reflecting its mixed voter registration. In 2010, almost 51 percent of registered voters in Louisiana were Democrats, 26 percent were Republicans, and 23 percent were independents or another third party.[105] Rather, their anger was focused on the moratorium, even though few people I met that day were either fishers or oil workers. Most people who live in Morgan City proper are employed as professionals or as service or government workers. Nonetheless, they were mad.

I was told that Obama was a socialist or a communist and that "Pelosi replaced Hillary [Clinton] as the most hated woman in America." After repeatedly hearing the comparison that the airline industry is not shut down every time a plane crashes, I ask whether the fact that the airline industry is badly regulated is really an argument for why we should allow the oil industry to be poorly regulated as well. I don't get a reply. When I then point out that when an airplane crashes, the disaster does not spread to numerous other industries, livelihoods, and entire ecosystems, I realize that I am probably getting a little punchy.

So I went off in search of people who I thought might hold a different opinion.

I sought out the handful of black faces at the festival and was able to talk to just one family, the Delcos. Not only were they among the only people I encountered that day who worked in the oil industry, they were also the only people who had anything positive to say about President Obama.

Raymond and Myra Delco have been coming to the Shrimp and Petroleum festival for thirty years. They live in Burwick, "across the water." Myra tells me that "Obama has done his best. He's done well. Really done well. Problems he got, best he could do in the face of it."

Her son, Jayson, with whom I had coffee the next morning, works offshore. Her husband, Raymond, paints the Christmas trees that are part of the offshore drilling equipment. Raymond tells me that his work has picked up since the explosion. Jayson tells me the same thing the next day. Jayson is a pipe inspector and inspections have significantly increased on all the rigs. Raymond and Myra say that the spill has not really affected them but that they worry for the "folks in Dulac. The folks in Grand Isle. It's really hurt them. We feel for them." Although

Oil worker Jayson Delco in September 2010, Morgan City, Louisiana.

everyone I spoke with that day except the Delcos had little positive to
say about the president, people could not say enough positive things
about their governor, Bobby Jindal.

In 2008, Governor Jindal—a young, attractive man of color—gave
the Republican response to President Obama's first State of the Union
address. Where Obama was the first African American president, Jindal
was the first Indian American governor. There was a great deal of buzz
that Jindal would be on the presidential ticket in 2012 and that this
speech was his national debut. Unfortunately for Jindal and the Republicans, he bombed, giving a stilted awkward speech that many believed
was career-ending. It may well have been, had it not been for the Gulf
oil disaster.

While I traveled the Gulf coast in the months after the disaster, I
repeatedly asked, "Who, in the political sphere, is a 'hero' in this disaster?
Who is doing the right thing?" There was only one frequent answer:
Bobby Jindal.

Jindal's family immigrated to the United States from India while his
mother was still pregnant with him so that she could complete her

doctorate in nuclear physics at Louisiana State University. Born Piyush Jindal in 1971 in Baton Rouge, Jindal took the nickname Bobby at the age of four from the TV show *The Brady Bunch*, being a big fan of the youngest male Brady. Although his parents are Hindus and registered Democrats, as a young man, Jindal sought out Roman Catholicism and conservative Republicanism. A fiscal and social conservative (Jindal is against abortion, even in cases of rape and incest, is against stem-cell research, and is for teaching intelligent design in schools), Jindal has been called by Rush Limbaugh "the next Ronald Reagan." After serving as a high-level George W. Bush appointee, he was elected to Congress in 2004 at age thirty-three and served as House assistant majority whip. In 2007, he was elected governor of Louisiana.

Since April 20, Jindal has been a constant presence in the Gulf coast media, railing equally against BP and President Obama. While Obama's "timidity" was a topic of national discussion, Jindal was in constant action and "mad as hell" at just about everyone. In a refrain I heard regularly throughout the Gulf and from across the political spectrum, out-of-work fishermen Wade Fruge and Kirk Chaisson of St. Mary Parish, Louisiana, tell me, "At least he's doing something. He's aggressive. TV-wise, he's great." They may not be able to point to concrete accomplishments by Jindal, but they are thrilled to see him doing *something.*

For many in the Gulf—conservative, progressive, or libertarian—a common thread stated repeatedly was a feeling of being abandoned by the federal government. The causes people routinely cited were the disinformation and lack of transparency, the use of dispersants that harmed their health and livelihoods, the disastrous claims process that Kenneth Feinberg failed to solve, the lack of preparation in advance of the spill, the lack of boom and skimmers once it happened, the early announcement that the oil was gone followed by the premature removal of cleanup workers and boom, the inability to get ahead and on top of BP, and a general sense that the administration was trying to put the spill out of sight and out of mind.

There was none of the hostility toward the administration that emerged from Katrina—a far more deadly event (sixteen hundred lives immediately lost compared to eleven) in which people were not merely abandoned, they were actively turned away from escape routes; forced to die in the Superdome, on roofs, and in homes and hospitals; and not just pushed away from beaches but actively hunted by private security forces in their own streets. And President Obama did make several trips to the Gulf that were welcomed, but they were not enough, and certainly not enough to erase the negative actions the administration was also taking.

Thus, the mere fact that Jindal was constantly on beaches, in town halls, and at restaurants talking to people, reporters, and the administration, almost regardless of what he had to say, made him widely popular. Then, among those who were additionally angry about the moratorium, his increasing focus on it garnered a great deal of support. In a June 2010 poll of Louisiana residents, 77 percent said that they support drilling for oil off the shore of Louisiana, with 28 percent saying that the oil spill made them *more* supportive of offshore drilling. The governor's approval rating was 63 percent overall and 65 percent on his handling of the spill. By contrast, President Obama's overall approval rating in Louisiana was only 37 percent and on the spill just 32 percent.[106] By the end of the year, Jindal polled as the most popular governor in his home state and, with the release of his book, *Leadership and Crisis*, and a national publicity tour, a renewed buzz circulated that he could once again be a contender for a spot on a Republican presidential ticket if not in 2012, then in 2016, with JINDAL-PALIN 2012 bumper stickers and buttons becoming increasingly popular across the Gulf.

Dead End

Back in Washington, D.C., the rest of the policy responses to address the oil spill were about as popular as was the moratorium in the Gulf. Over the course of the summer, as oil continued to spew from the wellhead of the Macondo, public attention across the country was gripped and legislation moved. Protests took place calling for action, including the posting of giant WANTED signs depicting the "BP Ten," the ten members of Congress wanted for taking more campaign contributions from BP than any of their peers in the last two campaign cycles. The posters were part of a campaign by Friends of Earth, Oil Change International, and others calling on these, and other elected officials, to donate all the "dirty oil dollars they received" to the Gulf Coast Restoration Network, a network of environmental protection organizations across the Gulf coast.

Keith Jones went to Washington eight times over the summer and fall. He testified in one of the dozens of hearings held by Congress, lobbied the president, met with every elected representative he could—and every staff person of every representative he could not—and talked endlessly to reporters. His life was dominated by his crusade against BP, the oil industry, and Congress. "These companies care about one thing: profits," he tells me. "So that's the only place we can affect them."

"I've been involved truly in almost nothing else. My interest has been in nothing else. The two things that I have effectively done are lobby these bills and deal with the press. It's defined my life," Jones told me.

Sandra, his second wife of barely a year and a real estate agent, was "happy to carry the weight and cover our expenses." But she worried about him and so did his daughter and his daughter-in-law. I ask Jones if the lobbying helped with his grief. "It puts it off. I don't know what I'll do when I have to stop and feel it."

It did not go well. In what may turn out to be the most successful lobbying effort of all time by the oil industry, the 110th Congress ended without turning a single piece of legislation written to address the Gulf oil disaster into law, including Jones's bill. "I'm disgusted," Jones says.

Congress went on its summer recess (extended by a week this year) in August. By the time it returned in September, not only was the Macondo well capped but also the White House had announced the oil from the well was virtually all gone. Every oil bill that had passed the House in the summer died in the Senate by the fall. Not even the $75 million liability cap was lifted. There were several reasons for this. One was that the Obama administration was not actively lobbying for passage while the oil industry was vehemently working against any changes. Another reason was a concerted effort by Republicans to stonewall *any* legislation.

As Anna Aurilio of Environment America told me, on the Senate's plate in September was funding for the Department of Defense, "and they couldn't even get that done." Aurilio tells the interns who come to her Washington, D.C., office, "It's like a tar pit, not a black hole because sometimes things do come out of it. But a tar pit. You have to pull and pull and pull to get something out." Reflecting on the upcoming November 2010 elections, Aurilio says, "But, if one House or the other flips [to Republican control], then I don't know what we're going to do."

"Of course, so few politicians are willing to do what is right if it offers the slightest political risk to them," Jones tells me. "The question is, of course, what a Republican House will do with the bill [in the 111th Congress]? Nevertheless, I am determined to get a bill passed and I won't stop fighting until I do."

It was not just the Republicans, however. There was also enormous pressure from Democrats in oil states resisting any new legislation and aggressively fighting the moratorium. Senator Mary Landrieu refused to allow the confirmation of President Obama's pick to serve as head of the Office of Management and Budget to go forward until the moratorium was lifted. "Although Mr. Lew clearly possesses the expertise necessary to serve as one of the President's most important economic advisers, I found that he lacked sufficient concern for the host of economic challenges confronting the Gulf Coast," Landrieu wrote of her decision.[107]

The counterweight to the industry and its representatives in Congress was simply not strong enough.

A rare exception took place on August 30 when San Franciscans marked the fifth anniversary of Hurricane Katrina with a series of protests at BP's and Chevron's offices in the city's downtown. The activists chanted, "Make Big Oil pay!" while blockading BP's building and the street corners surrounding it. A mass "die-in" organized by Rainforest Action Network on a giant black tarp mimicking a toxic oil stain took place at Chevron's offices. The Mobilization for Climate-Justice West organized the day's events as a call not only to hold the oil industry accountable for its crimes in the Gulf but also for meaningful action to address climate change in Washington.

Back in Washington, meanwhile, there was the ongoing fight over the climate bill that continued throughout the summer and fall. Many organizations involved in the fight had long pursued a strategy that involved eschewing direct conflict with corporations, which left many unprepared to take on BP or the rest of the oil industry in response to the disaster. There was also a failure at the grassroots level to mobilize people. "There was incredible energy, intensity, interest, and spirit in the grassroots to take action," but the movement was not directed at one target, Phil Radford, president of Greenpeace, told me. "The interesting experience with the spill was you couldn't tell the difference between BP and Obama. Who should the target be? BP? The administration? Who was in charge and who was to blame?"

Aurilio also cited the lack of images to mobilize even more people. "Keeping people away from the hardest hit areas meant that those shocking images just stopped coming," she says. "These were so critical in the wake of the *Valdez* to tell the story outside of Alaska."

Both organizers and congressional staff people I spoke to also continually cited the calming effect on the public's sense of the need for action created by the administration's downplaying of the amount of oil spilled and particularly by its premature declaration that the oil was gone, which created the false sense that the problem was resolved and no more action was required.

Finally, the "isolate BP" strategy worked: the problem was not seen as an industry-wide failure but rather the failure of just one company.

On October 12, one month before the looming midterm congressional elections, Secretary Salazar announced that the administration would lift the Gulf of Mexico moratorium a full forty-eight days before the six-month time line was reached and before most of the major investigations into the oil spill, including the National Oil Spill Commission and the Marine Safety Board investigations, were complete. The announcement was made shortly after Salazar introduced new rules on offshore drilling, creating the impression that a deal with the oil industry had been struck. In exchange for new rules on offshore drilling operations, the moratorium would be lifted.

Less than two months later, the work of those who mobilized and organized for meaningful policy change in response to the disaster received a significant partial victory. The administration announced that no new drilling permits would be issued off the western coast of Florida or off the Atlantic coast through 2017—effectively reimplementing the moratorium in those areas. Drilling would be permitted off Alaska (including in the sensitive Chukchi and Beaufort seas), the Pacific coast, and the rest of the Gulf of Mexico.

But then in January, the oil industry was handed another "win" when the Interior Department exempted thirteen companies, including

Chevron and Shell, with existing operations in the Gulf that were stalled by the moratorium, from the environmental reviews Kieran Suckling had helped to make so well known and which the administration had vowed to conduct on all new requests for drilling permits.[108]

The new rules imposed by Salazar included some of the changes that had been sought through legislation. All the new rules are both good and necessary, such as the requirement for maintaining a remotely operated vehicle (ROV) and having a trained ROV crew on each floating drilling rig. The rules also include the new safety and environmental measure sought by the Interior Department in June 2009 (discussed in chapter 6) but previously blocked by the industry. Salazar also sought new funding from Congress for more personnel and increased training for offshore oil rig inspectors. Thus far, Congress has not appropriated the money.

Professor Bea called the new regulations a step in the right direction. But he warns that "there are major systemic problems embedded in our current situation" in offshore drilling. "Thus, one set of guidelines or regulations will not 'solve' all of the major systemic problems," he says. "Guidelines and regulations must be translated to effective practice by the industry. They must also be enforceable and enforced by the regulatory and legislative responsibilities."

Thus far, neither the industry nor the government's track record is good on either count. Moreover, while the new rules and staff (if they are hired and trained) will undoubtedly make offshore drilling *safer*, the question is, can it ever be *safe*?

BP's Future

The BP president said yesterday that the company would survive. That's like someone running over your dog and saying, "Don't worry, my car is fine."

—Jimmy Fallon

Within days of the explosion of the *Deepwater Horizon*, a reporter asked if I thought BP would survive, as if that were ever a serious question. One of the most important issues to arise for me personally from this disaster is an even clearer sense that the public lacks a true appreciation for the sheer magnitude of the wealth and influence of the world's major oil companies. They are larger than any corporations have ever been in history. The largest, ExxonMobil, only became the biggest and wealthiest corporation in the world after the *Valdez* disaster when the Clinton administration allowed Exxon to merge with Mobil, permitting the two largest pieces from the breakup of the Standard Oil Company to be brought back together after a nearly hundred-year hiatus. ExxonMobil earned the highest profits of any corporation in world history in 2003 and then topped those profits every subsequent year until 2009.

Exxon walked away virtually unscathed from the *Valdez* disaster. In 1990, just under a year after the spill, a federal grand jury indicted Exxon on five criminal violations, with potential penalties totaling $5 billion. The Justice Department of George H. W. Bush then unveiled a plea bargain on the eve of the second anniversary of the spill, giving Exxon a get-out-of-jail-free card: Exxon pleaded guilty to just three counts and agreed to a fine of a mere $25 million, or less than 1 percent of the total potential criminal fine, plus $900 million in civil fines to be paid over a 10-year period. Exxon was made to spend an additional mere $2.1 billion on cleanup, recovering just 14 percent of the spilled oil. Two hundred million dollars was also paid to fishermen for one summer's missed catch. In total, Exxon paid $3.4 billion, an amount equivalent to approximately 68 percent of the company's 1989 profits. As discussed earlier, although a 1994 court ruling on the civil suits against Exxon imposed a $5 billion judgment against the company, the U.S. Supreme Court reduced the damages to just $507.5 million in 2008. Exxon's total *Valdez* payouts were therefore less than $3.5 billion.

"We got really screwed," Alaska fisherman Doug Pettit, fifty-eight and a former Marine, told the *New York Daily News* in June 2010 while breaking down in tears. "Twenty years later, it's still a big mess. I just wanted to fish, like my father and grandfather did."[109]

Initially, there were takeover rumors around BP following the spill. ExxonMobil and Chevron were discussed as potential purchasers. Any corporation that purchased BP, however, would also take on its debts and would have to pay out all judgments against BP. These could yet be very large.

In December, U.S. attorney general Eric Holder announced that the

Department of Justice had filed a civil lawsuit against BP, Anadarko, Triton, Transocean, and QBE Underwriting Ltd./Lloyd's Syndicate 1036. Oddly, the list does not include Halliburton. The government is charging the companies with violations under the Oil Pollution Act, including: "Failure to take necessary precautions to secure the Macondo Well prior to the April 20th explosion; Failure to utilize the safest drilling technology to monitor the well's condition; Failure to maintain continuous surveillance of the well; and Failure to utilize and maintain equipment and materials that were available and necessary to ensure the safety and protection of personnel, property, natural resources, and the environment."

Holder said that the Justice Department is also seeking civil penalties under the Clean Water Act for "the unauthorized discharge of oil into the nation's waters," keeping open the possibility of charging the companies for either simple fault or willful negligence. Holder added that while the civil lawsuits "marks a critical step forward, it is not a final step," as both the criminal and civil investigations are ongoing.[110]

In December, BP surprised no one when it submitted a letter to Justice, NOAA, and the National Oil Spill Commission, arguing that the government overstated the size of the spill by as much as 50 percent. If successful, that would cut the government's potential judgment under the Clean Water Act by half.

Nonetheless, estimates of BP's potential legal liability have eclipsed $60 billion.[111] This is separate from BP's expenses for claims, cleanup, shutting down the well, and so on. Thus, the total price tag could be well over $100 billion.

BP could yet, of course, settle the charges with the federal government and the many other legal challenges against it by individuals, businesses, local and state governments, and even the government of Mexico, and ride out those it does not settle, as did Exxon.

To address its economic losses, BP announced in July that it would divest up to $30 billion of assets by the end of 2011. To accomplish this goal it has already sold off some $22 billion of its far-flung corporate assets, including those in Pakistan, South Africa, Argentina, Vietnam, and Venezuela (although it essentially sold the latter two to itself by selling them to TNK-BP, its Russian partnership). The major oil companies have thus far stayed away from these assets, and most of the sales have been outside the United States, with a few exceptions. BP sold interests in four of its Gulf of Mexico deepwater operations to Marubeni Oil and Gas, a Japanese company. The only sale to a U.S. company was to the relatively small Houston-based Apache oil company and included assets in Texas, western Canada, and Egypt.

At the same time, BP continued to gain new leases and production opportunities, including seven new exploration licenses in the UK North Sea, such that, by November 2010, the company reported that it was once again in the black, earning third-quarter profits of $1.8 billion. If BP does settle the charges against it and governments around the world continue to support its operations, its solid economic fate is all but guaranteed.

BP's economic fate would be threatened if the federal government withdrew its U.S. leases, as its U.S. operations account for more than 50 percent of BP's oil and gas reserves. Legislation in Congress has sought to do this. BP has demonstrated that it is not to be trusted and that it will put cost savings and risk above safety, the environment, and public health. It may yet be determined, moreover, that BP acted with criminal negligence in the Gulf. It would certainly be in the public's interest, therefore, to deny BP access to these operations. The U.S. government could revoke BP's leases, including all those in the Gulf of Mexico, sell them, and use the money to pay claims and judgments against BP. Moreover, given the history of oil companies, BP would likely still survive such a move, given that at least 50 percent of its reserves would remain intact.

The problem, however, goes much beyond BP, as I do not believe that we would be in much better hands if Chevron, Exxon, or Shell, for example, purchased BP's U.S. holdings. Thus, we must look to broader change.

While Exxon emerged economically unscathed from the *Valdez* disaster, a powerful citizens' movement rose up in the wake of the disaster that achieved several policy changes, including passage of the Oil Pollution Act. An even more powerful movement emerged in 1969 in the wake of the Union Oil blowout in California, which ultimately yielded a full moratorium on new offshore drilling for much of the United States. The question is whether we have it in us to repeat, and even surpass, these events.

A Better Future for All

On the six-month anniversary of the explosion of the *Deepwater Horizon*, Casi Callaway of Mobile Baykeeper wrote in her blog, "I've put off writing this article" hoping "that the oil would actually be gone. But it's not." As oil continued to wash up on Alabama's beaches, "there are days when you can smell it on the beach or even in downtown Mobile (some 30 miles north of the Gulf)," she wrote. "So to those of you who ask if things have settled down now that the oil is 'gone,' please forgive me when my response is so passionate. We want our lives back! But we can't wish the oil away." Of the impact on her personally, Callaway wrote, "The last six months stretched the limits of family, friendships, health and certainly my knowledge" but "the reason I think we have been able to get out of bed in the morning and tackle this overwhelming challenge again and again is because we are inspired by the possibility for real change. We can use this disaster to see that, just as there are consequences to our more wasteful actions—oil consumption and irresponsible industrial oversight—there can be great value in our collective work to restore Coastal Alabama."

John Hocevar, the Oceans Campaigns director of Greenpeace, has been fighting for the protection of our oceans for twenty years without gaining much broad attention from the public. "The whole world is paying attention now," he tells me. "And the question is, what can you make out of that? Is there something positive that can come out of it? Can you build on the momentum it creates? Even though the administration and BP are clearly hoping this all fades from memory as quickly as possible, there is still a policy window where we can get something done that we couldn't earlier."

There are many challenges to be overcome. One is that the general public and activists alike have largely ignored oversight of the oil industry, either finding it too complex or too big to tackle. But, as Professor Bea colorfully explains of the oil industry, "They are cowboys. When they don't have adequate parentage, by which I mean the government and the public, they just get spoiled and run wild. The lack of parentage has been an extremely important part of this story. We turned our back collectively, which is part of why Macondo became such a catastrophe. I hope we don't do it again."

Or, as Stan Jones of Alaska, now a director with the Prince William Sound Regional Citizens' Advisory Council (another post-*Valdez* victory), has said even more simply, "You absolutely cannot trust government and industry when they are out of your sight."

The problem, however, is that even when they are in our sight, neither the government nor the public has the knowledge to regulate the industry, while the industry clearly lacks the ability to regulate itself. Even if the government gained the knowledge, moreover, if it lacks the will to do so, due to the enormous financial strength of the industry, than should it even be considered "capable"? And, if an industry that carries this much risk is beyond our capacity to regulate, should it be allowed to continue in these operations? I firmly believe the answer is no.

If an industry cannot operate within the bounds of basic human rights and ecological survivability, its risks and costs are too great. Offshore oil drilling is an operation we can live without, and one which far too many cannot live with. We need to move past it and beyond it, as we move past and beyond oil as a resource altogether.

As this book has demonstrated throughout, there are people, communities, activists, organizations, policy makers, and leaders ready to help us fight for new policies, better regulation, and meaningful alternatives. They need our attention, support, motivation, creativity, and participation.

We will only be secure in our energy needs when we are no longer reliant on the oil industry to provide them. We have already begun to make the transition. We now need public policies that help us get there faster. At a time when the United States faces its highest and most prolonged unemployment since the Great Depression, we need a national jobs program that puts people to work building a public transportation system even grander than our national highways. Rather than subsidize the oil industry to the tune of nearly $40 billion a year, we need to let those companies operate on their own steam and instead turn this money over to help both ourselves and our international allies transition to alternative energy sources. We need to sign international agreements that force everyone collectively to move off of dirty energy sources, including oil, so we not only stand a chance of saving our planet, but we also reduce our need to fight over the last of this ever-dwindling non-renewable resource.

As we transition off of oil, we need to pay much closer attention to the actions of the oil industry wherever it operates and on its political activities as it does so. We cannot wean ourselves off of oil overnight. Thus, we must better regulate the industry, restrict where it can operate, and under what conditions. As my colleagues at Oil Change International have proclaimed for years, we must demand a "separation of oil and state" and force our elected officials to wean themselves off of oil's money and influence. We must remain ever vigilant in this practice. We need to support those communities that live on the front lines of oil operations and heighten their voices so that it is not only oiled birds that move us to action, but the people who daily sacrifice their lives and livelihoods at the doorstep of our collective addiction.

As I write I cannot help but think of the children I encountered along this journey and the sacrifices they have been forced to make. Jamie Billiot's daughter, who may or may not be able to call her mother's home her own. Kieran Suckling's children forced to withstand their parents' divorce as "the emotional intensity and spectacular work load of the Gulf crisis pushed our relationship over the edge." Kindra Arnesen's children made ill by oil and dispersants. Kermit Duck's children suffering as their father remains unemployed, unable to fish as did his father, and his father before him. Casi Callaway's son hoping his mother has protected the beach, and Mandy Joye's daughter hoping her mother has saved the ocean.

Keith Jones's grandson spoke his first words the other day. The first question he can summon up because he's finally developed the language skills to express what is in his mind, what he's been wanting to ask all along, is, "Where is my daddy?"

Acknowledgments

This book was a monumental undertaking, written in a relatively minuscule amount of time. It would not have been possible without the incredible generosity of thousands of people who opened their lives, hearts, homes, and homelands to me. This is their story. It is also the accomplishment of an incredible team of which I am proud to be a part.

The book is based simultaneously on observation, lived experiences, and deep investigation. As I traversed beaches, bayous, and the halls of Congress, research fellow Amit Srivastava sat at his computer, never more than an e-mail away, providing the material I needed to ask the right questions, head in the right direction (often literally), and reach the truth. His research provides the backbone on which this book is built.

Legal intern Lindsey Ingraham broke down, with clarity and meticulous accuracy, the complexities of hundreds of lawsuits from workers, states and the federal government, and thousands of pages of Freedom of Information Act requests.

I also could not have written this book without the extreme dedication and great skill of research assistant Saif Rahman, the incredibly talented and devoted transcription powerhouses Crystal Redford and Erika Steeves, and the journalistic skills of Celia Alario, who conducted vital interviews.

My further gratitude to:

Diana Finch, my agent and friend, who also acts, when needed, as my editor, researcher, cheerleader, and coach. Diana has guided my writing career every incredible step of the way.

Editor Stephen Power, whose vision of and belief in this project pushed us all the way through (incredibly!) to the end, and the entire fabulous team at John Wiley & Sons, Inc., especially Kimberly Monroe-Hill and Ellen Wright.

My colleagues and friends at Global Exchange, who supported me in every possible way throughout this project—in particular, Energy Program associate T. J. Buonomo, who not only held down the office but also provided critical work on the book.

Sheri Revette and Keith and Sandra Jones, for your honesty, kindness, generosity, and trust. Your love and strength for your families are an inspiration and a guide. Keith, Gordon's memory is honored with every step you've taken in justice's name.

And the following people whose experiences fill these pages and whose support made this book possible: Samantha Joye, Christof Meile, and Sophie Joye Meile; Casi, Jarrett, and Coleman Callaway; Jamie, James, and Camille Billiot; Vinh Tran, Rot Thi Lam, and Dom Van Tran; David Pham, Boat People SOS; Ta Taponpanh; Chanty Prak; Sharon and David Gauthe, Bayou Interfaith Shared Community Organization; Wilma Subra, Subra Associates and Louisiana Environmental Action Network; the entire staff of Mobile Baykeeper, especially Tammy Herrington; Annie Ducmanis and LaTosha Brown, Gulf Coast Fund; Darryl Malek-Wiley, Sierra Club; Riki Ott; Lise Marie Jacobs; Derrick Evans and Karen Savage, Turkey Creek Community Initiatives; Kermit and Stephanie Duck; Chuck Frelich; Sean McQuiddy; Sherrill Boega; Rocky Kistner and Bob Deans, Natural Resources Defense Council; Dr. Joseph Montoya; Dr. Robert Bea; Pat Carrigan; Misty Ellis; Chief Michael Amos, Atakapa-Ishak Nation; Thomas Dardar Jr., United Houma Nation; Aaron Viles and Jason Henderson, Gulf Restoration Network; Steve Shepard, Ryan Vandermeulen, and everyone with Gulf Coast Sierra Club; John Wathen, Hurricane Creekkeeper; Tracy Kuhns, Louisiana Bayoukeeper; Zach Carter, Mobile South Bay Community Alliance; Raymond, Myra, and Jayson Delco; Kieran Suckling and Sarah Bergman, Center for Biological Diversity; Anna Aurilio, Environment

America; John Hocevar and Phil Radford, Greenpeace; Eric Picah and Nick Berning, Friends of the Earth; Diane Wilson, Medea Benjamin, Rae Abileah, Dana Balicki, and all the women of Code Pink Women for Peace; Tyson Slocum, Public Citizen; Sandy Cioffi; Greg Westhoff; Craig Stratton; Susanne Richter, Volker Barth, and Gernot Bayer, Anthro Media; Josh Gilbert; Bryan and Juan Parras and Liana Lopez, TEJAS; John Parras; Stan Jones; Lance Ryberg and Caroline Douglas, Southwings; Dean Blanchard; Clint Guidry, Louisiana Shrimp Association; Dr. Michael Ilana Freedhoff and Jeff Duncan, office of Congressman Edward Markey; Erin Allweiss, office of Congressman Earl Blumenauer; Sam Ricketts, office of Congressman Jay Inslee; Michael Shank and Gloria Chan; office of Congressman Michael Honda; Mayor Ron Davis; Dahr Jamail; Amy Goodman; Gwyn Miller, U.S. Career Institute; Rachel Miller, intern; Iraq Veterans Against the War; Jyri Kidwell, who provided daily doses of support; and everyone at Church Street Cafe, who, for the second book in a row, gave me a home away from home at which to write.

Notes

Epigraph

1. Joel Achenbach, "In the Gulf of Mexico, What Went Wrong with the Deepwater Horizon Oil Drilling Rig?", *Washington Post*, May 9, 2010, www.washingtonpost.com/wp-dyn/content/article/2010/05/08/AR2010050803429.html.

1. The Explosion of the *Deepwater Horizon*

1. Stephen Lane Stone, testimony before the House Judiciary Committee, May 27, 2010, http://judiciary.house.gov/hearings/hear_100527.html.
2. Allen Breed and Kevin McGill, "The Real Deal: Survivors Recall the *Deepwater Horizon* Explosion," Associated Press, May 9, 2010, http://blog.al.com/live/2010/05/the_real_deal_survivors_recall.html.
3. This fact is revealed in multiple lines of questioning at the United States Coast Guard/Bureau of Oceans and Environmental Management (USCG/BOEM) Marine Board of Investigation into the Marine Casualty, Explosion, Fire, Pollution, and Sinking of Mobile Offshore Drilling Unit *Deepwater Horizon*, with Loss of Life in the Gulf of Mexico, 21–22 April 2010. See, for example, the questions during the testimonies of Daun Winslow, August 23, 2010, David Young, May 27, 2010, and Captain Curt Kuchta, May 27, 2010, www.deepwaterjointinvestigation.com.
4. E-mail exchange between the author and Professor Robert Bea, head of the *Deepwater Horizon* Study Group (DHSG), January 6, 2011. Findings are the analysis of the DHSG to be included in its not-yet-released final report.

5. Transocean Ltd., U.S. Securities and Exchange Commission, form 10-K, annual report, filed on February 24, 2010.

6. Patrick O'Bryan, testimony at the USCG/BOEM Marine Board of Investigation, August 26, 2010, www.deepwaterinvestigation.com/external/content/document/3043/903591/1/USCGHEARING%2026_Aug_10.pdf.

7. Captain Curt Kuchta, testimony at the USCG/BOEM Marine Board of Investigation, May 27, 2010, www.deepwaterinvestigation.com/external/content/document/3043/670139/1/May%2027%20PDF.pdf.

8. Joel Achenbach, "In the Gulf of Mexico, What Went Wrong with the Deepwater Horizon Oil Drilling Rig?" *Washington Post*, May 9, 2010, www.washingtonpost.com/wpdyn/content/article/2010/05/08/AR2010050803429.html.

9. Daun Winslow, testimony at the USCG/BOEM Marine Board of Investigation, August 23, 2010, www.deepwaterinvestigation.com/external/content/document/3043/903575/1/USCGHEARING%2023_Aug_10.pdf.

10. O'Bryan, testimony.

11. Bob Cavnar, *Disaster on the Horizon* (White River Junction, VT: Chelsea Green, 2010), 33.

12. Michael K. Williams, testimony at the USCG/BOEM Marine Board of Investigation, July 23, 2010, www.deepwaterinvestigation.com/external/content/document/3043/856507/1/7–23–10.pdf.

13. U.S. Department of the Interior, Bureau of Ocean Energy Management, "Loss of Well Control—Statistics and Summaries, 2006–2010," accessed November 5, 2010, www.boemre.gov/incidents/blowouts.htm.

14. David Izon, E. P. Danenberger, and Melinda Mayes, "Absence of Fatalities in Blowouts Encouraging in MMS Study of OCS Incidents, 1992–2006," *Drilling Contractor*, July/August 2007, http://drillingcontractor.org/dcpi/dc-julyaug07/DC_July07_MMSBlowouts.pdf.

15. Geoff Goodman, original analysis of U.S. Department of the Interior, Mineral Management Service, Offshore Minerals Management data, using Gom-Explorer, Global Exchange, May 2010.

16. Brett Clanton, "New Tactic Might Seal Leaking Well Sooner, BP CEO Says," *Houston Chronicle*, May 5, 2010, www.chron.com/disp/story.mpl/business/6991278.html; Robin Pagnamenta, "Macondo Well May Contain 1bn Barrels of Oil—and May Flow for a Decade," *Times of London*, June 19, 2010, www.thetimes.co.uk/tto/news/world/americas/article2563253.ece#.

17. Jeff Sattler, "Pull Your BOP Stack—or Not? A Systematic Method to Making This Multi-Million Dollar Decision," WEST Engineering Services, SPE/IADG Drilling Conference and Exhibition, Amsterdam, March 17–19, 2009, http://s3.amazonaws.com/nytdocs/docs/396/396.pdf.

18. Mark Hafle, testimony at the USCG/BOEM Marine Board of Investigation,

May 28, 2010, www.deepwaterinvestigation.com/external/content/document/3043/670171/1/May%2028%20PDF.pdf.

19. Ben Casselman, "Rig Owner Had Rising Tally of Accidents," *Wall Street Journal*, May 10, 2010, http://online.wsj.com/article/SB10001424052748704307804575234471807539054.html.

20. "Blowout: The Deepwater Horizon Disaster," *60 Minutes*, CBS, May 16, 2010, www.cbsnews.com/stories/2010/05/16/60minutes/main6490197.shtml.

21. Williams, testimony.

22. Dane Schiller, "Relatives Remember the 11 Lost in the Oil Rig Blast," *Houston Chronicle*, May 24, 2010, www.chron.com/disp/story.mpl/business/deepwater-horizon/7020616.html.

23. Natalie Roshto, testimony at the USCG/BOEM Marine Board of Investigation, July 22, 2010, www.deepwaterinvestigation.com/external/content/document/3043/856503/1/7–22–10.pdf.

24. Paul Schneider, "The Well from Hell," *Men's Journal*, August 6, 2010, www.mensjournal.com/deepwater-horizon.

25. Tom Junod, "Eleven Lives," *Esquire*, August 16, 2010, www.esquire.com/features/Gulf-oil-spill-lives-0910.

26. Brian P. Morel, e-mail to Richard A. Miller and Mark E. Hafle, April 14, 2010, http://democrats.energycommerce.house.gov/documents/20100614/BP-April14.Email.calling.Macondo.a.nightmare.well.pdf.

27. Alexander John Guide, testimony at the USCG/BOEM Marine Board of Investigation, July 22, 2010, www.deepwaterinvestigation.com/external/content/document/3043/856503/1/7–22–10.pdf.

28. David Hammer, "5 Key Human Errors, Colossal Mechanical Failure Led to Fatal Gulf Oil Rig Blowout," *Times-Picayune*, September 5, 2010, www.nola.com/news/Gulf-oil-spill/index.ssf/2010/09/5_key_human_errors_colossal_me.html.

29. Guide, testimony.

30. Winslow, testimony.

31. Paul Parsons, "The Macondo Well," Energy Training Resources, July 15, 2010, www.EnergyTrainingResources.com.

32. Fred H. Bartlit Jr., letter to the National Commission on the BP Deepwater Horizon Oil Spill and Offshore Drilling, October 28, 2010.

33. Douglas Brown, testimony at the USCG/BOEM Marine Board of Investigation, May 26, 2010, www.deepwaterinvestigation.com/external/content/document/3043/670067/1/May%2026%20PDF.pdf.

34. Hammer, "5 Key Human Errors."

35. Brown, testimony.

36. Junod, "Eleven Lives."

37. Hammer, "5 Key Human Errors."

38. Ibid.

39. Ibid.

40. Guide, testimony.

41. Parsons, "The Macondo Well."

42. Miles Ezell, testimony at the USCG/BOEM Marine Board of Investigation, May 28, 2010, www.deepwaterinvestigation.com/external/content/document/3043/670171/1/May%2028%20PDF.pdf.

43. Ibid.

44. Parsons, "The Macondo Well."

45. Williams, testimony.

46. Winslow, testimony.

47. Brown, testimony,

48. David Sims, testimony at the USCG/BOEM Marine Board of Investigation, August 26, 2010, www.deepwaterinvestigation.com/external/content/document/3043/903591/1/USCGHEARING%2026_Aug_10.pdf.

49. Alwin Landry, testimony at the USCG/BOEM Marine Board of Investigation, May 11, 2010, www.deepwaterinvestigation.com/external/content/document/3043/621903/1/Deepwater%20Horizon%20Joint%20Investigation%20Transcript%20-%20May%2011,%202010.pdf.

50. Bill Lodge, "Joyous Anticipation Turns to Tragedy," Advocate, May 8, 2010, www.2theadvocate.com/news/93175714.html?showAll=y&c=y.

51. Junod, "Eleven Lives."

52. Anna Tinsley, "Two Texas Families Mourn Deaths of Rig Workers," Star Telegram, May 22, 2010, www.star-telegram.com/2010/05/21/2208529/two-texas-families-mourn-deaths.html.

53. Young's Funeral Home, "In Memory of Roy Wyatt Kemp," accessed November 5, 2010, http://youngsfuneral.frontrunnerpro.com/runtime/17947/runtime.php?SiteId=17947&NavigatorId=113940&op=tributeObituary&viewOpt=dpaneOnly&ItemId=459376.

54. Junod, "Eleven Lives."

55. Alberto Cervantes, Jovi Juan, and Susan McGregar, "The Final Moments," Wall Street Journal, May 27, 2010, http://online.wsj.com/article/SB10001424052748704113504575264721101985024.html#articleTabs_interactive%3D%26articleTabs%3Dinteractive; Junod, "Eleven Lives."

56. Tinsley, "Two Texas Families."

57. Scott Bronstein and Wayne Drash, "The Night the Oil Rig Exploded: Escape from a Fire-Breathing Monster," CNN, June 18, 2010, http://articles.cnn.com/2010–06–18/us/oil.rig.explosion_1_rig-drill-fight-fires?_s=PM:US.

58. Joseph Shapiro, "Rig Blast Survivor: 'I Thought I Was Going to Die,'" Morning Edition, NPR, May 10, 2010, www.npr.org/templates/transcript/transcript.php?storyId=126650691.

59. Breed and McGill, "The Real Deal."

60. BP Exploration and Production, "Initial Exploration Plan, Mississippi Canyon Block 252, OSC-G 32306," February 2009, www.gomr.boemre.gov/PI/PDF Images/PLANS/29/29977.pdf.

61. Brown, testimony.

62. Chad Murray, testimony at the USCG/BOEM Marine Board of Investigation, May 27, 2010, www.deepwaterinvestigation.com/external/content/document/3043/670139/1/May%2027%20PDF.pdf.

63. Jimmy W. Harrell, testimony at the USCG/BOEM Marine Board of Investigation, May 27, 2010, www.deepwaterinvestigation.com/external/content/document/3043/670139/1/May%2027%20PDF.pdf.

64. Douglas Blackmon, Vanessa O'Connell, Alexandra Berzon, and Ana Campoy, "There Was 'Nobody in Charge,'" *Wall Street Journal*, May 27, 2010, http://online.wsj.com/article/SB10001424052748704113504575264721101985024.html.

65. Breed and McGill, "The Real Deal."

66. Stephen Bertone, testimony at the USCG/BOEM Marine Board of Investigation, July 19, 2010, www.deepwaterinvestigation.com/external/content/document/3043/856483/1/7–19–10.pdf.

67. Ezell, testimony.

68. Laurel Brubaker Calkins, "Rig Survivor Blames BP's 'Screwed-Up Plan' for Gulf Oil Blowout," Bloomberg, August 23, 2010, www.bloomberg.com/news/2010–08–23/bp-s-flawed-oil-well-design-caused-blowout-transocean-rig-supervisor-says.html.

69. Ezell, testimony.

70. Brown, testimony.

71. Bertone, testimony.

72. Blackmon et al., "There Was 'Nobody in Charge.'"

73. Bertone, testimony.

74. Ibid.

75. Hammer, "5 Key Human Errors."

76. Parsons, "The Macondo Well."

77. Ronnie W. Sepulvado, testimony at the USCG/BOEM Marine Board of Investigation, July 20, 2010, www.deepwaterinvestigation.com/external/content/document/3043/856499/1/7–20–10.pdf.

78. Guide, testimony.

79. "Blowout," *60 Minutes*.

80. BP Exploration and Production, "Deepwater Horizon Accident Investigation Report," September 8, 2010, www.bp.com/liveassets/bp_internet/globalbp/globalbp_uk_english/incident_response/STAGING/local_assets/downloads_pdfs/Deepwater_Horizon_Accident_Investigation_Report.pdf.

81. See the testimonies of Yancy Keplinger (October 5, 2010), Stephen Bertone

(July 19, 2010), and Chad Murray (May 27, 2010) at the USCG/BOEM Marine Board of Investigation, www.deepwaterinvestigation.com.

82. Blackmon et al., "There Was 'Nobody in Charge.'"

83. Brown, testimony.

84. Micah Joseph Sandell, testimony at the USCG/BOEM Marine Board of Investigation, May 29, 2010, www.deepwaterinvestigation.com/external/content/document/3043/856499/1/7–20–10.pdf.

85. Gregory Meche, testimony at the USCG/BOEM Marine Board of Investigation, May 28, 2010, www.deepwaterinvestigation.com/external/content/document/3043/670171/1/May%2028%20PDF.pdf.

86. Shapiro, "Rig Blast Survivor."

87. Calkins, "Rig Survivor Blames BP's 'Screwed-Up Plan.'"

88. Bertone, testimony.

89. Ezell, testimony.

90. "Blowout," *60 Minutes.*

91. Blackmon et al., "There Was 'Nobody in Charge.'"

92. Bertone, testimony.

93. "Blowout," *60 Minutes.*

94. Breed and McGill, "The Real Deal."

2. What You Can't See Can Kill You: The BP Macondo Oil Monster Escapes

1. Alwin Landry, testimony at the USCG/BOEM Marine Board of Investigation, May 11, 2010, www.deepwaterinvestigation.com/external/content/document/3043/621903/1/Deepwater%20Horizon%20Joint%20Investigation%20Transcript%20-%20May%2011,%202010.pdf.

2. Joseph Shapiro, "Rig Blast Survivor: 'I Thought I Was Going to Die,'" *Morning Edition*, NPR, May 10, 2010, www.npr.org/templates/story/story.php?storyId=126650691.

3. Transcript, CNN Newsroom, April 23, 2010, www-cgi.cnn.com/TRANSCRIPTS/1004/23/cnr.03.html.

4. Ian Urbina and Justin Gillis, "Workers on Oil Rig Recall a Terrible Night of Blasts," *New York Times*, May 7, 2010, www.nytimes.com/2010/05/08/us/08rig.html?_r=1&pagewanted=1.

5. Josh Harkinson, "The Rig's on Fire! I Told You This Was Gonna Happen!" *Mother Jones*, June 7, 2010, http://motherjones.com/blue-marble/2010/06/rigs-fire-i-told-you-was-gonna-happen.

6. Joseph Shapiro, "Blast Survivors Kept Isolated on Gulf for Hours," *All Things Considered*, NPR, May 10, 2010, www.npr.org/templates/story/story.php?storyId=126667241.

7. U.S. Coast Guard Log, "Case History Report—April 20–May 5, 2010," Center

for Public Integrity, May 5, 2010, www.publicintegrity.org/documents/entry/
2124/?docID=1; Landry, testimony.

8. Barbara Wilk, testimony before the USCG/BOEM Marine Board of Investigation, May 12, 2010, www.deepwaterinvestigation.com/external/content/
document/3043/621931/1/Deepwater%20Horizon%20Joint%20Investiga
tion%20Transcript%20-%20May%2012,%202010.pdf.

9. Urbina and Gillis, "Workers on Oil Rig."

10. Joseph Shapiro, "Rig Survivors Coerced to Sign Waivers," *All Things Considered*, NPR, May 6, 2010, www.npr.org/templates/story/story.php?storyId=
126565283.

11. Kevin Robb, testimony at the USCG/BOEM Marine Board of Investigation,
May 11, 2010, www.deepwaterinvestigation.com/external/content/document/
3043/621903/1/Deepwater%20Horizon%20Joint%20Investigation%20Tran
script%20-%20May%2011,%202010.pdf; Aaron Mehta and John Solomon,
"Haphazard Firefighting Might Have Sunk BP Oil Rig," Center for Public
Integrity, July 27, 2010, www.publicintegrity.org/articles/entry/2286.

12. Mehta and Solomon, "Haphazard Firefighting."

13. Robb, testimony.

14. Ibid.

15. Mehta and Solomon, "Haphazard Firefighting."

16. David Barstow, Laura Dodd, James Glanz, Stephanie Saul, and Ian Urbina,
"Regulators Failed to Address Risks in Oil Rig Fail-Safe Device," *New York
Times*, June 20, 2010, www.nytimes.com/2010/06/21/us/21blowout.html.

17. U.S. Coast Guard Log, "Case History Report."

18. Ibid.

19. Barstow et al., "Regulators Failed to Address Risks."

20. Ibid.

21. *Deepwater Horizon* Study Group, "Progress Report 2," July 15, 2010,
http://ccrm.berkeley.edu/pdfs_papers/bea_pdfs/DHSG_July_Report-Final.pdf.

22. For figures on natural gas release, see Samantha B. Joye, Ian R. MacDonald, Ira
Leifer, and Vernon Asper, "Magnitude and Oxidation Potential of Hydrocarbon
Gases Released from the BP Blowout," *Nature Geoscience*, December 10, 2010.

23. BP Exploration and Production, "Initial Exploration Plan, Mississippi Canyon
Block 252, OSC-G 32306," February 2009, www.gomr.boemre.gov/PI/PDF
Images/PLANS/29/29977.pdf.

24. Juliet Eilperin, "U.S. Exempted BP's Gulf of Mexico Drilling from Environmental Impact Study," *Washington Post*, May 5, 2010, www.washington
post.com/wp-dyn/content/article/2010/05/04/AR2010050404118.html.

25. Campbell Robertson, a *New York Times* reporter who attended the press conference, in an e-mail to Saif Rahman, a *Black Tide* research assistant, November 17, 2010.

26. *This Just In*, "Coast Guard: Oil Rig That Exploded Has Sunk," CNN, April 22, 2010, http://news.blogs.cnn.com/2010/04/22/coast-guard-oil-rig-that-exploded-has-sunk.

27. CNN, "Oil Slick Spreads from Sunken Rig," April 22, 2010, www.cnn.com/2010/US/04/22/oil.rig.explosion/index.html?hpt=T1.

28. Robertson, e-mail.

29. Henry Fountain, "Notes from Wake of Blowout Outline Obstacles and Frustration," *New York Times*, June 21, 2010, www.nytimes.com/2010/06/22/science/earth/22blowout.html.

30. John Solomon and Aaron Mehta, "Coast Guard Logs Reveal Early Spill Estimate of 8,000 Barrels a Day," Center for Public Integrity, June 3, 2010, www.publicintegrity.org/articles/entry/2123/?utm_source=publicintegrity&utm_medium=related_heds&utm_campaign=side_v1.

31. National Commission on the BP Deepwater Horizon Oil Spill and Offshore Drilling, "Draft—the Amount and Fate of the Oil," Staff Working Paper No. 3, October 6, 2010, www.oilspillcommission.gov/sites/default/files/documents/Amount%20and%20Fate%20of%20the%20Oil%20Working%20Paper%2010%206%2010.pdf.

32. "Coast Guard: Oil Not Leaking from Sunken Rig," CBS News, April 23, 2010, www.cbsnews.com/video/watch/?id=6424647n.

33. WDSU, "RAW: Interview with Rear Adm. Mary Landry," New Orleans, April 23, 2010, www.wdsu.com/r-video/23247455/detail.html.

34. Fountain, "Notes from Wake of Blowout."

35. National Commission, "The Amount and Fate of the Oil."

36. Dahr Jamail, "Despite Heavy Oil, Louisiana Keeps Fisheries Open," Inter Press Service, October 26, 2010, http://ipsnews.net/news.asp?idnews=53297.

37. National Commission, "The Amount and Fate of the Oil."

38. Edward J. Markey, "On Flow Rate, Ask What Did BP Know and When Did They Know It," letter to BP Commission, September 28, 2010, http://globalwarming.house.gov/files/LTTR/2010–09–28_timelineDocuments.pdf.

39. Campbell Robertson and Leslie Kaufman, "Size of Spill in Gulf of Mexico Is Larger Than Thought," *New York Times*, April 28, 2010, www.nytimes.com/2010/04/29/us/29spill.html.

40. Adam Gabbatt, "Gulf Oil Spill 'Five Times' Larger Than Estimated," *Guardian*, April 29, 2010, www.guardian.co.uk/world/2010/apr/29/Gulf-oil-spill-larger-estimated.

41. National Commission, "The Amount and Fate of the Oil."

42. Ibid.

43. Barstow et al., "Regulators Failed to Address Risks."

44. Craig Guillot, "Pictures: Gulf Oil Spill Hits Land—and Wildlife," *National Geographic*, April 30, 2010, http://news.nationalgeographic.com/news/2010/04/

photogalleries/100430-oil-spill-Gulf-mexico-pictures/; *NewsHour*, PBS, May 1, 2010, www.pbs.org/newshour/multimedia/Gulfoil/?photos.

45. Ned Potter, Ryan Owens, and Bradley Blackburn, "Gulf of Mexico Oil Hits Coast," ABC News, April 29, 2010.

46. BP, "BP Steps Up Shoreline Protection Plans on US Gulf Coast," press release, April 29, 2010, www.bp.com/genericarticle.do?categoryId=2012968&con tentId=7061565.

47. "President Obama Honors 2010 National Teacher of the Year," remarks by the president in the Rose Garden, White House, April 29, 2010, www.whitehouse .gov/the-press-office/remarks-president-rose-garden.

48. Anna Fifield and Ed Crooks, "BP Will Pay Oil Spill Claims," *Financial Times*, April 30, 2010, www.ft.com/cms/s/0/5eebc5da-546c-11df-8bef-00144feab49a .html#axzz1AgDoj72s.

49. Fareed Zakaria, "Presidential Pony Show," *Newsweek*, June 13, 2010, www .newsweek.com/2010/06/13/presidential-pony-show.html.

50. National Commission on the BP Deepwater Horizon Oil Spill and Offshore Drilling, "Draft—Decision-Making Within the Unified Command," Staff Working Paper No. 2, October 6, 2010, www.oilspillcommission.gov/sites/ default/files/documents/Unified%20Command%20Working%20Paper%2010% 206%2010.pdf.

51. Ibid.

52. Ibid.

53. Ibid.

54. Ibid.

55. Suzanne Goldenberg, "*Deepwater Horizon* Oil Spill: Underwater Robots Trying to Seal Well," *Guardian*, April 26, 2010, www.guardian.co.uk/environment/ 2010/apr/26/deepwater-horizon-spill-underwater-robots; David Byrd, "Long-Term Effects Feared After Gulf of Mexico Oil Rig Explosion," Voice of America, April 27, 2010, www.voanews.com/english/news/usa/Long-Term-Effects-Feared-AfterGulf-of-Mexico-Oil-Rig-Explosion—92213014.html.

56. Byrd, "Long-Term Effects Feared."

57. John Amos, "Gulf Oil Spill—New Spill Calculation—Exxon Valdez Surpassed Today," SkyTruth, May 1, 2010, http://blog.skytruth.org/2010/05/gulf-oil-spill-new-spill-rate.html.

58. "Admiral: Impossible to Estimate Size of Oil Leak," *Newsday*, May 1, 2010, www.newsday.com/news/admiral-impossible-to-estimate-size-of-oil-leak-1 .1891150.

59. Justin Gillis, "Doubts Are Raised on Accuracy of Government's Spill Estimate," *New York Times*, May 14, 2010, http://query.nytimes.com/gst/fullpage.html?res= 9E05EED6143DF937A25756C0A9669D8B63.

60. National Commission, "The Amount and Fate of the Oil."

61. BP, "Work Begins to Drill Relief Well to Stop Oil Spill," press release, May 4, 2010, www.bp.com/genericarticle.do?categoryId=2012968&contentId=7061778.

62. John M. Broder, Campbell Robertson, and Clifford Krauss, "Amount of Spill Could Escalate, Company Admits," *New York Times*, May 4, 2010, www.nytimes.com/2010/05/05/us/05spill.html.

63. National Commission, "The Amount and Fate of the Oil."

64. Gillis, "Doubts Are Raised."

65. Guy Raz, "Scientists Find Huge Oil Plumes in Gulf," *All Things Considered*, NPR, May 16, 2010, www.npr.org/templates/story/story.php?storyId=126870185.

66. Committee on Oil in the Sea: Inputs, Fates, and Effects, National Research Council, "Oil in the Sea III: Inputs, Fates, and Effects," *National Academies Press*, 2003, www.nap.edu/catalog.php?record_id=10388#orgs; Oistein Johansen and Cortis Cooper, "Project 'Deep Spill,'" Minerals Management Service, March 24, 2005, www.boemre.gov/tarprojects/377.htm.

67. Senate Committee on Environment and Public Works, "BP Agrees to Comply with Boxer and Nelson Request to Turn Over Oil Spill Videos," press release, May 18, 2010, http://epw.senate.gov/public/index.cfm?FuseAction=Press Room.PressReleases&ContentRecord_id=ac4455bd-802a-23ad-4e41–31d53b47bfd0&Designation=Majority.

68. "Gulf Spill May Far Exceed Official Estimates," *Morning Edition*, NPR, May 14, 2010, www.npr.org/templates/transcript/transcript.php?storyId=126809525.

69. National Commission, "The Amount and Fate of the Oil."

70. BP, "Update on Gulf of Mexico Oil Spill Response," press release, May 5, 2010, www.bp.com/genericarticle.do?categoryId=2012968&contentId=7061856.

71. National Commission, "The Amount and Fate of the Oil."

72. "Gulf Spill May Far Exceed Official Estimates."

73. Barstow et al., "Regulators Failed to Address Risks."

74. Justin Gillis, "Size of Oil Spill Underestimated, Scientists Say," *New York Times*, May 13, 2010, www.nytimes.com/2010/05/14/us/14oil.html.

75. ———, "Giant Plumes of Oil Forming Under the Gulf," *New York Times*, May 15, 2010, www.nytimes.com/2010/05/16/us/16oil.html.

76. Edward Markey, "Markey: BP Burying Heads in Sand on Underwater Plumes, Oil Flow," press release, May 16, 2010, http://markey.house.gov/index.php?option=com_content&task=view&id=3982&Itemid=141.

77. Gillis, "Giant Plumes of Oil Forming."

78. Mark Schrope, "*Deepwater Horizon*: A Scientist at the Centre of the Spill," *Nature*, August 4, 2010, www.nature.com/news/2010/100804/full/466680a.html.

79. *Deepwater Horizon* Incident Joint Information Center, "Statement from NOAA Administrator Jane Lubchenco on Ongoing Efforts to Monitor Subsea Impacts of the BP Oil Spill," May 17, 2010, www.restorethegulf.gov/release/2010/05/

17/statement-noaa-administrator-jane-lubchenco-ongoing-efforts-monitor-subsea-impact.

80. National Oceanic and Atmospheric Administration, "Update on NOAA's Oil Spill Research Missions," June 13, 2010, www.noaanews.noaa.gov/stories 2010/20100613_missions.html.

81. Richard Camilli et al., "Tracking Hydrocarbon Plume Transport and Biodegradation at Deepwater Horizon," Science, August 19, 2010, www.sciencemag.org/content/330/6001/201.abstract.

82. "Major Study Charts Long-Lasting Oil Plume in Gulf," Associated Press, August 20, 2010, www.aolnews.com/2010/08/19/major-study-charts-long-lasting-oil-plume-in-gulf.

83. Craig Pittman, "USF Says Government Tried to Squelch Their Oil Plume Findings," St. Petersburg Times, August 10, 2010, www.tampabay.com/news/environment/article1114225.ece.

84. National Commission, "The Amount and Fate of the Oil."

85. The White House Blog, May 19, 2010, www.whitehouse.gov/blog/2010/05/05/ongoing-administration-wide-response-deepwater-bp-oil-spill.

86. National Commission, "The Amount and Fate of the Oil."

87. Ibid.

88. Richard Harris, "Scientist: BP's Oil Spill Estimates Improbable," Morning Edition, NPR, May 20, 2010, www.npr.org/templates/story/story.php?storyId=126975907.

89. Deepwater Horizon Incident Joint Information Center, "U.S. Scientific Teams Refine Estimates of Oil Flow from BP's Well Prior to Capping," press release, August 2, 2010, http://app.restorethegulf.gov/release/2010/08/02/us-scientific-teams-refine-estimates-oil-flow-bps-well-prior-capping.

90. National Commission, "The Amount and Fate of the Oil."

3. Body Count: Oil, Gas, and Dispersant Attack the People, Wildlife, and Wild Places of the Gulf

1. Jason Berry, "BP Storm: Tulane Prof Oliver Houck Warned for Decades of Peril of Lax Energy Regulation," PoliticsDaily.com, June 13, 2010, www.politicsdaily.com/2010/06/13/bp-storm-tulane-prof-oliver-houck-warned-for-decades-of-peril-o.

2. Brenda Dardar Robichaux, "Our Natural Resources at Risk: The Short- and Long-Term Impacts of the Deepwater Horizon Oil Spill," testimony before the Subcommittee on Insular Affairs, Oceans and Wildlife, June 10, 2010, http://resourcescommittee.house.gov/images/Documents/20100610/2010_06_10_oceans/testimony_robichaux.pdf.

3. Agency for Toxic Substances and Disease Registry, "Toxicological Profile for Toluene," September 2000, www.atsdr.cdc.gov/ToxProfiles/tp.asp?id=161&

tid=29; U.S. Department of Labor, Occupational Safety and Health Administration, "Occupational Safety and Health Guideline for Xylene," www.osha.gov/SLTC/healthguidelines/xylene/recognition.html, accessed November 27, 2010.

4. Aubrey Keith Miller, "NIEHS Activities Related to the Gulf Oil Spill," testimony before the Committee on Energy and Commerce, June 16, 2010, www.hhs.gov/asl/testify/2010/06/t20100616d.html.

5. Samantha B. Joye, "Beneath the Surface of the BP Spill: What's Happening Now, What's Needed Next," testimony before the House Committee on Energy and Commerce, Subcommittee on Energy and Environment, June 9, 2010, http://energycommerce.house.gov/documents/20100609/Joye.Statement.06.09.2010.pdf.

6. National Oceanic and Atmospheric Administration, "*Deepwater Horizon* Oil: Characteristics and Concerns," May 15, 2010, http://sero.nmfs.noaa.gov/sf/deepwater_horizon/OilCharacteristics.pdf.

7. Agency for Toxic Substances and Disease Registry, "Toxic Substances Portal: Nickel," www.atsdr.cdc.gov/substances/toxsubstance.asp?toxid=44.

8. Agency for Toxic Substances and Disease Registry, "Toxic Substances Portal: Lead," www.atsdr.cdc.gov/substances/toxsubstance.asp?toxid=22.

9. Ellycia Harrould-Kolieb, Jacqueline Savitz, Jeffrey Short, and Marianne Veach, "Toxic Legacy: Long-Term Effects of Offshore Oil on Wildlife and Public Health," Oceana, www.oceana.org/index.php?id=2894.

10. Joye, "Beneath the Surface of the BP Spill."

11. Ibid.

12. Jane Lubchenco et al., "BP *Deepwater Horizon* Oil Budget: What Happened to the Oil?" National Oceanic and Atmospheric Administration, August 4, 2010, www.noaanews.noaa.gov/stories2010/PDFs/OilBudget_description_%2083final.pdf.

13. National Commission on the BP *Deepwater Horizon* Oil Spill and Offshore Drilling, "Draft Staff Working Paper: Response/Clean-Up Technology Research & Development and the BP *Deepwater Horizon* Oil Spill," November 22, 2010, www.oilspillcommission.gov.

14. Jacqui Goddard, "BP to Try Golf Long Shot to Stop Gulf of Mexico Oil Leak," *Sunday Times*, May 10, 2010, www.timesonline.co.uk/tol/news/world/us_and_americas/article7121185.ece.

15. *Deepwater Horizon* Incident Joint Information Center, "The Ongoing Administration-Wide Response to the Deepwater BP Oil Spill," August 14, 2010, www.restorethegulf.gov/release/2010/08/14/ongoing-administration-wide-response-deepwater-bp-oil-spill.

16. National Commission on the BP *Deepwater Horizon* Oil Spill and Offshore Drilling, "The Use of Surface and Subsea Dispersants during the BP *Deepwater*

Horizon Oil Spill," Staff Working Paper No. 4, October 6, 2010, www.oilspill commission.gov.

17. Antonio Vargas, "Oil Is Leaking from Well at *Deepwater Horizon* Explosion Site," *Times-Picayune*, April 29, 2010, www.nola.com/business/index.ssf/ 2010/04/deepwater_horizon_oil_rig_expl.html.

18. U.S. Coast Guard Log, "Case History Report—April 20–May 5, 2010," Center for Public Integrity, May 5, 2010, www.publicintegrity.org/documents/entry/ 2124/?docID=1.

19. Environmental Protection Agency, "Government Response to the BP Oil Spill," July 2010, www.epa.gov/bpspill/factsheets/dispersants-factsheet.pdf.

20. Anne Casselman, "10 Biggest Oil Spills in History," *Popular Mechanics*, 2010, www.popularmechanics.com/science/energy/coal-oil-gas/biggest-oil-spills-in-history.

21. Edward J. Markey, "Coast Guard Allowed BP, Spill Responders to Excessively Use Dispersants," press release, Office of Congressman Markey, July 31, 2010, http://markey.house.gov/index.php?option=content&task=view&id=4073& Itemid=125.

22. Deepwater Horizon Incident Joint Information Center, "Operations and Ongoing Response," August 28, 2010, www.restorethegulf.gov/release/2010/08/28/ operations-and-ongoing-response-august-28-2010.

23. National Commission, "The Use of Surface and Subsea Dispersants."

24. "Robot Vessels Used to Cap Gulf of Mexico Oil Leak," BBC News, April 26, 2010, http://news.bbc.co.uk/2/hi/8643782.stm.

25. Campbell Robertson, "White House Takes a Bigger Role in the Oil Spill Cleanup," *New York Times*, April 30, 2010, www.nytimes.com/2010/04/ 30/us/30Gulf.html.

26. Jeff Goodell, "The Poisoning: It's the Biggest Environmental Disaster in American History—and BP Is Making It Worse," *Rolling Stone*, July 21, 2010, www.rollingstone.com/politics/news/17390/183349?RS_show_page=3.

27. National Commission, "The Use of Surface and Subsea Dispersants."

28. Nalco Company, "Safety Data Sheet Corexit EC9527A," May 11, 2010, www.restorethegulf.gov/sites/default/files/imported_pdfs/external/content/ document/2931/539295/1/Corexit%20EC9527A%20MSDS.pdf.

29. Elana Schor, "Ingredients of Controversial Dispersants Used on Gulf Spill Are Secrets No More," *New York Times*, June 9, 2010, www.nytimes.com/gwire/2010/ 06/09/09greenwire-ingredients-of-controversial-dispersants-used-42891.html; "EPA Whistleblower Accuses Agency of Covering Up Effects of Dispersant in BP Oil Spill Cleanup," Democracy Now!, July 20, 2010, www.democracynow.org/ 2010/7/20/epa_whistleblower_accuses_agency_of_covering.

30. Wilma Subra, "Human Health and Environmental Impacts Associated with the *Deepwater Horizon* Crude Oil Spill Disaster," testimony before the House

Energy and Commerce Committee, Subcommittee on Oversight and Investigations, June 7, 2010, http://democrats.energycommerce.house.gov/docu
ments/20100607/Subra.Testimony.06.07.2010.pdf; "Material Safety Data
Sheet," Fisher Scientific, March 18, 2003, http://fscimage.fishersci.com/msds/
89683.htm.

31. Subra, testimony.

32. Jacqueline Savitz, "Oversight Hearing on the Use of Oil Dispersants in the
Deepwater Horizon Spill," written testimony for the Senate Committee on
Environment and Public Works, August 4, 2010, http://epw.senate.gov/
public/index.cfm?FuseAction=Files.View&FileStore_id=b26e4d69–6472–
4c0c-a0f1–865f71c60303.

33. Barbara A. Mikulski, "Mikulski Chairs Hearing on Environmental Effects of
Dispersants Used in Gulf Oil Spill Clean Up," press release, July 15, 2010,
http://mikulski.senate.gov/media/pressrelease/07–15–10–1.cfm.

34. "Mobile Baykeeper Says BP Spill Could Destroy Most Productive Fishery
in the World," Mobile Baykeeper, April 30, 2010, www.redorbit.com/news/
business/1858098/mobile_baykeeper_says_bp_spill_could_destroy_most_pro
ductive_ fishery.

35. "No More Shrimp," CNN, May 1, 2010, www.cnn.com/video/#/video/us/2010/
05/01/holmes.seafood.oil.damage.intv.cnn?iref=videosearch.

36. Ibid.

37. Goodell, "The Poisoning."

38. Rose Aguilar, "Experts: Health Hazards in Gulf Warrant Evacuations,"
Truthout, July 22, 2010, www.truth-out.org/toxic-dispersants-causing-wide-
spread-illness61604.

39. Kate Sheppard, "Is the EPA Playing Dumb on Dispersants?" *Mother Jones*, July
20, 2010, http://motherjones.com/environment/2010/07/epa-whistleblower-
bp-dispersants.

40. Suzanne Goldenberg, "BP Oil Spill: Obama Administration's Scientists Admit
Alarm Over Chemicals," *Guardian*, August 3, 2010, www.guardian.co.uk/
environment/2010/aug/03/Gulf-oil-spill-chemicals-epa.

41. "EPA Whistleblower Accuses Agency."

42. Ibid.

43. Joye, "Beneath the Surface of the BP Spill."

44. Miles O'Brien and Marsha Walton, "Hidden Oil and Gas Plumes in the Gulf,"
Science Nation, September 20, 2010, www.nsf.gov/news/special reports/
science_nation/hiddenoilplumes.jsp.

45. David C. Smith, testimony before the Committee on Environment and Public
Works Oversight Hearing on the Use of Oil Dispersants in the *Deepwater Horizon*
Oil Spill, August 4, 2010, http://epw.senate.gov/public/index.cfm?FuseAction=
Files.View&FileStore_id=1f9c46fb-8075–4d69-a5ffb7b2d438a1ff.

46. Environmental Protection Agency, "BP Must Use Less Toxic Dispersant," press release, May 20, 2010, http://yosemite.epa.gov/opa/admpress.nsf/d0cf661 8525a9efb85257359003fb69d/0897f55bc6d9a3ba852577290067f67f!Open Document.

47. Lisa P. Jackson, letter to David Rainey of BP, Environmental Protection Agency, May 26, 2010, www.epa.gov/bpspill/dispersants/Rainey-letter-052610.pdf.

48. Environmental Protection Agency, "Dispersant Monitoring and Assessment Directive—Addendum 3," May 26, 2010, www.epa.gov/bpspill/dispersants/directive-addendum3.pdf.

49. Markey, "Coast Guard Allowed BP."

50. Goodell, "The Poisoning."

51. National Commission, "The Use of Surface and Subsea Dispersants."

52. Lawrence Berkeley National Laboratory, "Study Shows Deepwater Oil Plume in Gulf Degraded by Microbes," press release, August 24, 2010, http://newscenter.lbl.gov/news-releases/2010/08/24/deepwater-oil-plume-microbes.

53. Energy Biosciences Institute, "EBI At-A-Glance," 2010, www.energybio sciencesinstitute.org/index.php?option=com_content&task=blogsection&id=2&Itemid=3.

54. Robert Lee Hotz, "New Study Sees Dissipating Oil Spill," Wall Street Journal, August 25, 2010, http://online.wsj.com/article/SB1000142405274870412560 4575449690602451532.html.

55. Lawrence Berkeley National Laboratory, "Study Shows Deepwater Oil Plume in Gulf Degraded."

56. Matt Gutman and Kevin Dolak, "Oil From the BP Spill Found at Bottom of Gulf," ABC World News with Diane Sawyer, September 12, 2010, http://abcnews.go.com/WN/oil-bp-spill-found-bottom-gulf/story?id=11618039.

57. Seth Borenstein, "Where's the Oil? On the Gulf Floor, Scientists Say," Associated Press, September 13, 2010, www.ajc.com/news/nation-world/wheres-the-oil-on-612509.html.

58. "Verbatim," Time, September 27, 2010, www.time.com/time/magazine/article/0,9171,2019632,00.html.

59. Berry Yeoman, "Where the Oil Went? Gulf Scientists Find Layer of 'Slime Snot' on Sea Floor," One Earth, November 4, 2010, www.onearth.org/article/where-the-oil-went-gulf-scientists-find-layer-of-slime-snot-on-sea-floor.

60. "Oil Plume Approaches Mobile Bay," Free Speech Radio News, May 28, 2010, http://fsrn.org/audio/oil-plume-approaches-mobile-bay/6822.

61. Jeremy Peters, "Efforts to Limit the Flow of Spill News," New York Times, June 9, 2010, www.nytimes.com/2010/06/10/us/10access.html?pagewanted=2&r=1.

62. Matthew Philips, "BP's Photo Blockade of the Gulf Oil Spill," Newsweek, May 26, 2010, www.newsweek.com/2010/05/26/the-missing-oil-spill-photos.html.

63. Jeffrey Kofman, *Nightline*, ABC News, June 28, 2010, http://abcnews.go.com/nightline.

64. *Deepwater Horizon* Incident Joint Information Center, "Update 11—Controlled Burn Scheduled to Begin," April 28, 2010, www.restorethegulf.gov/release/2010/04/27/update-11-controlled-burn-scheduled-begin; Environmental Protection Agency, "EPA Releases Reports on Dioxin Emitted During Deepwater Horizon BP Spill," November 11, 2010, www.epa.gov/research/dioxin.

65. *Deepwater Horizon* Incident Joint Information Center, "The Ongoing Administration-Wide Response to the Deepwater BP Oil Spill," September 15, 2010, www.restorethegulf.gov/release/2010/09/15/ongoing-administration-wide-response-deepwater-bp-oil-spill.

66. National Commission, "The Amount and Fate of the Oil."

67. "EPA Whistleblower Accuses Agency."

68. Paul Rioux, "Giant Oil Skimmer 'A Whale' Deemed a Bust for Gulf of Mexico Spill," *Times-Picayune*, July 16, 2010, www.nola.com/news/gulf-oil-spill/index.ssf/2010/07/giant_oil_skimmer_a_whale_deem.html.

69. National Oceanic and Atmospheric Administration, "NOAA Releases Data Report on Air Quality Measurements Near the Deepwater Horizon/BP Oil Spill Area," July 21, 2010, www.noaanews.noaa.gov/stories2010/20100721_p3_oilspill.html.

70. Greg Hall, "WTF!!! Surf on Pensacola Beach Florida Boiling Like Acid!!!.flv," YouTube, June 24, 2010, www.youtube.com/watch?v=BTCs-cmyHSs&feature=related. See also "Bubbling Sh*t in the Water! Pensacola Beach Oil Disaster," YouTube, June 24, 2010, www.youtube.com/watch?v=inP69itgczc&feature=related; Dahr Jamail and Erika Blumenfeld, "Uncovering the Lies That Are Sinking the Oil," Truthout, August 16, 2010, www.truth-out.org/uncovering-lies-that-are-sinking-oil62345; Ed Cake, "Fisherman Documents Underwater Oil Plumes," Save Our Gulf, August 23, 2010, http://saveourgulf.org/updates/fisherman-documents-underwater-oil-plumes; "Bubble, Bubble, Toil and Trouble," YouTube, September 1, 2010, www.youtube.com/watch?v=DciiIQiZcIo.

71. Hall, "WTF!!!"; see also "Bubbling Sh*t in the Water!"

72. Cary Chow, "Gov. Riley Urging People to Visit Coast," Fox 10 TV, June 19, 2010, www.fox10tv.com/dpp/news/gulf_oil_spill/gov-riley-urging-people-to-visit-coast.

73. Jessica Taloney, "WKRG investigates: Testing the Water," WKRG, July 18, 2010, www.wkrg.com/gulf_oil_spill/article/news-5-investigates-testing-the-water/906545/Jul-18-2010_6-40-pm.

74. Louisiana Environmental Action Network, "Fishermen Win Another Round," May 5, 2010, http://leanweb.org/news/latest/fishermen-win-another-round.html.

75. Kindra Arnesen, speech at the Gulf Emergency Summit, June 19, 2010, www.youtube.com/watch?v=jkYJDI8pK9Y.

76. Elizabeth Cohen, "Fisherman's Wife Breaks the Silence," CNN, June 3, 2010, http://articles.cnn.com/2010–06–03/health/Gulf.fishermans.wife_1_ shrimping-exxon-valdez-oil-spill-cell-phone?_s=PM:HEALTH.

77. "Dispute Over Oil Worker Illness; Residents Coping with Major Oil," CNN Newsroom, May 30, 2010, http://edition.cnn.com/TRANSCRIPTS/1005/ 30/cnr.06.html.

78. Arnesen, speech at the Gulf Emergency Summit.

79. Subra, testimony.

80. Janet Kwak, "Mysterious Illness Plagues Gulf Oil Disaster Workers," WOAI, June 15, 2010, www.woai.com/content/health/story/Mysterious-illness-plagues-Gulf-oil-disaster/PNcpQeot20qXs_L5nfSR4w.cspx.

81. Anne Rolfes, "CJS Hearing on Use of Dispersants in Response to the Oil Spill," testimony before the Subcommittee on Commerce, Justice, Science and Related Agencies, July 15, 2010, www.labucketbrigade.org/downloads/ ROLFES%20Testimony%20to%20Appropriations%20Subcommittee%207.15 .10.pdf.

82. Callie Casstevens, "CTEH? Don't Know Them, Actually We Do Use Their Results," Louisiana Bucket Brigade Blog, July 9, 2010, http://labucket brigade.wordpress.com/2010/07/09/%E2%80%9Ccteh-don%E2%80%99t-know-them-actually-we-do-use-their-results%E2%80%9D.

83. Associated Press, "Gulf Oil Spill Workers Complaining of Flu-like Symptoms," June 3, 2010, http://blog.al.com/live/2010/06/gulf_spill_workers_complaining .html.

84. Ibid.

85. Louisiana Department of Health and Hospitals, MS Canyon 252 Oil Spill Surveillance Report, July 4–10, 2010, www.dhh.louisiana.gov/offices/ publications/pubs-378/_OilSpillSurveillance2010_06.pdf.

86. Ibid.

87. Anne Rolfes, "Our Natural Resources at Risk: The Short and Long Term Impacts of the Deepwater Horizon Oil Spill," testimony before the House Natural Resources Subcommittee on Insular Affairs, Oceans and Wildlife, June 10, 2010, http://resourcescommittee.house.gov/UploadedFiles/Rolfes Testimony06.10.09.pdf.

88. Associated Press, "Gulf Oil Spill Workers Complaining of Flu-like Symptoms."

89. Environmental Protection Agency, "Odors from the BP Oil Spill," June 2010, www.epa.gov/bpspill/reports/odorfactsheet.pdf.

90. Anne Rolfes, "Review of EPA's Air Sampling Data in Louisiana," Louisiana Bucket Brigade, July 20, 2010, www.labucketbrigade.org/downloads/ Review%20of%20EPA%20Sampling%20Data%207.20.10_1.pdf.

91. Ibid.

92. Centers for Disease Control and Prevention, "CDC Response to the Gulf of

Mexico Oil Spill," CDC Emergency Preparedness and Response, August 31, 2010, http://emergency.cdc.gov/gulfoilspill2010/cdcresponds.asp.

93. "BP Gives $10 million to National Institutes of Health to Study Health Effects of Oil Spill," *Times-Picayune*, September 7, 2010, www.nola.com/news/gulf-oil-spill/index.ssf/2010/09/bp_gives_10_million_to_nationa.html.

94. John Howard, "Evaluating the Health Impacts of the Gulf of Mexico Oil Spill," testimony before the Senate Committee on Health, Education, Labor, and Pensions, June 15, 2010, www.hhs.gov/gulfoilspill/Documents/061510niosh_help.pdf.

95. "BP, Coast Guard Officers Block Journalists from Filming Oil-Covered Beach," *Huffington Post*, May 19, 2010, www.huffingtonpost.com/2010/05/19/bp-coast-guard-officers-b_n_581779.html.

96. *Deepwater Horizon* Incident Joint Information Center, "Coast Guard Establishes 20-Meter Safety Zone around All *Deepwater Horizon* Protective Boom Operations," press release, June 30, 2010, www.restorethegulf.gov/release/2010/06/30/coast-guard-establishes-20-meter-safety-zone-around-all-deepwater-horizon-protect.

97. *Anderson Cooper 360*, CNN, July 1, 2010, www.youtube.com/watch?v=WpJBsjKhRTo.

98. Philips, "BP's Photo Blockade of the Gulf Oil Spill."

99. Peters, "Efforts to Limit the Flow of Spill News."

100. Ibid.

101. Marjorie R. Esman, "Open Letter Concerning Media and Public Access to BP Oil Spill," American Civil Liberties Union of Louisiana, June 28, 2010, www.laaclu.org/PDF_documents/Media_Public_Access_Oil_Spill_Letter_062810.pdf.

102. Christy George, letter to U.S. Coast Guard, Society of Environmental Journalists, June 4, 2010, www.sej.org/initiatives/freedom-information/sej-letter-coast-guard-media-access-spill-response-operations.

103. "Gulf Oil Spill Containment Attempt Fails," Environment News Service, May 10, 2010, www.ens-newswire.com/ens/may2010/2010–05–10–02.html.

104. U.S. Fish and Wildlife Service, "Wildlife Threatened on the Gulf Coast," June 2010, www.fws.gov/home/dhoilspill/pdfs/NewWildlifeOfGulf.pdf.

105. Michael Gravitz, "Oceans under the Gun: Living Seas or Drilling Seas?" Environment America Research and Policy Center, October 28, 2009, www.environmentamerica.org/home/reports/report-archives/ocean-conservation/ocean-conservation/oceans-under-the-gun-living-seas-or-drilling-seas#idDE88aglFnpScAOxy9Q9YtA.

106. Jane Lyder, "Our Natural Resources at Risk: The Short- and Long-Term Impacts of the *Deepwater Horizon* Spill," testimony before the House Natural Resources Subcommittee on Insular Affairs, Oceans and Wildlife, June 10,

2010, http://resourcescommittee.house.gov/images/Documents/20100610/
2010_06_10_oceans/testimony_lyder.pdf.

107. *Exxon Valdez* Oil Spill Trustee Council, "Legacy of an Oil Spill—20 Years
After the Exxon Valdez," 2009, www.evostc.state.ak.us/Universal/Docu
ments/Publications/20th%20Anniversary%20Report/2009%20Status%20
Report%20(Low-Res.pdf).

108. April Reese, "Gulf Spill: Wildlife Toll Mounts as BP Oil Inundates Coastal
Marshes," E&E, June 3, 2010, www.eenews.net/public/Landletter/2010/
06/03/1.

109. Oregon State University, "OSU Researchers Find Heightened Levels of
Known Carcinogens in Gulf," press release, September 30, 2010, http://
oregonstate.edu/ua/ncs/archives/2010/sep/osu-researchers-find-heightened-
levels-known-carcinogens-gulf.

110. Riki Ott, "From the Ground: BP Censoring Media, Destroying Evidence,"
Huffington Post, June 11, 2010, www.huffingtonpost.com/riki-ott/from-the-
ground-bp-censor_b_608724.html.

111. Timothy J. Ragen, "The *Deepwater Horizon* Oil Spill and Its Effects on
Marine Mammals," testimony before the House Subcommittee on Insular
Affairs, Oceans and Wildlife, June 10, 2010, http://resourcescommittee
.house.gov/images/Documents/20100610/2010_06_10_oceans/testimony_
ragen.pdf.

112. National Oceanic and Atmospheric Administration, National Marine Fisheries Service, "Sea Turtles, Dolphins, and Whales and the Gulf of Mexico
Oil Spill," November 2, 2010, www.nmfs.noaa.gov/pr/health/oilspill.htm.

113. Sea Turtle Restoration Project, "Sea Turtle Fact Sheet: Kemp's Ridley," March
2003, www.seaturtles.org/downloads/Kemps.pdf.

114. Center for Biological Diversity, "More Than 150,000 Call on BP and Federal
Officials to Stop Burning Endangered Sea Turtles Alive," press release, June
28, 2010, www.biologicaldiversity.org/news/press_releases/2010/sea-turtles-
06–28–2010.html.

115. Sea Turtle Restoration Project, "BP and Coast Guard Halt Burning of Endangered Sea Turtles in Gulf Oil Spill Clean-Up," press release, July 2, 2010,
www.seaturtles.org/article.php?id=1685.

116. Based on "Consolidated Fish and Wildlife Collection Reports" issued by the
Deepwater Horizon Incident Joint Information Center from May 28 to November 2, 2010. The reports are available at www.restorethegulf.gov/fish-wildlife.

117. National Oceanic and Atmospheric Administration, National Marine Fisheries Service, "Endangered Species Act of 1973 (ESA) Penalties and Enforcement," www.nmfs.noaa.gov/pr/pdfs/laws/esa_section11.pdf.

118. National Wildlife Federation and others, Freedom of Information Act request,
fax to Johnny Hunt, U.S. Fish and Wildlife Services FOIA officer, August 3,

2010, www.nwf.org/News-and-Magazines/Media-Center/News-by-Topic/Wildlife/2010/~/media/PDFs/Wildlife/FOIA-Bird-Oil-Spill-Data.ashx.

119. National Wildlife Federation and others, letter to Eric H. Holder and Robert W. Dudley, August 3, 2010, www.nwf.org/News-and-Magazines/Media-Center/News-by-Topic/Wildlife/2010/~/media/PDFs/Wildlife/Oil-Spill-Transparency-Letter.ashx.

120. Bob Serata, "Scientist Finds Deadly Effects of Dispersant Used in Gulf Oil Disaster," National Wildlife Federation, October 8, 2010, http://blog.nwf.org/wildlifepromise/2010/10/scientist-finds-deadly-effects-of-dispersant-used-in-Gulf-oil-disaster.

121. Ibid.

122. Subra, testimony.

123. Reese, "Gulf Spill."

124. World Wildlife Fund, "Gulf of Mexico Oil Disaster: Species Impacts," October 4, 2010, www.worldwildlife.org/what/howwedoit/policy/oil-disaster-species-impacts.html.

125. Ibid.

126. United Houma Nation, "Existence of United Houma Nation in Peril," press release, May 17, 2010, http://unitedhoumanation.org/node/1184.

4. When the Oil Kills the Fish, Can the Fishers Survive?

1. National Oceanic and Atmospheric Administration, "NOAA Strategy for Future Reopenings: NOAA Seafood Sampling and Testing Priorities for Federal Closed Area," September 27, 2010, www.noaanews.noaa.gov/stories2010/20100927_reopening.html.

2. Phone conversations with each state's respective state government fisheries representative with Lindsey Ingraham, Black Tide research assistant, December 2010 and January 2011.

3. National Oceanic and Atmospheric Administration, "NOAA Closes 4,200 Square Miles of Gulf Waters to Royal Red Shrimping," November 24, 2010, www.noaanews.noaa.gov/stories2010/20101124_closing.html.

4. National Oceanic and Atmospheric Administration, "Fish Stocks in the Gulf of Mexico," May 12, 2010, http://response.restoration.noaa.gov/book_shelf/1886_Fish-Stocks-Gulf-fact-sheetv2.pdf; National Marine Fisheries Service, "Fisheries of the United States 2009," September 2010, www.st.nmfs.noaa.gov/st1/fus/fus09/fus_2009.pdf.

5. National Marine Fisheries Service, "Fisheries of the United States 2009."

6. City-Data, "Bayou La Batre, Alabama," 2010, www.city-data.com/city/Bayou-La-Batre-Alabama.html.

7. Miya Saika Chen and Audrey Buehring, "The Impact of the Gulf Oil Spill on AAPI [Asian American/Pacific Islander] Communities," White House Office

of Public Engagement, June 17, 2010, www.whitehouse.gov/blog/2010/06/17/impact-gulf-oil-spill-aapi-communities-0.

8. Ibid.

9. Russ Henderson, "Bayou Residents Lose Homes," *Mobile Register*, September 1, 2005, www.al.com/mobileregister/090105/4mr0501b0901.pdf.

10. Gina Pace, "Rice: Race Not an Issue in Efforts," CBS News, September 4, 2005, www.cbsnews.com/stories/2005/09/04/katrina/main815171.shtml.

11. City-Data, "Bayou La Batre."

12. Irwin Redlener, David Abramson, Tasha Stehling-Ariza, Jonathan Sury, Akilah Banister, and Yoon Soo Park, "Impact on Children and Families of the *Deepwater Horizon* Oil Spill: Preliminary Findings of the Coastal Population Impact Study," Columbia University Mailman School of Public Health, National Center for Disaster Preparedness, August 3, 2010, www.ncdp.mailman.columbia.edu/files/NCDP_Oil_Impact_Report.pdf.

13. Matthew R. Lee and Troy C. Blanchard, "Health Impacts of *Deepwater Horizon* Oil Disaster on Coastal Louisiana Residents," Louisiana State University, Public Policy Research Lab, July 2010, www.lsu.edu/pa/mediacenter/tipsheets/spill/publichealthreport_2.pdf.

14. Louisiana Department of Health and Hospitals, "DHH Secretary Requests $10 Million from BP to Provide Mental Health Services to Residents Affected by the Oil Spill," press release, June 28, 2010, http://emergency.louisiana.gov/Releases/06282010-health.html.

15. Louisiana Department of Health and Hospitals, "Louisiana Oysters Are Safe to Eat," 2010, www.dhh.louisiana.gov/offices/page.asp?id=214&detail=9468.

16. National Marine Fisheries Service, "Fisheries of the United States 2009."

17. Arnold and Itkin (law firm), "Gulf Coast Oil Spill / Commercial Fishermen Economic Damage Claims," 2010, http://transoceanlawsuits.com/those-affected/commercial-fishermen.

18. Molly Hennessy-Fiske, "Gulf Oil Spill: Boat Captain, Despondent over Spill, Commits Suicide," *Los Angeles Times*, June 23, 2010, http://latimesblogs.latimes.com/greenspace/2010/06/gulf-oil-spill-boat-captain-despondent-over-spill-commits-suicide.html.

19. Chuck Hopkinson, "Outcome/Guidance from Georgia Sea Grant Program: Current Status of BP Oil Spill," Georgia Sea Grant, August 17, 2010, http://uga.edu/aboutUGA/joye_pkit/GeorgiaSeaGrant_OilSpillReport 8–16.pdf.

20. Public Employees for Environmental Responsibility, "FDA Should Test Gulf Seafood for Dispersant Contamination—Legal Petition Demands FDA Move beyond Sensory Test to Look at Chemicals," August 4, 2010, www.peer.org/news/news_id.php?row_id=1385.

21. Donald W. Kraemer, "The BP Oil Spill: Accounting for the Spilled Oil and

Ensuring the Safety of Seafood from the Gulf," testimony before the House Committee on Energy and Commerce, Subcommittee on Energy and Environment, August 19, 2010, www.fda.gov/NewsEvents/Testimony/ucm223246 .htm.

22. White House Office of the Press Secretary, "Press Briefing by Press Secretary Robert Gibbs, Admiral Thad Allen, Carol Browner, and Dr. Lubchenco," August 4, 2010, www.whitehouse.gov/the-press-office/press-briefing-press-secretary-robert-gibbs-admiral-thad-allen-carol-browner-and-dr.

23. Cook Inletkeeper, "How Toxic Are Oil Dispersants? Groups Press EPA to Find Out before Next Spill, Shrimpers, Community Groups Petition Agency for Info, Clear Rules Before OK'ing Future Use," press release, October 13, 2010, www.inletkeeper.org/energy/dispersants.htm.

5. Making BP Pay: Cleanup Workers, Vessels of Opportunity, Claims, and Protests

1. Casandra Andrews, "Local Fishermen Block Mouth of Bayou to Protest BP Hiring, Payment Practices," *Press-Register*, June 2, 2010, http://blog.al.com/ live/2010/06/local_fishermen_block_mouth_of.html.

2. Susan A. Fleming, "Cost of Major Spills May Impact Viability of Oil Spill Liability Trust Fund," testimony before the Subcommittee on Federal Financial Management, Government Information, Federal Services, and International Security, June 16, 2010, http://hsgac.senate.gov/public/index.cfm?FuseAc tion=Files.View&FileStore_id=32035873–9a84–496a-8f56–00b6f08c426b.

3. Thomas Morrison, letter to BP Exploration and Production, U.S. Coast Guard, April 28, 2010, http://oilspill.ago.alabama.gov/April%2028%20Responsible %20Party%20Letter%20from%20USCG%20to%20BP.pdf; Thomas Morrison, letter to Transocean Holdings Incorporated, U.S. Coast Guard, April 28, 2010, http://oilspill.ago.alabama.gov/April%2028%20Responsible%20Party %20Letter%20from%20USCG%20to%20Transocean.pdf.

4. Thomas S. Morrison, e-mail to Terry L. Stoltz, U.S. Coast Guard, May 11, 2010, http://oilspill.ago.alabama.gov/May%2011%20Email%20from%20US CG%20to%20Transocean%20re%20BP%20Advertising%20Adequate.pdf.

5. Henry Waxman and Bart Stupack, "Committee Releases Details on BP's Advertising Expenditures Related to the Gulf Oil Spill," letter to Kathy Cator, Committee on Energy and Commerce, September 1, 2010, http://democrats .energycommerce.house.gov/index.php?q=news/committee-releases-details-on-bps-advertising-expenditures-related-to-the-gulf-oil-spill.

6. Phuong Le and John Flesher, "Oil-Cleanup Technology Is Lacking," Associated Press, June 26, 2010, http://seattletimes.nwsource.com/html/nationworld/ 2012217841_oilcleanup27.html.

7. Caroline McCarthy, "Fake BP Twitter Account Remains Shrouded in Mystery,"

CNET News, May 27, 2010, http://news.cnet.com/8301–13577_3–20006199–36.html; Rupal Parekh and Michael Bush, "Why BP Isn't Fretting Over Its Twitter Impostor," *Advertising Age*, May 24, 2010, http://adage.com/article?article_id=144062.

8. NBC, *Late Night with Jimmy Fallon*, May 25, 2010, www.latenightwithjimmy fallon.com/video/dana-white-52510/1230968.

9. Mail Foreign Service, "BP Buys Google Search Term Oil Spill to Repair Reputation," *Daily Mail*, June 10, 2010, www.dailymail.co.uk/news/worldnews/article-1285190/BP-buys-Google-search-term-oil-spill-help-repair-reputa tion.html.

10. Daren Beaudo, phone conversation with Saif Rahman, *Black Tide* research assistant, December 14, 2010.

11. Lower Mississippi Riverkeeper, "Gulf Fishermen Win First Legal Battle Against BP," May 3, 2010, http://lmrk.org/issues/bp-s-deep-water-drilling-disaster/gulf-fishermen-win-first-legal-battle-against-bp.html.

12. Wilma Subra, "Human Health and Environmental Impacts Associated with the Deepwater Horizon Crude Oil Spill Disaster," testimony before the House Energy and Commerce Committee, June 7, 2010, http://energycommerce .house.gov/documents/20100607/Subra.Testimony.06.07.2010.pdf.

13. Tom Brown, "BP Admits 'Misstep' Over Oil Spill Claims Waivers," Reuters, May 3, 2010, www.reuters.com/article/idUSN0315696620100503.

14. "Vessels of Opportunity Lawsuit Filed Against BP Oil," PRWeb, November 29, 2010, www.prweb.com/releases/VOOLawsuitAgainstBP/TaylorMartino/prweb4762304.htm.

15. Harlan Kirgan, "BP Official Says 800 Vessels of Opportunity Deployed in Mississippi," *Mississippi Press*, July 8, 2010, http://blog.Gulflive.com/mississippi-press-news/2010/07/bp_official_says_800_vessels_o.html; Kimberly Kindy, "Early Cleanup Efforts of Gulf Oil Spill Marred by Communication Woes, Scammers," *Washington Post*, August 14, 2010, www.washingtonpost.com/wp-dyn/content/article/2010/08/13/AR2010081306236.html.

16. BP, "Vessels of Opportunity Sample Master Vessel Charter Agreement," Sea Grant, May 6, 2010, www.flseagrant.org/images/stories/oil_spill/vessels%20of %20opportunity-charter-checklist2–1.pdf.

17. Phone interview with Celia Alario, *Black Tide* interview assistant, October 7, 2010.

18. Andrews, "Local Fishermen Block Mouth of Bayou."

19. Kindy, "Early Cleanup Efforts of Gulf Oil Spill Marred by Communication."

20. National Commission on the BP Deepwater Horizon Oil Spill and Offshore Drilling, "Decision-Making within the Unified Command-Draft," Staff Working Paper No. 2, October 6, 2010, www.oilspillcommission.gov.

21. Mac McClelland, "Official Government Stats = BP Spin," *Mother Jones*, June

9, 2010, http://motherjones.com/rights-stuff/2010/06/official-government-stats-bp-spin; Brad Johnson, "BP's Secret Army of Oil Disaster," Think Progress: Wonk Room, June 18, 2010, http://wonkroom.thinkprogress.org/2010/06/18/bp-contractor-army.

22. Beaudo, phone conversation with Saif Rahman.

23. Catherine Clifford, "BP Hires the Unemployed for Clean-Up," CNN, June 8, 2010, http://money.cnn.com/2010/06/08/smallbusiness/bp_hiring_unemployed/index.htm.

24. Spencer Michels, "On the Gulf Coast, Media Access Can Be Hard to Come By," *The Rundown Blog*, PBS NewsHour, June 9, 2010, www.pbs.org/newshour/rundown/2010/06/on-the-gulf-coast-media-access.html.

25. ProPublica, letter from Doug Suttles, June 9, 2010, www.propublica.org/documents/item/bp-clarification-of-media-access-june-9–2010.

26. "BP Buses in 400 Workers During Obama's Visit," WDSU, May 28, 2010, www.wdsu.com/r/23711711/detail.html.

27. Brett Michael Dykes, "BP Bused in 100s of Temp Workers for Obama Visit, State Official Says," Yahoo! News, May 28, 2010, http://news.yahoo.com/s/ynews/ynews_ts2320.

28. BP, "Total Claims Payments Top $256 Million," press release, July 28, 2010, www.bp.com/genericarticle.do?categoryId=2012968&contentId=7064024.

29. BP, "Claims and Government Payments, Gulf of Mexico Oil Spill, Public Report—9/30/10," September 30, 2010, www.bp.com/liveassets/bp_internet/globalbp/globalbp_uk_english/incident_response/STAGING/local_assets/downloads_pdfs/Public_Report_9_30_2010.pdf.

30. Amy Schoenfeld, "Where BP's Money Is Landing," *New York Times*, July 3, 2010, www.nytimes.com/2010/07/04/business/04metricstext.html?_r=1&fta=y.

31. Darryl Willis, "Liability Issues Surrounding the Gulf Coast Oil Disaster," testimony before the House Committee on the Judiciary, May 27, 2010, http://judiciary.house.gov/hearings/pdf/Willis100527.pdf.

32. Ibid.

33. ESIS, "ESIS' Services," 2010, http://www2.esis.com/ESISRoot/ESIS/Services/Recovery+Services.

34. Monique Harden and Nathalie Walker, letter to Thad Allen, Advocates for Environmental Human Rights, June 9, 2010, www.ehumanrights.org/docs/AEHR_letter_to_Thad_Allen.pdf.

35. BP, "BP to Appoint Independent Mediator to Ensure Timely, Fair Claims Process," press release, May 26, 2010, www.bp.com/genericarticle.do?categoryId=2012968&contentId=7062448.

36. Kevin McGill, "Some Laborers Lack Papers for BP Claims," Associated Press, July 7, 2010, www.msnbc.msn.com/id/38125441/ns/business-us_business.

37. Thad Allen, letter to Tony Hayward, National Incident Command, June 8,

2010, www.restorethegulf.gov/sites/default/files/imported_pdfs/posted/2931/NIC_Letter_to_BP_CEO.621247.pdf.

38. Office of the Press Secretary, "FACT SHEET: Claims and Escrow," press release, White House, June 16, 2010, www.whitehouse.gov/the-press-office/fact-sheet-claims-and-escrow.

39. "Governor Jindal: BP Claims System Is Broken," press release, Office of the Governor of Louisiana, June 22, 2010, http://gov.louisiana.gov/index.cfm?md=newsroom&tmp=detail&catID=2&articleID=2255.

40. Sasha Chavkin, "BP Confirms That Thousands of Claims Decisions Will Be Deferred," *ProPublica*, August 3, 2010, www.propublica.org/article/bp-confirms-that-thousands-of-claims-decisions-will-be-deferred.

41. Phone interview with Celia Alario, *Black Tide* interview assistant, October 18, 2010.

42. Jayne Clark, "What Are the Best/Worst Beaches? National Geographic Traveler Names Names," *USA TODAY*, November 30, 2010, http://travel.usatoday.com/destinations/dispatches/post/2010/11/what-are-the-bestworst-beaches-national-geographic-traveler-names-names/132436/1.

43. "Potential Impact of the Gulf Oil Spill on Tourism," *Oxford Economics*, July 22, 2010, www.ustravel.org/news/press-releases/bp-oil-spill-impact-gulf-travel-likely-last-3-years-and-cost-227-billion.

44. Charisse Jones and Rick Jervis, "Oil Spill Takes Toll on Tourism on Gulf Coast," *USA Today*, June 25, 2010, http://travel.usatoday.com/destinations/2010–06–25–1Aspill25_CV_N.htm.

45. Matt Gutman, "BP Oil Spill: Tourism Industry Suffering on July 4th Weekend," ABC World News, July 2, 2010, http://abcnews.go.com/WN/bp-oil-spill-tourism-industry-suffering-july-weekend/story?id=11077632.

46. "Potential Impact of the Gulf Oil Spill on Tourism."

47. Maureen Mackey, "BP Oil Spill: Gulf Tourism Takes a Huge Hit," *Fiscal Times*, June 24, 2010, www.thefiscaltimes.com/Issues/The-Economy/2010/06/24/BP-Oil-Spill-Gulf-Tourism-Takes-a-Huge-Hit.aspx.

48. Steve Huettel, "Oil Spill Sent Florida Visitors to East Coast, Statistics Show," *St. Petersburg Times*, September 9, 2010, www.tampabay.com/news/business/tourism/oil-spill-sent-florida-visitors-to-east-coast-statistics-show/1120326.

49. Carlton Proctor, "The Top 10 Business Stories of 2010," *Pensacola Business Journal*, December 2, 2010, www.pnj.com/article/20101202/BUSINESSJOURNAL/11190343/1003/The-Top-10-Business-Stories-of-2010.

50. Office of the Press Secretary, "FACT SHEET."

51. Frances Romero, "Compensation Czar Kenneth Feinberg," *Time*, October 23, 2009, www.time.com/time/nation/article/0,8599,1903547,00.html.

52. Gulf Coast Claims Facility, "Gulf Coast Claims Facility Now Processing Oil

Spill Claims," press release, August 23, 2010, www.gulfcoastclaimsfacility
.com/press1.php.

53. Gulf Coast Claims Facility, "General Information About the Gulf Coast Claims Facility," 2010, www.gulfcoastclaimsfacility.com/faq#Q1.

54. Gulf Coast Claims Facility, "Gulf Coast Claims Facility Now Processing Oil Spill Claims."

55. Thomas Perrelli, letter to Kenneth Feinberg, U.S. Department of Justice, Office of the Associate Attorney General, September 17, 2010, http://blogs .tampabay.com/files/2010–9–17-perrelli-to-feinberg-letter.pdf.

56. Phone interview with Celia Alario, *Black Tide* interview assistant, October 7, 2010.

57. Thomas Perrelli, letter to Kenneth Feinberg, U.S. Department of Justice, Office of the Associate Attorney General, November 19, 2010, http://media .al.com/live/other/2010–11–19%20Perrelli%20Letter%20to%20Feinberg.pdf.

58. Gulf Coast Claims Facility, "GCCF Program Statistics: Overall Summary," December 4, 2010, www.gulfcoastclaimsfacility.com/GCCF_Overall_Status_ Report.pdf.

59. Dionne Searcey, "BP Oil-Spill Claims Get Fast Track," *Wall Street Journal*, December 13, 2010, http://online.wsj.com/article/SB10001424052748704058 704576015591156318386.html.

6. Big Oil Plays Defense: BP and the Oil Industry Respond to Disaster

1. Clifford Krauss, "Oil Spill's Blow to BP's Image May Eclipse Costs," *New York Times*, April 29, 2010, www.nytimes.com/2010/04/30/business/30bp.html.

2. "Profile: Tony Hayward," *Sunday Times*, June 6, 2010, http://business.timeson line.co.uk/tol/business/industry_sectors/natural_resources/article7144758.ece.

3. World Bank, "Gross Domestic Product 2009," World Development Indicators database, December 15, 2010, http://siteresources.worldbank.org/DATA STATISTICS/Resources/GDP.pdf.

4. Kenny Bruno, "BP: Beyond Petroleum or Beyond Preposterous?" CorpWatch, December 14, 2000, www.corpwatch.org/article.php?id=219.

5. Terry Macalister, "BP Shuts Alternative Energy HQ," *Guardian*, June 29, 2009, www.guardian.co.uk/business/2009/jun/28/bp-alternative-energy.

6. Anthony DiPaola and Maher Chmaytelli, "BP, CNPC Beat Exxon to Win First Iraqi Oil Contract," Bloomberg, June 20, 2010, www.bloomberg.com/apps/ news?pid=newsarchive&sid=aZ6zxJbJdRK8.

7. Tim Webb, "WikiLeaks Cables: BP Suffered Blowout on Azerbaijan Gas Plat-form," *Guardian*, December 16, 2010, www.guardian.co.uk/world/2010/dec/ 15/wikileaks-bp-azerbaijan-Gulf-spill?CMP=twt_gu.

8. Ibid.

9. Greg Palast, "Palast Arrested Busted by BP in Azerbaijan," December 20, 2010, http://groups.yahoo.com/group/Babel/message/21482.

10. Ian Urbina, "BP Is Pursuing Alaska Drilling Some Call Risky," *New York Times*, June 23, 2010, www.nytimes.com/2010/06/24/us/24rig.html.

11. Kenneth W. Abbott, "The Deepwater Horizon Incident: Are the Minerals Management Service Regulations Doing the Job?" testimony before the House Subcommittee on Energy and Mineral Resources, June 17, 2010, http://resources committee.house.gov/UploadedFiles/AbbottTestimony06.17.10.pdf.

12. David S. Hilzenrath, "BP Atlantis Rig Plans Never Received Proper Approvals, Former Contractor Says," *Washington Post*, June 17, 2010, www.washington post.com/wp-dyn/content/article/2010/06/17/AR2010061703872_2.html.

13. Jason Leopold, "Whistleblower: BP Risks More Massive Catastrophes in Gulf," Truthout.org, April 30, 2010, www.truth-out.org/whistleblower-bps-other-offshore-drilling-project-gulf-vulnerable-catastrophe59027.

14. Laurel Brubaker Calkins, "BP Sued by Watchdog Group Over Atlantis Platform," Bloomberg, September 13, 2010, www.bloomberg.com/news/2010–09–10/bp-sued-over-alleged-safety-gaps-at-atlantis-production-platform.html.

15. Elizabeth Douglas, "Concern Grows on Refinery Safety," *Los Angeles Times*, March 23, 2007.

16. Tyson Slocum, "Cost of Doing Business: BP's $730 Million in Fines/Settlement + 2 Criminal Convictions," *Public Citizen*, May 5, 2010, http://public citizenenergy.org/2010/05/05/cost-of-doing-business-bps-550-million-in-fines-2-criminal-convictions.

17. Investigation Report, "Refinery Explosion and Fire (15 Killed, 180 Injured)," U.S. Chemical Safety and Hazard Investigation Board, report no. 2005-04-1-TX, March 2007, www.csb.gov/assets/document/CSBFinalReportBP.pdf.

18. Ibid.

19. Jeannette Lee, "BP Subsidiary Pleads Guilty to Alaska Oil Spill," Associated Press, November 29, 2007, www.chron.com/disp/story.mpl/front/5339374 .html.

20. Slocum, "Cost of Doing Business."

21. Terry Macalister, "BP Boss Warns of Shake-Up After Dreadful Results," *Guardian*, September 26, 2007, www.guardian.co.uk/business/2007/sep/26/oilandpetrol.news.

22. *Deepwater Horizon* Study Group, "Letter to the National Commission on the BP Deepwater Horizon Oil Spill and Offshore Drilling," November 24, 2010, http://ccrm.berkeley.edu/pdfs_papers/bea_pdfs/DHSG_lettertoNationalCom mission.pdf.

23. BP, "BP Offers Full Support to Transocean After Drilling Rig Fire," press release, April 21, 2010, www.bp.com/genericarticle.do?categoryId=2012968& contentId=7061458.

24. BP, "BP Initiates Response to Gulf of Mexico Oil Spill," press release, April 22, 2010, www.bp.com/genericarticle.do?categoryId=2012968&contentId=7061490.

25. National Commission on the BP *Deepwater Horizon* Oil Spill and Offshore Drilling, "Stopping the Spill: The Five-Month Effort to Kill the Macondo Well," Staff Working Paper No. 6, November 22, 2010, www.oilspillcommission.gov.

26. Joel Achenbach, "In BP 'War Room,' Small Victories, Many Uncertainties," *Washington Post*, July 3, 2010, www.washingtonpost.com/wp-dyn/content/article/2010/07/02/AR2010070205570.html.

27. National Commission, "Stopping the Spill."

28. Juliet Eilperin, "Seeking Answers in MMS' Flawed Culture," *Washington Post*, August 25, 2010, www.washingtonpost.com/wp-dyn/content/article/2010/08/24/AR2010082406771.html.

29. Joseph A. Pratt, Tyler Priest, and Christopher J. Castaneda, *Offshore Pioneers: Brown & Root and the History of Offshore Oil and Gas* (Houston: Gulf, 1997); "FACTBOX-Offshore Increasingly Important to Oil Industry," Reuters, July 6, 2010, http://uk.reuters.com/article/idUKLDE6640YV20100706.

30. Pratt, Priest, and Castaneda, *Offshore Pioneers*.

31. Ibid.; Kevin G. Hall, "Despite Disaster, U.S. Has Little Choice but to Drill Offshore," McClatchy Newspapers, May 5, 2010, www.mcclatchydc.com/2010/05/05/93609/crucial-ultra-deepwater-drilling.html; Alex Chakhmakhchev and Peter Rushworth, "Global Overview of Offshore Oil and Gas Operations for 2005–2009," *Offshore*, May 1, 2010, www.offshore-mag.com/index/article-display/7580142997/articles/offshore/volume-70/issue-50/international-e_p/global-overview_of.html.

32. Chevron, "Deepwater Drilling: How It Works," 2011, www.chevron.com/stories/#/allstories/deepwaterdrilling.

33. Joe Carroll, "Rig Shortage Slows Chevron Bid to Tap Offshore Fields," Reuters, December 6, 2006, www.bloomberg.com/apps/news?pid=newsarchive&sid=asDAIInNRKIg.

34. Eilperin, "Seeking Answers."

35. Dan Van Natta Jr. with Neela Banerjee, "Bush Policies Have Been Good to Energy Industry," *New York Times*, April 21, 2002, www.nytimes.com/2002/04/21/us/bush-policies-have-been-good-to-energy-industry.html.

36. Dana Milbank, "Bush Energy Order Wording Mirrors Oil Lobby's Proposal," *Washington Post*, March 28, 2002, www.washingtonpost.com/ac2/wp-dyn/A28281-2002Mar27.

37. Eilperin, "Seeking Answers."

38. Jeffrey St. Clair, "Meet Steven Griles: Big Oil's Inside Man," CounterPunch.org, June 28, 2003, www.counterpunch.org/stclair06282003.html; Antonia Juhasz, *The Tyranny of Oil: The World's Most Powerful Industry—and What We Must Do to Stop It* (New York: HarperCollins, 2008), 259.

39. St. Clair, "Meet Steven Griles."

40. John Helprin, "Ex-Deputy Pleads Guilty in Abramoff Case," Associated Press, March 23, 2007, www.11alive.com/news/article_news.aspx?storyid=94300.

41. Earl E. Devaney, memorandum from Inspector General Earl E. Devaney to Secretary Kempthorne, U.S. Department of the Interior, Office of Inspector General, September 9, 2008, www.doioig.gov/images/stories/reports/pdf/RIKinvestigation.pdf.

42. Eric Lipton and John Broder, "Regulator Deferred to Oil Industry on Rig Safety," *New York Times*, May 8, 2010, www.nytimes.com/2010/05/08/us/08agency.html?hpw.

43. Eilperin, "Seeking Answers."

44. Lipton and Broder, "Regulator Deferred to Oil Industry."

45. Jason DeParle, "Minerals Service Had a Mandate to Produce Results," *New York Times*, August 7, 2010, www.nytimes.com/2010/08/08/us/08mms.html?pagewanted=all.

46. Larry West, "U.S. House Votes to End Subsidies and Tax Breaks for Oil Companies," About.com Environmental Issues, January 19, 2007, http://environment.about.com/od/environmentallawpolicy/a/end_oil_subsidy.htm.

47. Carolyn Maloney, "Rep. Maloney Reacts to Promotion of Official Responsible for Oil and Gas Royalty Rip-Off," press release, Office of Representative Carolyn B. Maloney, February 7, 2007, http://maloney.house.gov/index.php?Itemid=61&id=1283&option=content&task=view.

48. Ibid.

49. U.S. Department of the Interior, Office of Inspector General, "Investigative Report, Island Operating Company et al.," Case No. PI-GA-09–0102-I, March 31, 2010, www.doioig.gov/images/stories/reports/pdf/IslandOperatingCo.pdf.

50. Mark Schleifstein, "Gulf Region MMS Employees Accepted Gifts, Food, Tickets at Oil and Gas Company Expense," *Times-Picayune*, May 25, 2010, www.nola.com/news/gulf-oil-spill/index.ssf/2010/05/gulf_region_minerals_management.html.

51. Sandi M. Fury, Gulfletter to Department of Interior, *Federal Register* 74, no. 115, September 14, 2009, www.boemre.gov/federalregister/PublicComments/AD15SafetyEnvMgmtSysforOCSOilGasOperations/ChevronSept142009.pdf.

52. Jonathan W. Armstrong, letter to Department of Interior, *Federal Register* 74, no. 115, September 14, 2009, www.boemre.gov/federalregister/PublicComments/AD15SafetyEnvMgmtSysforOCSOilGasOperations /ExxonMobil9–14–09.pdf.

53. Eilperin, "Seeking Answers."

54. Michael Saucier, testimony at the USCG/BOEM Marine Board of Investigation, May 12, 2010, www.deepwaterinvestigation.com/external/content/document/3043/621931/1/Deepwater%20Horizon%20Joint%20Investigation%20Transcript%20-%20May%2012,%202010.pdf.

55. National Commission, "Stopping the Spill."

56. Deborah Zabarenko, "Walruses in Louisiana? Eyebrow-Raising Details of BP's Spill Response Plan," Reuters, May 27, 2010, http://blogs.reuters.com/envi ronment/2010/05/27/walruses-in-louisiana-eyebrow-raising-details-of-bps- spill-response-plan.

57. Steven Mufson and Juliet Eilperin, "Lawmakers Attack Plans Oil Companies Had in Place to Deal with a Spill," *Washington Post*, June 16, 2010, www.wash ingtonpost.com/wp-dyn/content/article/2010/06/15/AR2010061501700.html.

58. Ibid.

59. David Schneider, "How to Drill a Relief Well," *IEEE Spectrum*, August 2010, http://spectrum.ieee.org/energy/fossil-fuels/how-to-drill-a-relief-well.

60. National Commission, "Stopping the Spill."

61. Ibid.

62. BP, "BP Gulf of Mexico Spill Response Accelerating," press release, April 26, 2010, www.bp.com/genericarticle.do?categoryId=2012968&contentId= 7061537.

63. Sam Dolnick and Henry Fountain, "Unable to Stanch Oil, BP Will Try to Gather It," *New York Times*, May 6, 2010, www.nytimes.com/2010/05/06/ science/06container.html.

64. David Mattingly, "Interview with BP CEO Tony Hayward," CNN, May 7, 2010, http://edition.cnn.tv/TRANSCRIPTS/1005/07/cnr.04.html.

65. National Commission, "Stopping the Spill."

66. Ibid.

67. Tim Webb, "BP Boss Admits Job on the Line over Gulf Oil Spill," *Guardian*, May 13, 2010, www.guardian.co.uk/business/2010/may/13/bp-boss-admits- mistakes-gulf-oil-spill.

68. Greg Milam, "BP Chief: Oil Spill Impact 'Very Modest,'" *Sky News*, May 18, 2010, http://news.sky.com/skynews/Home/World-News/BP-Oil-Spill-In-Gulf- Of-Mexico-Will-Have-Very-Modest-Environmental-Impact-Says-Firms- CEO/Article/201005315633987.

69. Christopher Helman, "In His Own Words: Forbes Q&A with BP's Tony Hay- ward," *Forbes*, May 18, 2010, www.forbes.com/2010/05/18/oil-tony-hayward- business-energy-hayward.html.

70. National Commission, "Stopping the Spill."

71. "Exxon Valdez Oil Spill Disaster Fact Sheet," Davis Wright Tremaine LLP, June 2008, www.dwt.com/portalresource/06–08_Exxon_Valdez_Fact_Sheet.

72. Simon Usborne, "Tony Hayward: Right in the Thick of It," *Telegraph*, May 29, 2010, www.independent.co.uk/news/people/profiles/tony-hayward-right-in- the-thick-of-it-1986232.html.

73. National Commission, "Stopping the Spill."

74. Simon Usborne, "Tony Hayward."

75. "BP Chief to Gulf Residents: 'I'm Sorry,'" CNN, May 30, 2010, http://articles
.cnn.com/2010–05–30/us/gulf.oil.spill_1_oil-spill-heavy-oil-dudley?_s=
PM:US.

76. Louie Ludwig, "Hey Tony," YouTube, June 8, 2010, www.youtube.com/watch?
v=Gsa29VhnZeM.

77. "Profile: Tony Hayward."

78. Kate Pickert, "Hershey's, Hip Hop, and Hayward," *Time*, June 4, 2010,
http://swampland.blogs.time.com/2010/06/04/hershey%E2%80%99s-hip-hop-
and-hayward.

79. Henry A. Waxman, "The Role of BP in the *Deepwater Horizon* Explosion and
Oil Spill," opening statement before the House Committee on Energy and
Commerce, Subcommittee on Oversight and Investigations, June 17, 2010,
http://energycommerce.house.gov/documents/20100617/Waxman.Statement
.oi.06.17.2010.pdf.

80. Brian Montopoll, "Rep. Joe Barton Apologizes to BP's Tony Hayward," CBS
News, June 17, 2010, www.cbsnews.com/8301-503544_162-20008020-
503544.html.

81. Transcript, "The Role of BP in the *Deepwater Horizon* Explosion and Oil Spill,"
House Committee on Energy and Commerce, Subcommittee on Oversight
and Investigations, June 17, 2010, http://energycommerce.house.gov/
documents/20100617/transcript.06.17.2010.oi.pdf.

82. CNN Newsroom, transcript, June 18, 2010, http://edition.cnn.com/TRAN
SCRIPTS/1006/18/cnr.06.html.

83. National Commission, "Stopping the Spill."

84. Steven Mufson, "Altered BP Photo Comes into Question," *Washington Post*,
July 20, 2010, www.washingtonpost.com/wp-dyn/content/article/2010/07/19/
AR2010071905256.html.

85. BP, "BP CEO Tony Hayward to Step Down and Be Succeeded by Robert Dudley," press release, July 27, 2010, www.bp.com/genericarticle.do?categoryId=
2012968&contentId=7063976.

86. Ed Markey, "Markey: No Golden Parachute for Hayward Until Gulf Costs
Paid," press release, House Select Committee on Energy Independence and
Global Warming, July 26, 2010, http://globalwarming.house.gov/mediacenter/
pressreleases_2008?id=0298#main_content.

87. BP, "Board and Executive Management—Robert Dudley," 2010, www.bp
.com/sectiongenericarticle.do?categoryId=9028860&contentId=7052605.

88. BP, "BP CEO Tony Hayward to Step Down."

89. Richard Wray, "BP Makes Record Loss as Tony Hayward Quits," *Guardian*,
July 27, 2010, www.guardian.co.uk/business/2010/jul/27/tony-hayward-leaves-
bp-1m-payoff.

90. Monica Langley, "Hayward Defends Tenure, BP's Spill Response," *Wall Street*

Journal, July 30, 2010, http://online.wsj.com/article/SB1000142405274870 35781045753974832561880088.html?mod=wsjcrmain.

91. Tim Webb, "Tony Hayward on BP Oil Crisis: 'I'd Have Done Better with an Acting Degree,'" *Guardian*, November 9, 2010, www.guardian.co.uk/business/2010/nov/09/bp-tony-hayward-bbc-interview.

92. *Deepwater Horizon* Incident Joint Information Center, "Statement from Admiral Allen on the Successful Completion of the Relief Well," September 19, 2010, www.restorethegulf.gov/release/2010/09/19/statement-admiral-allen-successful-completion-relief-well.

93. "Allen: Well Is Dead, But Much Gulf Coast Work Remains," CNN, September 20, 2010, http://articles.cnn.com/2010–09–20/us/Gulf.oil.disaster_1_worst-oil-disaster-deepwater-horizon-rig-louisiana-coast?z_s=PM:US.

7. Obama Steps Up: But Is the Disaster Over and Could It Happen Again?

1. Office of the Press Secretary, "Remarks by the President on Oil Spill," press release, White House, May 2, 2010, www.whitehouse.gov/the-press-office/remarks-president-oil-spill.

2. "BP Official—Open Heart Surgery at 5,000 Feet; Believes Cause Is Failed Equipment," ABC News, May 2, 2010, http://blogs.abcnews.com/political punch/2010/05/bp-official-open-heart-surgery-at-5000-feet-believes-cause-is-failed-equipment.html.

3. John M. Broder, "Panel Is Unlikely to End Deepwater Drilling Ban Early," *New York Times*, June 21, 2010, www.nytimes.com/2010/06/22/us/politics/22panel.html?_r=4.

4. Campbell Robertson, "Gulf of Mexico Has Long Been Dumping Site," *New York Times*, June 29, 2010, www.nytimes.com/2010/07/30/us/30gulf.html?partner=rss&emc=rss&pagewanted=all.

5. Steven Mufson, "Oil Spills. Poverty. Corruption. Why Louisiana Is America's Petro-State," *Washington Post*, July 18, 2010, www.washingtonpost.com/wp-dyn/content/article/2010/07/16/AR2010071602721.html.

6. David Ivanovich and Kristen Hays, "Offshore Drilling Safer, But Small Spills Routine," *Houston Chronicle*, July 28, 2008, www.chron.com/disp/story.mpl/business/5897424.html.

7. Ibid.

8. Steve Mufson, "Federal Records Show Steady Stream of Oil Spills in Gulf Since 1964," *Washington Post*, July 24, 2010, www.washingtonpost.com/wp-dyn/content/article/2010/07/23/AR2010072305603.html.

9. Ibid.

10. Sierra Club, "The Threat of Offshore Drilling: America's Coasts in Peril," May 24, 2006, http://www.sierraclub.org/wildlands/coasts.

11. Debbie Boger, testimony before the Senate Committee on Energy and Natural Resources, April 19, 2005, http://energy.senate.gov/hearings/testimony.cfm?id=1463&wit_id=4188.

12. Gulf Restoration Network, "Natural Defenses," 2010, http://healthygulf.org/our-work/natural-defenses/natural-defenses-home.

13. Lionel D. Lyles and Fulbert Namwamba, "Louisiana Coastal Zone Erosion: 100+ Years of Landuse and Land Loss Using GIS and Remote Sensing," ESRI Education User Conference Proceedings, July 2005, http://proceedings.esri.com/library/userconf/educ05/papers/pap1222.pdf.

14. U.S. Geological Survey, "America's Wetland: Historical and Projected Land Change in Coastal Louisiana (1932–1950)," April 23, 2004.

15. Geoff Colvin, "Chevron's CEO: The Price of Oil," *Fortune*, November 28, 2007, http://money.cnn.com/2007/11/27/news/newsmakers/101644366.fortune.

16. "Barack Obama on Offshore Drilling," *World News*, June 20, 2008, http://wn.com/barack_obama_on_offshore_oil_drilling.

17. Matthew Mosk, "Oil Dollars Gush after McCain Backs Drilling," *Washington Post*, July 28, 2010, http://articles.sfgate.com/2008–07–28/news/17172448_1_sen-john-mccain-oil-industry-mccain-last-month.

18. "Barack Obama on Offshore Drilling."

19. Michael C. Bender, "Obama Would Consider Off-Shore Drilling as Part of Comprehensive Energy Plan," *Palm Beach Post*, August 1, 2008, www.palmbeachpost.com/state/content/state/epaper/2008/08/01/0801obama1.html.

20. Josiah Ryan, "Democrat Leader: Restoring Offshore Drilling Ban a 'Top Priority' for Next Year," CNS News, September 24, 2008, www.cnsnews.com/news/article/36268.

21. Nick Snow, "House Democrats Won't Try to Restore OCS Moratoriums, Majority Leader Says," *Oil and Gas Journal*, November 21, 2008.

22. "Fortune 500," *Fortune*, 2009, http://money.cnn.com/magazines/fortune/fortune500/2009/index.html.

23. Office of the Press Secretary, "Remarks by the President on Energy Security at Andrews Air Force Base," press release, White House, March 31, 2010, www.whitehouse.gov/the-press-office/remarks-president-energy-security-andrews-air-force-base-3312010.

24. Office of the Press Secretary, "Remarks by the President in a Discussion on Jobs and the Economy in Charlotte, North Carolina," press release, White House, April 2, 2010, www.whitehouse.gov/the-press-office/remarks-president-a-discussion-jobs-and-economy-charlotte-north-carolina.

25. Sierra Club, "The Threat of Offshore Drilling."

26. The White House Blog, May 5, 2010, www.whitehouse.gov/blog/2010/05/05/ongoing-administration-wide-response-deepwater-bp-oil-spill.

27. Office of the Press Secretary, "Statement by the Press Secretary on the President's

Oval Office Meeting to Discuss the Situation in the Gulf of Mexico," press release, White House, April 22, 2010, www.whitehouse.gov/the-press-office/state ment-press-secretary-presidents-oval-office-meeting-discuss-situation-gulf-mex.

28. U.S. Government Printing Office, "National Oil and Hazardous Substances Pollution Contingency Plan," September 15, 1994, http://ecfr.gpoaccess .gov/cgi/t/text/text-idx?c=ecfr&sid=5c520868602917ce7125e06d3da5b91a &rgn=div5&view=text&node=40%3A27.0.1.1.1&idno=40#40:27.0.1.1.1.4.1.2.

29. National Commission on the BP Deepwater Horizon Oil Spill and Offshore Drilling, "Decision-Making Within the Unified Command," Staff Working Paper No. 2, October 6, 2010, www.oilspillcommission.gov.

30. James L. Jones, "Transition Plan for Securing America's Energy Future," November 17, 2008, www.energyxxi.org/reports/Transition_Plan.pdf.

31. Michael J. Keegan, "Profiles in Leadership: Admiral Thad W. Allen," *Business of Government*, Spring 2007, www.businessofgovernment.org/sites/default/ files/profiles07.pdf.

32. "BP Has Finished Pouring Cement into Blown-Out Well, Thad Allen Tells NPR," *All Things Considered*, NPR, August 5, 2010, www.npr.org/templates/ transcript/transcript.php?storyId=129005047.

33. National Commission, "Decision-Making Within the Unified Command."

34. Joel Achenbach, "With BP's Know-How and U.S. Authority, the Macondo Well Was Plugged," *Washington Post*, August 21, 2010, www.washingtonpost .com/wp-dyn/content/article/2010/08/20/AR2010082005747.html?sid= ST2010082005804.

35. "100 Days of Gushing Oil—Media Analysis and Quiz: A Different Kind of Disaster Story," Pew Research Center's Project for Excellence in Journalism, August 25, 2010, www.journalism.org/analysis_report/oil_spill_was_very_ different_kind_disaster_story.

36. TPM PollTracker, "National: U.S.-Obama Job Approval (BP Spill)," September 13, 2010, http://polltracker.talkingpointsmemo.com/contests/us-obama-job- approval-bp-spill#page=1.

37. Rasmussen Reports, "Bush Katrina Ratings Fall After Speech," September 18, 2005, http://legacy.rasmussenreports.com/2005/Katrina_September%2018.htm.

38. "Presidential Pony Show: Obama Needs to Lead, Not Emote," *Newsweek*, June 13, 2010, www.newsweek.com/2010/06/13/presidential-pony-show.html; Frank Rich, "Don't Get Mad, Mr. President. Get Even," *New York Times*, June 6, 2010, www.nytimes.com/2010/06/06/opinion/06rich.html.

39. Maureen Dowd, "A Storyteller Loses the Story Line," *New York Times*, June 1, 2010, www.nytimes.com/2010/06/02/opinion/02dowd.html.

40. Sam Youngman, "Obama in Gulf: 'The Buck Stops with Me,'" *The Hill*, May 28, 2010, http://thehill.com/homenews/administration/100553-obama-in-gulf- the-buck-stops-with-me.

41. "Interview with President Barack Obama," *Larry King Live*, CNN, June 3, 2010, http://transcripts.cnn.com/TRANSCRIPTS/1006/03/lkl.01.html.

42. Ruth King, "President Obama, Race, and Rage," Ezine Articles, June 30, 2010, http://ezinearticles.com/?President-Obama,-Race-and-Rage&id=4561679.

43. Mike Allen, "Gulf Commander to Begin Solo Briefings," *Politico*, May 13, 2010, www.politico.com/news/stories/0510/37965.html.

44. Deepwater Horizon Incident Joint Information Center, "Rear Adm. Landry to Resume Her Role as Coast Guard Eighth District Commander to Focus on Hurricane Readiness," June 1, 2010, www.restorethegulf.gov/release/2010/06/01/rear-adm-landry-resume-her-role-coast-guard-eighth-district-commander-focus-hurri.

45. Helene Cooper and Peter Baker, "U.S. Opens Criminal Inquiry into Oil Spill," *New York Times*, June 1, 2010, www.nytimes.com/2010/06/02/us/02spill.html.

46. Heidi Avery, "The Ongoing Administration-Wide Response to the Deepwater BP Oil Spill," White House Blog, May 5, 2010, www.whitehouse.gov/blog/2010/05/05/ongoing-administration-wide-response-deepwater-bp-oil-spill.

47. Kieran Suckling, "Interior Department Exempted BP Drilling From Environmental Review," Center for Biological Diversity, May 5, 2010, www.biologicaldiversity.org/news/press_releases/2010/bp-exempted-05–05–2010.html.

48. Juliet Eilperin, "U.S. Exempted BP's Gulf of Mexico Drilling from Environmental Impact Study," *Washington Post*, May 5, 2010, www.washingtonpost.com/wp-dyn/content/article/2010/05/04/AR2010050404118.html.

49. Suzanne Goldenberg, "Oil Spill: US Failing to Tighten Ecological Oversight, Say Activists," *Guardian*, May 9, 2010, www.guardian.co.uk/environment/2010/may/09/oil-spill-ecological-review-environment.

50. Center for Biological Diversity, "MMS Approved 27 Gulf Drilling Operations After BP Disaster 26 Were Exempted From Environmental Review, Including Two to BP," press release, May 7, 2010, www.biologicaldiversity.org/news/press_releases/2010/post-disaster-permits-05–07–2010.html.

51. U.S. Department of the Interior, "Salazar Meets with BP Officials and Engineers at Houston Command Center to Review Response Efforts, Activities," press release, May 6, 2010, www.doi.gov/news/pressreleases/Salazar-Meets-with-BP-Officials-and-Engineers-at-Houston-Command-Center-to-Review-Response-Efforts-Activities.cfm.

52. Juliet Eilperin, "Firms Operating Offshore Will Have to Give Information about Risks, Precautions, MMS Says," *Washington Post*, June 2, 2010, www.washingtonpost.com/wp-dyn/content/article/2010/06/02/AR2010060204385.html.

53. "MMS Was Troubled Long Before Oil Spill," CNN, May 27, 2010, http://articles.cnn.com/2010–05–27/politics/mms.salazar_1_salazar-oil-spill-rig/2?_s=PM:POLITICS.

54. Gardiner Harris, "Crisis Places Focus on Beleaguered Agency's Chief," *New*

York Times, May 25, 2010, www.nytimes.com/2010/05/26/us/politics/26birn baum.html.

55. U.S. Department of the Interior, "Interior Issues Directive to Guide Safe, Six-Month Moratorium on Deepwater Drilling," press release, May 30, 2010, www.doi.gov/news/pressreleases/Interior-Issues-Directive-to-Guide-Safe-Six-Month-Moratorium-on-Deepwater-Drilling.cfm.

56. Seth Borenstein and Alan Fram, "Obama Gets Good Marks on Spill Response," Associated Press, May 13, 2010, www.chron.com/disp/story.mpl/business/deepwaterhorizon/7004066.html.

57. AP-GfK Poll, June 2010, www.ap-gfkpoll.com/pdf/AP-GfK_Poll_Politics%20Topline%20June%209–14–2010%20Topline.pdf.

58. Greenpeace USA, "Greenpeace Activists Paint Message in Oil on Arctic-Bound Drilling Supply Ship," press release, May 24, 2010, www.greenpeace.org/usa/en/media-center/news-releases/greenpeace-activists-paint-mes.

59. John M. Broder, "Oil Executives Break Ranks in Testimony," *New York Times*, June 16, 2010, www.nytimes.com/2010/06/16/business/16oil.html.

60. Center for Biological Diversity, "Lawsuit Seeks Release of Interior Secretary Salazar's Communications with Oil Industry about Drilling," press release, July 12, 2010, www.biologicaldiversity.org/news/press_releases/2010/salazar-07–12–2010.html.

61. Center for Responsive Politics, "Lobbying Spending Database Oil and Gas—2010," OpenSecrets, November 29, 2010, www.opensecrets.org/lobby/indusclient.php?year=2010&lname=E01&id=.

62. Peter H. Stone, "U.S. Chamber Boosts Election Budget to $75 Million," Center for Public Integrity, July 1, 2010, www.publicintegrity.org/blog/entry/2203.

63. David Prestipino, "PTTEP Confirms Oil Spill Cause," *WA Today*, January 14, 2010, www.watoday.com.au/environment/pttep-confirms-oil-spill-cause-20100114-m8x9.html.

64. U.S. Congress, Office of Technology Assessment, "Coping with an Oiled Sea: An Analysis of Oil Spill Response Technologies," March 1990, www.fas.org/ota/reports/9011.pdf.

65. "Gulf Oil Spill: New Orleans Judge Held Energy-Related Stocks," *Los Angeles Times*, June 22, 2010, http://latimesblogs.latimes.com/greenspace/2010/06/gulf-oil-spill-judge-martin-feldman-deepwater-drilling-moratorium.html.

66. Martin Feldman, "Injunction on Drilling Moratorium," *New York Times*, June 22, 2010, http://documents.nytimes.com/injunction-on-deepwater-drilling-moratorium#document/p1; John M. Broder, "Court Rejects Moratorium on Drilling in the Gulf," *New York Times*, July 8, 2010, www.nytimes.com/2010/07/09/us/09drill.html.

67. Bloomberg National Poll, "Study #1993, July 9–12, 2010," July 14, 2010, http://mediabugs.org/peoplepods/files/docs/57.original.pdf.

68. Center for Responsive Politics, "Lobbying Spending Database Oil and Gas."

69. Robert Walsh, "Balancing Family, Church and Work as Oil CEO," *Mormon Times*, December 15, 2008, www.mormontimes.com/article/11563/yourtimes/lifestyle/food%20&%20health.

70. American Petroleum Institute, "API Statement on Court's Decision to Lift the Moratorium," press release, June 22, 2010, www.api.org/Newsroom/reaction-lift-mor.cfm.

71. American Petroleum Institute, "API Opposes New Offshore Drilling Moratorium," press release, July 12, 2010, www.api.org/Newsroom/no-new-moratorium.cfm.

72. American Petroleum Institute, "House Resources Bill Threatens American Jobs, Economy," press release, July 15, 2010, www.api.org/Newsroom/hrbilljobseconomy.cfm.

73. Will Evans, "New Group Tied to Oil Industry Runs Ads Promoting Drilling, Attacking Democrat," NPR, September 22, 2008, www.npr.org/blogs/secretmoney/2008/09/udall_radio_ad.html.

74. Crossroads Media, www.crossroadsmedia.tv.

75. Save U.S. Energy Jobs, "Energy Group Launches Project to Promote Jobs, Affordability, Industry Safety," press release, July 5, 2010, www.saveusenergyjobs.com/2010/07/energy-group-launches-project-to-promote-jobs-affordability-safety.

76. United States Department of Labor, Occupational Safety and Health Administration, OSHA Enforcement Inspections by Name of Establishment, accessed January 2011, www.osha.gov/pls/imis/establishment.html.

77. Antonia Juhasz, ed., "The True Cost of Chevron: An Alternative Annual Report," p. 22, Global Exchange, May 2009, http://truecostofchevron.com/2009-alternative-annual-report-with-endnotes.pdf.

78. Lise Olsen and Eric Nadler, "Offshore Accidents Bring Few Penalties," *Houston Chronicle*, June 7, 2010, www.chron.com/disp/story.mpl/business/deepwaterhorizon/7039960.html.

79. Save U.S. Energy Jobs, "White House Drilling Moratorium: The Second Gulf Disaster?," accessed January 12, 2010, www.saveusenergyjobs.com/2010/09/white-house-drilling-moratorium-the-second-gulf-disaster.

80. "Astroturf Activism: Leaked Memo Reveals Oil Industry Effort to Stage Rallies Against Climate Legislation," Democracy Now! August 21, 2009, www.democracynow.org/2009/8/21/astroturf_activism_leaked_memo_reveals_oil.

81. Kate Sheppard, "API's Recycled Astroturf," *Mother Jones*, July, 21, 2010, http://motherjones.com/blue-marble/2010/07/american-petroleum-institute-astroturf-energy-citizens.

82. American Petroleum Institute, "America's Oil and Natural Gas Industry Supports Over 9 Million Jobs," October 2010, www.api.org/policy/americatowork/upload/JOBS_AMERICA_100410.pdf.

83. Mary L. Landrieu, "Landrieu Urges Obama Administration to Lift Moratorium on New Shallow Water Drilling," press release, Office of U.S. Senator Mary L. Landrieu, May 21, 2010, http://landrieu.senate.gov/mediacenter/press releases/05–21-2010-2.cfm.

84. Erwin Seba, "Talks Ongoing for New Refinery Workers Contract," Reuters, February 1, 2009, www.reuters.com/article/idUSTRE50T6DG20090201? pageNumber=1; Dean Corgey, "Lack of Union Workers Hurts Offshore Oil Industry," *Houston Chronicle*, June 10, 2010, www.chron.com/disp/story.mpl/editorial/outlook/7047234.html.

85. Bureau of Labor Statistics, "Union Members—2009," press release, January 22, 2010, www.bls.gov/news.release/pdf/union2.pdf; "Union Workers Have Better Health Care and Pensions," AFL-CIO, 2010, www.aflcio.org/joina union/why/uniondifference/uniondiff6.cfm; "Unions Making a Difference," American Rights at Work, 2010, www.americanrightsatwork.org/component/option,com_issues/Itemid,366/view,issue/id,12.

86. Tom Fowler, "Oil-Field Workers Striking It Rich," *Houston Chronicle*, May 2, 2007, www.chron.com/disp/story.mpl/business/4767136.html.

87. U.S. Department of Labor, Bureau of the Census, "Oil and Gas Extraction Industry—Establishments, Employees, and Payroll by State: 2007," 2010, www.census.gov/compendia/statab/2010/tables/10s0881.xls.

88. Rebecca M. Blank, "Understanding the Impact of the Drilling Moratorium on the Gulf Coast Economy," testimony before the Senate Committee on Small Business and Entrepreneurship, September 16, 2010, http://sbc.senate.gov/public/?a=Files.Serve&File_id=31647442-c186–48c8-b199-eea8fde7a0e1.

89. Tom Fowler, "Fund to Help Workers Has Few Takers," *Houston Chronicle*, September 14, 2010, www.chron.com/disp/story.mpl/business/energy/7201320.html.

90. Sue Sturgis, "Big Oil Rallies to Save Big Oil," Facing South, Institute for Southern Studies, September 2, 2010, www.southernstudies.org/2010/09/big-oil-rallies-to-save-big-oil.html.

91. "White House: Turning Point in Oil Containment," *Good Morning America*, ABC, August 4, 2010, http://abcnews.go.com/GMA/video/white-house-turning-point-oil-containment-11320458.

92. Jane Lubchenco et al., "BP Deepwater Horizon Oil Budget: What Happened to the Oil?" August 4, 2010, www.noaanews.noaa.gov/stories2010/PDFs/OilBudget_description_%2083final.pdf.

93. David Fahrenthold, "Scientists Question Government Team's Report of Shrinking Gulf Oil Spill," *Washington Post*, August 5, 2010, www.washington post.com/wp-dyn/content/article/2010/08/04/AR2010080407082.html?sid=ST2010080406374.

94. Ibid.

95. Justin Gillis and Leslie Kaufman, "Oil Spill Calculations Stir Debate on Damage," *New York Times*, August 4, 2010, www.nytimes.com/2010/08/05/us/05oil.html.

96. "Current Status of BP Oil Spill," Georgia Sea Grant, August 16, 2010, http://uga.edu/aboutUGA/joye_pkit/GeorgiaSeaGrant_OilSpillReport8–16.pdf.

97. Bob Marshall, "New Wave of Oil Comes Ashore West of Mississippi River," *Times-Picayune*, September 12, 2010, www.nola.com/outdoors/index.ssf/2010/09/new_wave_of_oil_comes_ashore_w.html.

98. ———, "Oil Spill Is Far From Over for Those Who Live, Work Along the Gulf," *Times-Picayune*, September 19, 2010, www.nola.com/news/gulf-oil-spill/index.ssf/2010/09/oil_spill_is_far_from_over_for.html.

99. U.S. Department of Labor, Bureau of the Census, "Oil and Gas Extraction Industry."

100. Louis Jacobson, "Landrieu Says Louisiana Doesn't Get 'One Single Penny' from Offshore Drilling," PolitiFact, *St. Petersburg Times*, May 13, 2010, www.politifact.com/truth-o-meter/statements/2010/may/13/mary-landrieu/landrieu-says-louisiana-doesnt-get-one-single-penny.

101. Marisa Di Natale, "The Economic Impact of the Gulf Oil Spill," *Moody's Analytics*, July 21, 2010, www.economy.com/dismal/article_free.asp?cid=191641&src=moodys.

102. "Crude Oil Production Table," U.S. Government Energy Information Administration, December 8, 2010, www.eia.gov/dnav/pet/pet_crd_crpdn_adc_mbblpd_a.htm.

103. "4-Week Average U.S. Exports of Crude Oil and Petroleum Products," U.S. Energy Information Administration, January 12, 2011, www.eia.gov/dnav/pet/hist/LeafHandler.ashx?n=PET&s=WTTEXUS2&f=4; U.S. Bureau of the Census: Foreign Trade Division, USA Trade Online, U.S. Import and Export Merchandise Trade Statistics, http://data.usatradeonline.gov/Dim/dimension.aspx?ReportId=3165.

104. Russell Gold and Ana Campoy, "Oil Industry Braces for Drop in U.S. Thirst for Gasoline," *Wall Street Journal*, April 13, 2009, http://online.wsj.com/article/SB123957686061311925.html.

105. Jeff David, "Louisiana's Electoral Map, Fall of 2010," *Mobile News*, October 14, 2010.

106. Eric Kleefeld, "Poll: Louisiana Loves Bobby Jindal—and Offshore Drilling," *Talking Points Memo*, June 15, 2010, http://tpmdc.talkingpointsmemo.com/2010/06/poll-louisiana-loves-bobby-jindal-and-offshore-drilling.php.

107. Ed O'Keefe, "Mary Landrieu to Block Jack Lew's Nomination to Serve as OMB Chief," *Washington Post*, September 23, 2010, http://voices.washingtonpost.com/federal-eye/2010/09/jack_lew_wins_senate_budget_pa.html.

108. Katie Howell, "Obama Admin Waives NEPA Reviews for 13 Gulf Projects," *E&E News*, January 3, 2011.
109. Nancy Dillon, "Future Looks Bleak for Gulf: Alaska STILL Hasn't Recovered from Exxon Valdez Disaster 21 Years Later," *New York Daily News*, June 20, 2010, www.nydailynews.com/news/national/2010/06/20/2010–06–20_it_wont_go_away_21_years_later_oil_still_stains_alaskan_coast.html.
110. "Attorney General Eric Holder Announces Civil Lawsuit Regarding Deepwater Horizon Oil Spill," speech by Eric Holder, U.S. Department of Justice, December 15, 2010, www.justice.gov/iso/opa/ag/speeches/2010/ag-speech-101215.html.
111. John Schwartz, "With Criminal Charges, Costs to BP Could Soar," *New York Times*, June 16, 2010, www.nytimes.com/2010/06/17/us/17liability.html.

Photo Credits

Index

NOTE: Page numbers in italics refer to photos.